Application and Theory of Multimedia Signal Processing Using Machine Learning or Advanced Methods

Application and Theory of Multimedia Signal Processing Using Machine Learning or Advanced Methods

Editor

Cheonshik Kim

MDPI • Basel • Beijing • Wuhan • Barcelona • Belgrade • Manchester • Tokyo • Cluj • Tianjin

Editor
Cheonshik Kim
Sejong University
Korea

Editorial Office
MDPI
St. Alban-Anlage 66
4052 Basel, Switzerland

This is a reprint of articles from the Special Issue published online in the open access journal *Applied Sciences* (ISSN 2076-3417) (available at: https://www.mdpi.com/journal/applsci/special_issues/Multimedia_Signal_Processing_Using_Machine_Learning_Advanced_Methods).

For citation purposes, cite each article independently as indicated on the article page online and as indicated below:

LastName, A.A.; LastName, B.B.; LastName, C.C. Article Title. *Journal Name* **Year**, *Volume Number*, Page Range.

ISBN 978-3-0365-5393-1 (Hbk)
ISBN 978-3-0365-5394-8 (PDF)

Contents

About the Editor

Cheonshik Kim

Cheonshik Kim is a professor in the Department of Computer Science, Sejong University, Korea. He is the Editor of the *Real-Time Image Processing Journal and ICACT Transaction on Advanced Communications Technology (TACT)*, and a Topical Advisory Panel Member of *Applied Sciences* (MDPI). His research interests include multimedia systems, data hiding, and watermarking. He was a subject of a biographical record in the *Marquis Who's Who in the World 2013–2015*.

applied sciences

Editorial

Application and Theory of Multimedia Signal Processing Using Machine Learning or Advanced Methods

Cheonshik Kim

Department of Computer Engineering, Sejong University, Seoul 05006, Korea; mipsan@sejong.ac.kr

1. Introduction

Machine learning (ML) uses algorithms to identify and predict useful patterns from data. Although it has found success in many areas, the results of multimedia mining are not satisfactory. ML in multimedia application extracts relevant data from multimedia files, such as audio, video, and still images, to perform similar searches, identify associations, and perform entity identification and classification. CNN emerged as a new breakthrough in the fields of data mining and AI, and has proven useful in both data analysis and application. In addition, CNN has made great progress in the area of multimedia. CNN is a field of machine learning that is applied in smart phones for face recognition and voice commands. Additionally, CNN technology contributes to the development of algorithms for the safety and security of multimedia data and the development of new applications.

This Special Issue will share the achievements of key researchers and practitioners in academia, as well as in the industry, dealing with a wide range of theoretical and applied problems in the field of multimedia.

2. Published Papers

In view of the above, this Special Issue is introduced to collect the latest research on the related subject and to solve the present challenging problems related to the various technologies based on digital imaging technology. In this feature, 10 papers have been published, and 21 papers have been received (i.e., 47% acceptance rate). Looking back at the special feature, various topics were covered with a focus on data hiding, encryption, object detection, image classification, and text recognition.

The first paper (Kim et al. (2020)) [1] shows an effective data hiding method for two quantization levels of each block of AMBTC using Hamming codes. Bai and Chang introduced a method of applying Hamming (7,4) to two quantization levels; however, the scheme is ineffective, and the image distortion error is relatively large. To solve the problem with the image distortion errors, this paper introduces a way of optimizing codewords and reducing pixel distortion by utilizing Hamming (7,4) and lookup tables.

The second paper provides another review of efficient inner product encryption approach by Tseng et al. (2020) [2]. The formal security proof and implementation result are also demonstrated. Compared with other state-of-the-art schemes, our scheme is the most efficient in terms of the number of pairing computations for decryption and the private key length.

The third paper proposed by Kim et al. (2021) [3] introduced a self-embedded watermarking technique based on Absolute Moment Block Truncation Coding (AMBTC) for reconstructing tampered images by cropping attacks and forgery. The watermark is embedded in the pixels of the cover image using 3LSB and 2LSB, and the checksum is hidden in the LSB. Through the recovering procedure, it is possible to recover the original marked image from the tampered marked image.

The fourth paper is a study on image reconstruction based on sparse constraints, which is an important research topic in compressed sensing. This paper is a constrained

Citation: Kim, C. Application and Theory of Multimedia Signal Processing Using Machine Learning or Advanced Methods. *Appl. Sci.* **2022**, *12*, 6426. https://doi.org/10.3390/app12136426

Received: 22 June 2022
Accepted: 22 June 2022
Published: 24 June 2022

Publisher's Note: MDPI stays neutral with regard to jurisdictional claims in published maps and institutional affiliations.

backtracking matching pursuit (CBMP) algorithm for image reconstruction, and is written by Bi et al. (2021) [4].

In the fifth paper, written by Kim et al. (2021) [5], they present a new method to detect foreground objects in video surveillance using multiple difference images as the input of convolutional neural networks, which guarantees improved generalization power compared to current deep learning-based methods.

The sixth paper (Xiang et al. (2021)) [6] is a study of an example for splitting the inference component of the YOLOv2 trained machine learning model between client, network, and service side processing to reduce the overall service latency. The approach of this research is not only applicable to object detection, but can also be applied in a broad variety of machine learning-based applications and services.

The seventh paper (Gu et al. (2021)) [7] proposes a method of power allocation for secrecy capacity optimization artificial-noise secure MIMO precoding systems under perfect and imperfect channel state information.

The eighth paper (Copiaco et al. (2021)) [8] presents a detailed study of the most apparent and widely-used cepstral and spectral features for multi-channel audio applications. Additionally, the paper details the development of a compact version of the AlexNet model for computationally limited platforms through studies of performances against various architectural and parameter modifications of the original network.

This ninth paper (Zhang et al. (2022)) [9] empirically evaluates two kinds of features, which are extracted, respectively, with traditional statistical methods and convolutional neural networks (CNNs), in order to improve the performance of seismic patch image classification.

The tenth paper (Tan et al. (2022)) [10] proposes a pipeline that locates texts on a page and recognizes the text types, as well as the context of the texts within the detected region.

3. Future Research Directions

Although the special feature has ended, more in-depth research on digital image security technology is expected. In order to support the basic technology of the 4th industrial revolution, it can be expected that more advanced research will occur in the future.

Funding: This research received no external funding.

Acknowledgments: This issue would not be possible without the contributions of various talented authors, hardworking and professional reviewers, and the dedicated editorial team of Applied Sciences. Congratulations to all authors—no matter what the final decisions of the submitted manuscripts were, the feedback, comments, and suggestions from the reviewers and editors helped the authors to improve their papers. We would like to take this opportunity to record our sincere gratefulness to all reviewers. Finally, we place on record our gratitude to the editorial team of Applied Sciences, and MDPI Branch Office, Beijing.

Conflicts of Interest: The authors declare no conflict of interest.

References

1. Kim, C.; Shin, D.; Yang, C.; Leng, L. Hybrid Data Hiding Based on AMBTC Using Enhanced Hamming Code. *Appl. Sci.* **2020**, *10*, 5336. [CrossRef]
2. Tseng, Y.; Liu, Z.; Tso, R. Practical Inner Product Encryption with Constant Private Key. *Appl. Sci.* **2020**, *10*, 8669. [CrossRef]
3. Kim, C.; Yang, C. Self-Embedding Fragile Watermarking Scheme to Detect Image Tampering Using AMBTC and OPAP Approaches. *Appl. Sci.* **2021**, *11*, 1146. [CrossRef]
4. Bi, X.; Leng, L.; Kim, C.; Liu, X.; Du, Y.; Liu, F. Constrained Backtracking Matching Pursuit Algorithm for Image Reconstruction in Compressed Sensing. *Appl. Sci.* **2021**, *11*, 1435. [CrossRef]
5. Kim, J.; Ha, J. Foreground Objects Detection by U-Net with Multiple Difference Images. *Appl. Sci.* **2021**, *11*, 1807. [CrossRef]
6. Xiang, Z.; Seeling, P.; Fitzek, F. You Only Look Once, But Compute Twice: Service Function Chaining for Low-Latency Object Detection in Softwarized Networks. *Appl. Sci.* **2021**, *11*, 2177. [CrossRef]
7. Gu, Y.; Huang, B.; Wu, Z. Power Allocation for Secrecy-Capacity-Optimization-Artificial-Noise Secure MIMO Precoding Systems under Perfect and Imperfect Channel State Information. *Appl. Sci.* **2021**, *11*, 4558. [CrossRef]
8. Copiaco, A.; Ritz, C.; Abdulaziz, N.; Fasciani, S. A Study of Features and Deep Neural Network Architectures and Hyper-Parameters for Domestic Audio Classification. *Appl. Sci.* **2021**, *11*, 4880. [CrossRef]

9. Zhang, C.; Wei, X.; Kim, S. Empirical Evaluation on Utilizing CNN-Features for Seismic Patch Classification. *Appl. Sci.* **2022**, 12, 197. [CrossRef]
10. Tan, Y.; Connie, T.; Goh, M.; Teoh, A. A Pipeline Approach to Context-Aware Handwritten Text Recognition. *Appl. Sci.* **2022**, 12, 1870. [CrossRef]

Article

A Study of Features and Deep Neural Network Architectures and Hyper-Parameters for Domestic Audio Classification

Abigail Copiaco [1], Christian Ritz [2,*], Nidhal Abdulaziz [1] and Stefano Fasciani [3]

[1] Faculty of Engineering and Information Sciences, University of Wollongong in Dubai, Dubai 20183, United Arab Emirates; abigailcopiaco@uowdubai.ac.ae (A.C.); nidhalabdulaziz@uowdubai.ac.ae (N.A.)

[2] School of Electrical, Computer and Telecommunications Engineering, University of Wollongong, Northfields Ave, Wollongong, NSW 2522, Australia

[3] Department of Musicology, University of Oslo, Sem Sælands vei 2, 0371 Oslo, Norway; stefano.fasciani@imv.uio.no

[*] Correspondence: critz@uow.edu.au

Featured Application: The algorithms explored in this research can be used for any multi-level classification applications.

Abstract: Recent methodologies for audio classification frequently involve cepstral and spectral features, applied to single channel recordings of acoustic scenes and events. Further, the concept of transfer learning has been widely used over the years, and has proven to provide an efficient alternative to training neural networks from scratch. The lower time and resource requirements when using pre-trained models allows for more versatility in developing system classification approaches. However, information on classification performance when using different features for multi-channel recordings is often limited. Furthermore, pre-trained networks are initially trained on bigger databases and are often unnecessarily large. This poses a challenge when developing systems for devices with limited computational resources, such as mobile or embedded devices. This paper presents a detailed study of the most apparent and widely-used cepstral and spectral features for multi-channel audio applications. Accordingly, we propose the use of spectro-temporal features. Additionally, the paper details the development of a compact version of the AlexNet model for computationally-limited platforms through studies of performances against various architectural and parameter modifications of the original network. The aim is to minimize the network size while maintaining the series network architecture and preserving the classification accuracy. Considering that other state-of-the-art compact networks present complex directed acyclic graphs, a series architecture proposes an advantage in customizability. Experimentation was carried out through Matlab, using a database that we have generated for this task, which composes of four-channel synthetic recordings of both sound events and scenes. The top performing methodology resulted in a weighted F1-score of 87.92% for scalogram features classified via the modified AlexNet-33 network, which has a size of 14.33 MB. The AlexNet network returned 86.24% at a size of 222.71 MB.

Keywords: neural network; transfer learning; scalograms; MFCC; Log-mel; pre-trained models

Citation: Copiaco, A.; Ritz, C.; Abdulaziz, N.; Fasciani, S. A Study of Features and Deep Neural Network Architectures and Hyper-Parameters for Domestic Audio Classification. *Appl. Sci.* **2021**, *11*, 4880. https://doi.org/10.3390/app11114880

Academic Editor: Cheonshik Kim

Received: 1 May 2021
Accepted: 25 May 2021
Published: 26 May 2021

Publisher's Note: MDPI stays neutral with regard to jurisdictional claims in published maps and institutional affiliations.

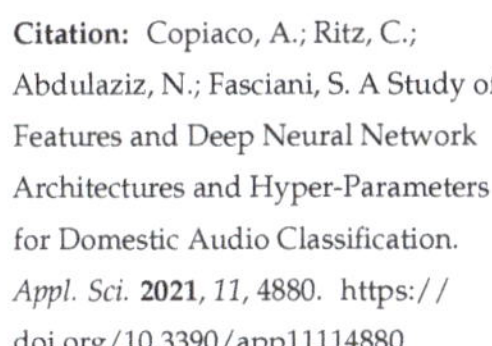

1. Introduction

The continuous research advances in the field of single and multi-channel audio classification suggests its importance and relevance in a broad range of real-world applications. In this work, we focus on domestic multi-channel audio classification, which can be applied to monitoring systems and assistive technology [1,2].

The majority of the existing works within this area are based on the classification of sound events found in single channel audio [3,4] rather than classifying multi-channel audio signals containing acoustic scenes, which is required to understand the continuous nature of daily domestic activities. Acoustic scenes refer to the sound scene recording of a certain activity over time, while sound events refer to more specific sound classes happening at short periods of time within a duration [5]. The detection of multi-channel audio was also found to be 10% more accurate when compared to single channel audio, considering the case of overlapping sounds that commonly occur in real-life [6]. Such overlapping sounds may be better detected through joint processing from different channels, reducing the effects of background noise and other interference. A similar concept to this work is the Detection and Classification of Acoustic Scenes and Events (DCASE) 2018 Task 5 challenge, which focuses on domestic multi-channel acoustic scene classification [7]. In this challenge, top performing methods often involve the use of Log-Mel energies and Mel-frequency Cepstral Coefficients (MFCC), while VGG-16 and VGG-ish pre-trained models are common choices for classification. The use of Log-Mel continues to be a popular choice for features in top performing methods of the DCASE 2019 and 2020 Task 4 challenges on sound event detection and classification [7]. Nonetheless, the utilization of spectro-temporal scalograms for multi-channel classification has not yet been thoroughly explored.

Log-Mel energies are a subset of spectral features, which consider the frequency components of a signal [8]. On the other hand, MFCCs are based on the cepstral representation of a signal, which results from the Inverse Fourier Transform (IFT) of the spectral components of the signal [8]. Although these algorithms are commonly used and are popular for noise-free environments, they have several challenges when faced in noisy acoustic environments [8,9].

Hence, this work aims to determine the optimum feature for domestic multi-channel acoustic scene classification, which takes into account real-life scenarios, such as the presence of different types of background noise. Although the DCASE 2018 Task 5 challenge had real recordings in real environments, the specific characteristics of the noise and reverberation were unknown. Hence, here we conduct a controlled study on these effects using a new database with known characteristics. Experimentation is done by conducting a thorough analysis and comparison of the classification performances and processing time of cepstral and spectral features for several pre-trained neural network and compact neural network models, using weight-sensitive metrics. It is important to note that the use of weight-sensitive metrics is important, in order to take into account the biasing that may be caused by imbalanced datasets. Further, a study on the effects of architectural and hyper-parameter modification on the optimum pre-trained network has also been looked into, in order to reduce the size of the network while maintaining its performance. In turn, we propose the use of spectro-temporal features in the form of scalograms, which are computed through a fast Fourier transform (FFT)-based continuous wavelet transform (CWT) [10]. These features possess excellent time and frequency localization, allowing a thorough representation of continuous signals with minimal loss of information [10]. This is coupled with a modified AlexNet Model, which consists of 33 layers instead of 25, and utilizes a leaky rectified linear unit (ReLU) activation function instead of a traditional ReLU function. Finally, we also synthesize an original database, which aims to recreate scenarios that could occur in real life, in order to test and verify the overall robustness of the system. In summary, the contributions described in this article include:

- A detailed performance comparison between different cepstral, spectral, and spectro-temporal features for audio classification.
- A direct performance comparison of pre-trained models and a detailed study of the effects of network modification on the optimum model.
- The development of a modified, compact AlexNet model that maintains the model's accuracy while reducing the network size by over 90%, allowing compatibility with mobile devices and applications.
- The development of a multi-channel synthetic domestic acoustic scene and event database to test the overall system robustness.

In this work, we focus on the classification and labelling of sound event and scenes, which are relevant for dementia patient monitoring systems. However, applications of the techniques explored in this work are not limited to acoustic scene classification and can be extended to other domains. For example, the compact network and the features examined can be modified to fit any image classification problem, such as emotion detection systems [11] and image-based diagnosis for healthcare applications [12]. Further, features explored in this work, as well as their combination, can also be used for regression problems, such as the estimation of characteristics of seismic waves [13], which is based on STFT features combined with CNN.

It is important to note that the compact neural network development is not a step towards an actual deployment in any specific resource-limited system. Rather, we explore and experiment the extent to which the system can be scaled down while maintaining high performance.

2. Audio Features and Pre-Trained Neural Networks

2.1. Audio Signal Features

Audio classification is typically achieved by extracting discriminative features that represent the underlying common characteristics of audio signals belonging to the same class. Similar to the DCASE challenge, it is assumed that the audio signals are recorded by microphone arrays placed at different locations (nodes) within a room. The recorded audio signals can then be represented as:

$$y_m(t) = \sum_{i=1}^{K} h_{m,i}(t) * S_i(t) + v_m(t) \tag{1}$$

where, $y_m(t)$ is the signal recorded at time t by microphone m in the array at each node, $S_i(t)$ is the ith sound source signal (where K is the total number of sounds), $h_{m,i}(t)$ is the room impulse response (RIR) from source i to microphone m, and $v_m(t)$ is additive background noise at microphone m. The audio recordings used in this work are four-channel and are time-aligned.

This section discusses several top performing features considered for multi-channel acoustic scenes and evaluates them in terms of their advantages and drawbacks according to the requirements of the system. The following subsections evaluate the possible features according to their relevant categories within the feature engineering process [8], as shown in Figure 1.

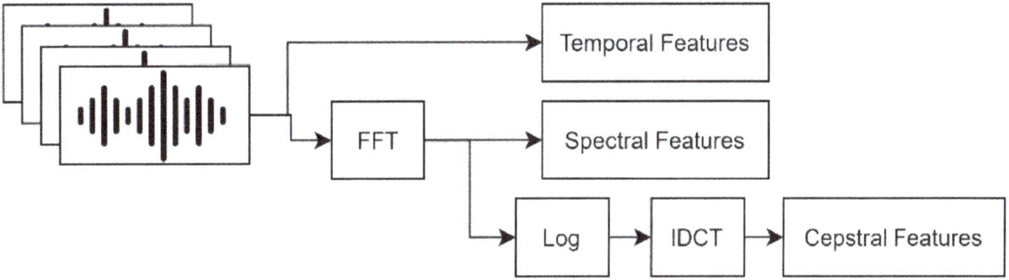

Figure 1. Taxonomy of features extracted from audio.

As observed, features are sub-divided into three main categories, namely: temporal features, spectral features, and cepstral features. Temporal features are computed in the time-domain and have the least computational complexity [8]. Spectral features, on the other hand, are extracted starting from the frequency representation of the signal [8]. Cepstral features then represent the rate of change within the different spectrum bands [8].

Finally, the fusion between spectral and temporal features results in spectro-temporal features, which combine both time and frequency attributes of a signal [8].

Since temporal features are directly extracted from the audio signal, they often deter from providing reliable descriptors for multi-channel audio classification, as they do not contain information about the frequency. Hence, in this work, we examine cepstral and spectral features only. Along with this, we also examine spectro-temporal features, which are a combination of temporal and spectral features.

2.1.1. Cepstral Features

Cepstral features represent the cepstrum, a depiction of acoustic signals that is commonly utilized in homomorphic signal processing, and is often characterized by the conversion of signals combined through convolution, into the sums of their specific cepstra [14]. Cepstral coefficients were found to be one of the most commonly utilized features for classification of acoustic scene and events.

The mel-frequency cepstral coefficients (MFCC) were the most widely apparent, and are based on a filter that models the behaviour of the human auditory system [14], making it advantageous in terms of sound identification. The MFCCs can be acquired through taking the log of the mel spectrum. Following this, the discrete cosine transform (DCT) of the log spectrum are obtained, with the MFCCs being the result of the DCT's amplitudes [15].

Calculation of the MFCC coefficient starts by dividing the time-aligned four-channel averaged audio signal $y_{avg}(t)$ into multiple segments. Windowing is then applied to each of these segments prior to being subject to the discrete Fourier transform (DFT), resulting in the short-term power spectrum $P(f)$ [16].

The power spectrum $P(f)$ is then warped along the frequency axis f, and into the mel-frequency axis M, resulting in a warped power spectrum $P(M)$. The warped power spectrum is then discretely convolved with a triangular bandpass filter with K filters, resulting in $\theta(M_k)$ [16]. The MFCC coefficients are calculated according to Equation (2) [16].

$$MFCC(d) = \sum_{k=1}^{K} X_k \cos\left[d(k-0.5)\frac{\pi}{K}\right], \, d = 1 \dots D \tag{2}$$

where $X_k = \ln(\theta(M_k))$, and $D << K$ due to the compression ability of the MFCC [16]. Nonetheless, these were also found to be prone to loss of substantial information due to its sensitivity to noise [17]. Similarly, its performance can be affected by the shape and spacing of the filters and the warping of the power spectrum [16]. Nevertheless, the MFCC approach has several advantages due to its simple computation, and flexibility with regards to integration with several other features [16].

2.1.2. Spectral Features

Spectral features are computed from the frequency components of the audio signal. The two-dimensional representation of the frequency components of an audio signal is called a spectrogram, which often results from the application of the short time discrete Fourier transform (STFT) to constantly compare the input signal with a sinusoidal analysis function [18]. Although this representation is known to work well with neural networks [19], the signal processing techniques used in order to display the representation can cause inconsistency within the structure of the spectrogram [18]. Further, the majority of the works concerning the spectrogram solely makes use of the magnitude component representation of the audio signal, omitting the phase information [20].

Although spectral features have several advantages, the information yielded may not be sufficient for the characterization of multi-channel audio scene acoustics. Often, they are combined with other features in order to produce a considerable representation of the signal magnitude [8]. However, since different audio scenes have different requirements in terms of temporal and frequency resolutions [21], the combination of several spectral features does not necessarily improve the accuracy of the classifier. A study by Chu, S. et al. [22] had shown that combining several spectral features, including centroid, bandwidth, flatness, and asymmetry for sound classification, does not really improve the accuracy. Instead, an increase in the computational complexity is observed due to the individual computation of multiple features that had to be combined.

Nonetheless, the log-Mel energy features are deemed beneficial for multi-channel acoustic scene classification and were utilized in notable related works mentioned in this research [23,24]. Log-Mel energy features had also been a well-received choice of features for DCASE challenge entries, as per the review of Mesaros, A. et al. [25], due to the two-dimensional matrix output that it yields, which is a suitable input for the CNN classifier. Log-Mel features are extracted through the application of a STFT applied to Hamming windowed audio segments [9]. A Mel-scale filter bank is then implemented after taking the square of the absolute value per bin, which are then processed to fit the requirements of the system [9].

2.1.3. Spectro-Temporal Features

Spectro-temporal features stem from the fusion of temporal and spectral features. Although not widely explored in the field of multi-channel audio classification, several works have devised algorithms that integrate the use of both temporal and spectral features for acoustic event detection [26,27]. Cotton, et al. proposed the use of a non-negative matrix factorization algorithm in order to detect a set of patches containing relevant spectral and temporal information that best describes the data [27]. The results achieved in their experiment suggest that their features provide more robustness in noisy environments as opposed to MFCCs as sole features. Schroder, et al. [26], on the other hand, devises a spectro-temporal feature extraction algorithm through two-dimensional Gabor functions for robust classification.

Nevertheless, these algorithms were tested solely on acoustic events as opposed to acoustic scenes. Similarly, the applicability of these algorithms to multi-channel audio scenes remains controversial; aside from not being widely utilized, comparison against other top performing feature combinations for the same application were not apparent.

However, one of the most notable works in the field of spectro-temporal features is scalogram features, which are computed through the continuous wavelet transform (CWT) [28]. Such methods consider both the time and frequency components of a signal. The time components represent the motion of the signal, and the frequency components symbolize the pixel positions in an image [28]. Taking a computer vision approach, the velocity vectors are first calculated through multi-scale wavelets, which are localized in time [29]. The CWT of a continuous signal is defined by Equation (3) [29].

$$CWT_c(s,t) = \int_{-\infty}^{\infty} y_{avg}(u) \frac{1}{\sqrt{s}} \psi^* \left(\frac{u-t}{s} \right) du \tag{3}$$

where ψ^* refers to the complex conjugate of the mother wavelet, t refers to the time domain, u signifies the signal segment, and s refers to the scale, which is a function of the frequency [29].

Separation of the audio channels is then performed via the low-dimensional models that reverberated from the firmness of the harmonic template models [28]. Such a process is beneficial for multi-channel audio classification due to its ability to separate mixed audio sources, which allows a thorough analysis for individual audio channels.

The scalogram is a visual representation of the absolute value of the CWT coefficients, represented by Equation (4) [30]:

$$E(s,t) = |CWT_c(s,t)|^2 \tag{4}$$

Nonetheless, despite its advantages, computation of CWT coefficients are often extensive and are subject to high computational time duration [31]. Wavelets are computed through comparing and inverting the DFT of the signal against the DFT of the wavelet, which can be computationally expensive. Thus, integration of other techniques in order to reduce this complexity must also be examined.

2.2. Pre-Trained Networks

Convolutional neural networks (CNN) have been commonly used for multi-channel sound scene classification in the recent years. CNNs are a sub-type of neural networks that utilize multiple convolution stages for classification [32]. Similar to the traditional neural network, CNNs are composed of three layers, namely: the convolutional layer, the pooling layer, and the fully connected layer [33]. Nonetheless, instead of a traditional fully connected layer, only a subset of the previous layer neurons is connected to the next ones. This suggests improvements in run time, computational complexity, and memory requirements.

There are various pre-trained convolutional neural network models for classification. This is achieved through the use of transfer learning, which allows the reuse of a previously trained network's weights to train a new network model [34], typically using new training data representing new classes. Several advantages of transfer learning include an improved efficiency both in time duration requirements of the model building process, training, and the learning workflow [35]. Further, several research works also report improved results by using transfer learning on pre-trained networks as opposed to training a network from scratch [36].

Various examples of pre-trained CNN models include AlexNet [37], GoogleNet [38], ResNet [39], Inception-ResNet [40], Xception [41], SqueezeNet [42], VGGNet [43], and LeNet [44]. These networks are trained with large datasets, and the weights are saved in order to be re-used for transfer learning. Table 1 provides a summary of the comparison between these pre-trained networks in terms of their basic characteristics, including the year of introduction, network size in MB, image input size, number of layers, number of parameters, and the 5% error rate. Nonetheless, as per our previous works, the AlexNet model returns the highest accuracy for domestic audio classification applications [45,46].

Table 1. General Comparison Summary between Pre-trained CNN Models.

Model	Year	Size (MB)	Input Size	Layers	Parameters	5% ER
AlexNet [37]	2012	227	227 × 227	8	62.3 million	16.4%
GoogleNet [38]	2014	27	224 × 224	22	4 million	6.70%
ResNet [39]	2015	167	224 × 224	101 *	25 million	3.57%
Inception-ResNet [40]	2017	209	299 × 299	164 *	55.9 million	
Xception [41]	2016	85	299 × 299	71	22.9 million	
SqueezeNet [42]	2016	5.2	227 × 227	18	1.25 million	
VGGNet [43]	2014	515	224 × 224	41*	138 million	7.30%
LeNet [44]	1998			7	60,000	28.2%

* Number of layers may vary depending on the version used.

3. Experimental Methodology

Based on the above discussion on the advantages and disadvantages of different feature and classification techniques, this section starts by explaining the dataset utilized and details the methodology and process we used to carry out this study.

3.1. Synthetic Domestic Acoustic Database

Synthesizing our own database allows the production of data that address issues commonly faced in a certain environment and recreates scenarios that could occur in real life. This includes noisy environments, as well as various source-to-receiver distances. Furthermore, this also provides the exact locations of the sound sources.

For this work, the generation of the synthetic database was done based on a 92.81 m^2 one-bedroom apartment modelled after the Hebrew Senior Life Facility [47], illustrated in Figure 2. We assumed a 3 m height for the ceiling. Multi-channel recordings were aimed for; hence, microphone arrays were placed on each of the four corners of the six rooms at 0.2 m below the ceiling. This produced four recordings, one from each of the receiver nodes.

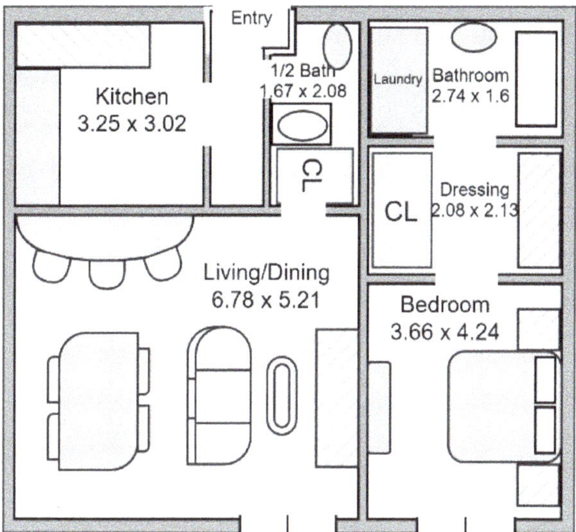

Figure 2. Floorplan of one-bedroom apartment used as acoustic environment for the synthetic database, dimensions in meters [47].

Accordingly, the microphone arrays were composed of four linearly arranged omnidirectional microphones with 5 cm inter-microphone spacing (n), as per the geometry provided in Figure 3, where d refers to the distance from the sound source to the microphones.

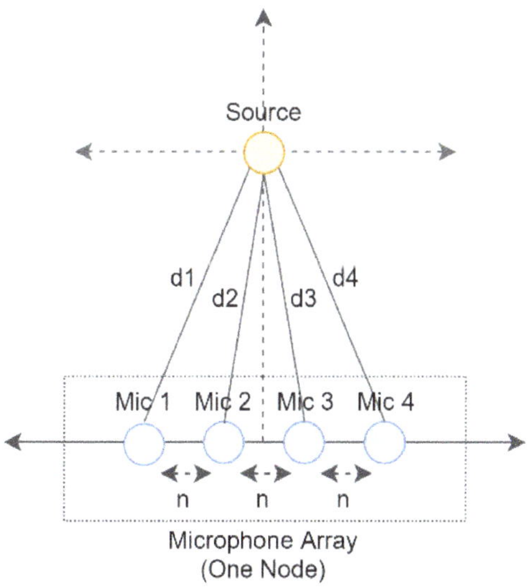

Figure 3. Microphone array geometry for a single node: four linearly spaced microphones.

Dry samples are taken from Freesound (FSD50K) [48], Kaggle [49], DESED Synthetic Soundscapes [50], and Open SLR [51], depending on the audio class. Due to the variations in sampling frequency, some of the audio signals were down sampled to 16 kHz for uniformity purposes. The room dimensions, source and receiver locations, wall reflectance, and other relevant information, were then used in order to calculate the impulse response for each room using the image method, incorporating source directivity [52]. This was then convolved with the sounds, specifying their location, in order to create the synthetic data. The data generated included clean signals, as well as different types of noisy signals, including: children playing, air conditioner, and street music, added at three different SNR levels: 15 dB, 20 dB, and 25 dB. The duration of each audio signal was uniformly kept at 5-s, as this was found to provide satisfactory time resolution for the sound scenes and events detected in this work.

Table 2 describes this dataset. This data was curated such that the testing data consisted of one noise level for each node. Any instances of the data contained in the test set were then removed from the training data. The testing set content is summarized for a specific sound being recorded at four nodes:

- Node 1: Clean Signal with 15 dB Noise
- Node 2: Clean Signal with 20 dB Noise
- Node 3: Clean Signal with 25 dB Noise
- Node 4: Clean Signal

Table 2. Summary of the Source Node Estimation Dataset.

Category	Training Data	Testing Data
Absence/Silence	11,286	876
Alarm	2765	260
Cat	11,724	1080
Dog	6673	792
Kitchen Activities	12,291	1062
Scream	4308	376
Shatter	2877	370
Shaver/toothbrush	11,231	1077
Slam	1565	268
Speech	30,113	2374
Water	6796	829
TOTAL	**101,629**	**9364**

This ensures that even when the same sound is being recorded by the four nodes present, it reduces the chances of biasing through the addition of different types of noise at different SNR levels. Further, this was also designed to reflect real life recordings, where the sound from different microphones may differ based on their distance to the source and other sounds present in their surroundings.

As observed, audio classes used in the generation of this database focus on sound events and scenes that often occur, or require an urgent response, in dementia patients' environment. Further, this was also generated through the room impulse responses of the HebrewLife Senior Facility [47], in order to reflect a realistic patient environment. This is because assistance monitoring systems are real-world applications of deep-learning audio classifiers, such as the work presented in this paper. Nonetheless, this can also be extended to other application domains as previously discussed.

3.2. Feature Extraction Using Fast CWT Scalograms

The CWT has several similarities to the Fourier transforms, such that it utilizes inner products in order to compute the similarity between the signal and an analysing function [53]. However, in the case of CWT, the analysing function is a wavelet, and the coefficients are the results of the comparison of the signal against shifted, scaled, and dilated versions of the wavelet, which are called constituent wavelets [53]. Compared with the STFT, wavelets provide better time-localization [30] and are more beneficial to non-stationary signals [53].

However, in order to reduce the computational requirements for deriving scalograms, this work proposes the use of the Fast Fourier Transform (FFT) algorithm for CWT coefficients computation [30]. Such that, if we define the mother wavelet (Ψ) to be [30], where t refers to continuous time:

$$\psi_{ts}(u) = \frac{1}{\sqrt{t}}\psi\left(\frac{u-s}{t}\right) \tag{5}$$

Then Equation (3), involving the *CWT* coefficients, can be rewritten as follows [30], where y_{avg} refers to the average of the four-channels of the audio signal:

$$CWT_c(s,t) = \int_{-\infty}^{\infty} y_{avg}(u)\psi_t^*(s-u)du \tag{6}$$

This shows that *CWT* coefficients can be expressed by the convolution of wavelets and signals. Thus, this can be written in the Fourier transform form domain, resulting in Equation (7) [30]:

$$CWT_c(s,t) = \frac{1}{2\pi}\int_{-\infty}^{\infty} y_{avg}(\omega)\psi_{s,t}^*(\omega)d\omega \tag{7}$$

where $\psi_{s,t}^*(\omega)$ specifies the Fourier transform of the mother wavelet at scale t:

$$\psi_{s,t}^*(\omega) = \sqrt{t}\psi^*(t\omega)e^{j\omega s} \tag{8}$$

Further, $y_{avg}(\omega)$ then denotes the Fourier transform of the analysed signal $y_{avg}(t)$:

$$y_{avg}(\omega) = \int_{-\infty}^{\infty} y_{avg}(t)e^{j\omega t}dt \tag{9}$$

Hence, the discrete versions of the convolutions can be represented as per Equation (10), where n is in discrete time domain:

$$W(s) = \sum_{n=0}^{N-1} y_{avg}(n)\psi^*(s-n) \tag{10}$$

From the sum in Equation (10), we can observe that CWT coefficients can be derived from the repetitive computation of the convolution of the signal, along with the wavelets, at every value of the scale per location [30]. This work follows this process in order to extract the DFT of the CWT coefficients at a faster rate compared to the traditional method.

In summary, CWT coefficients are calculated through obtaining both the DFT of the signal, as per Equation (9), and the Morlet analysing function, as per Equation (8), via the FFT. The products of these are then derived and integrated, as per Equation (6), in order to extract the wavelet coefficients. Accordingly, the discrete version of the integration can be represented as a summation, which is observed in Equation (10).

3.2.1. Feature Representation

Feature computation is carried out in MATLAB, exploiting functionalities provided in the Audio System and Data Communications toolboxes. A total of 20 filter bank channels with 12 cepstral coefficients are used for the cepstral feature extraction, as per the standard after DCT application [54]. An FFT size of 1024 is utilized, while the lower and upper filter bank frequency limits are set to 300 Hz and 3700 Hz. This frequency range includes the main components of speech signals (specifically, narrowband speech), while filtering out the humming sounds from the alternating current power, as well as high frequency noise [55]. Further, this range is relevant to the sound classes of speech and scream, and was found to also include the main components of the other classes. While larger frequency ranges could also be considered, this would require much larger FFT sizes to maintain the same frequency resolution, which in turn would increase the computational requirements. The extraction of the feature vectors is carried out by computing the average of the four time-aligned channels in the time domain, $y_{avg}(t)$. The coefficients are then extracted accordingly, from which single feature matrices are generated. The feature images are resized into 227 × 227 matrices using a bi-cubic interpolation algorithm with antialiasing [56], in order to match the input dimensionality of the AlexNet neural network model. Figure 4 shows samples of feature images for each of the three features compared, using the 'Speech' and 'Kitchen sound' classes.

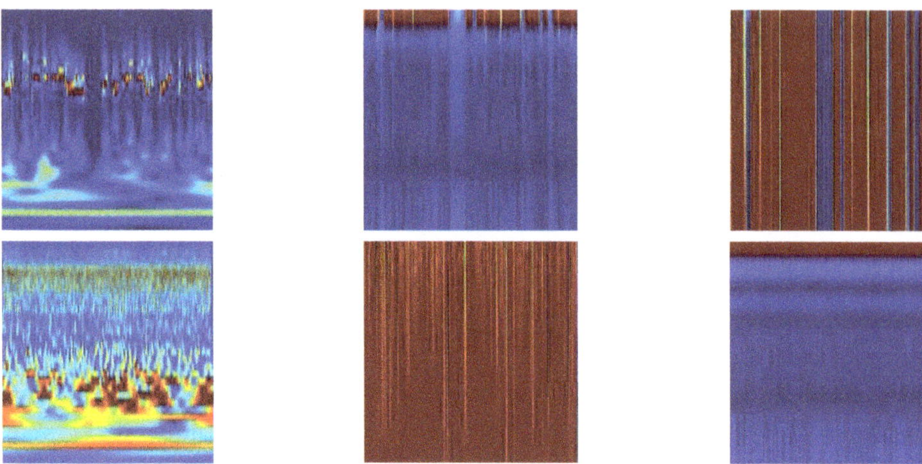

Figure 4. Feature representation samples using the 'Speech' (**top**) and 'Kitchen' (**bottom**) classes: Left to Right: CWT Scalograms, Log-Mel, and MFCC.

3.3. Modified AlexNet Network Model

Domestic multi-channel acoustic scenes consist of several signals that are captured with microphone arrays of different sizes and geometrical configurations. As discussed previously, CNNs have been widely popular for their advantage with regards to efficiency when used with data of spatial behaviour [57]. Thus, the experimentation part of this work compares different pre-trained network models for transfer learning. Modifications on the hyper-parameters are then made on the best performing network, the response being observed in three ways:

1. Effects of changing the network activation function.
2. Effects of fine-tuning the weight and bias factors, and parameter variation.
3. Effects of modifications in the network architecture.

Activation functions in neural networks are a very important aspect of deep learning. These functions heavily influence the performance and computational complexity of the deep learning model [58]. Further, such functions also affect the network in terms of its convergence speed and ability to perform the task. Aside from exploring different activation functions, we also look at fine-tuning the weights and bias factors of the convolutional layers, as well as investigating the effects of the presence of convolutional layers based on performance.

For the modified AlexNet model, we examine the traditional Rectified Linear Unit (ReLU) activation function, along with three of its variations. The ReLU offers advantages in solving the vanishing gradient problem [59], which is common with the traditional sigmoid and tanh activation functions. The gradients of neural networks are computed through backpropagation, which calculates the derivatives of the network through every layer. Hence, for activation functions such as the sigmoid, the multiplication of several small derivatives causes a very small gradient value. This, in turn, negatively affects the update of weights and biases across training sessions [59]. Provided that the ReLU function has a fixed gradient of either 1 or 0, aside from providing a solution to the vanishing gradient problem and overfitting, it also results in lower computational complexity, and therefore significantly faster training. Another benefit of ReLUs is the sparse representation, which is caused by the 0 gradient for negative values [60]. Over time, it has been proven that sparse representations are more beneficial compared to dense representations [61].

Nonetheless, despite the numerous advantages of the ReLU activation function, there are still a number of disadvantages. Because the ReLU function only considers positive components, the resulting gradient has a possibility to go towards 0. This is because the weights do not get adjusted during descent for the activations within that area. This means that the neurons that will go into that state would stop responding to any variations in the input or the error, causing several neurons to die, which makes a substantial part of the network passive. This phenomena is called the dying ReLU problem [62]. Another disadvantage of the ReLU activation function is that values may range from zero to infinity. This implies that the activation may continuously increase to a very large value, which is not an ideal condition for the network [63]. The following activations attempt to mitigate the disadvantages faced by the traditional ReLU function through modifications and will be explored in this work:

a. Leaky ReLU: The leaky ReLU is a variation of the traditional ReLU function that attempts to fix the dying ReLU problem by adding an alpha parameter, which creates a small negative slope when x is less than zero [64].

b. Clipped ReLU: The clipped ReLU activation function attempts to prevent the activation from continuously increasing to a large value. This is achieved cutting the gradient at a pre-defined ceiling value [63].

c. eLU: The exponential linear unit (eLU) is a similar activation function to ReLU. However, instead of sharply decreasing to zero for negative inputs, eLU smoothly decreases until the output is equivalent to the specified alpha value [65].

Aside from activation functions, variations in the convolutional and fully connected layers will also be examined. The study will be done in terms of both the number of parameters and the number of existing layers within the network.

For parameter modification, we explore the reduction of output variables in the fully connected layers. This method immensely reduces the overall network size [66]. However, it is important to note that recent works solely reduce the number of parameters from the first two fully connected layers. Hence, here we introduce the concept of uniform scaling, which is achieved by dividing the output parameters of fully connected layers by a common integer, based on the subsequent values.

Modification of the network architecture is also considered through examining the model's performance when the number of layers within the network is varied. These layers may include convolutional, fully-connected, and activation function layers. Nonetheless, throughout the layer variation process, the model architecture is maintained to be of a series network type. A series network contains layers that are arranged subsequent to one another, containing a single input, and output layer. Directed Acyclic Graph (DAG) networks, on the other hand, have a complex architecture, from which layers may have inputs from several layers, and the outputs of which may be used for multiple layers [67]. The higher number of hidden neurons and weights, which is apparent on DAG networks, could increase risks of overfitting. Hence, maintaining a series architecture allows for a more customizable and robust network. Further, as per the state-of-the-art, all other compact networks that currently exist present a DAG architecture. Thus, the development of a compact network with a more customizable format, and through using fewer layers, proposes advantages in designing sturdy custom networks.

3.4. Performance Evaluation Metrics

To evaluate the performance of the proposed systems, the following aspects are investigated:

1. Per class and overall comparison of different cepstral, temporal, and spectro-temporal features classified using various pre-trained neural network and machine learning models.

2. Effects of balancing the dataset

Aside from the standard accuracy, evaluations of the performances of different techniques were also compared and measured in terms of their F1-scores. This is defined to be a measure that takes into consideration both the recall and the precision, which are derived from the ratios of True Positives (TP), True Negatives (TN), False Positives (FP), and False Negatives (FN) [68], which can be extracted from confusion matrices.

The databases used for this research compose of unequal numbers of audio files per category. To account for the data imbalance, two different techniques are used:

1. Balancing the Dataset

Particularly used for the initial development and experiments conducted for this work, in this technique, the dataset was equalized across all levels in order to preserve a balanced dataset. This is done in order to avoid biasing in favour of specific categories with more samples. It is achieved by reducing the amount of data per level to match the minimum amount of data amongst the categories. Selection of the data was done randomly throughout the experiments.

2. Using Weight-sensitive Performance Metrics

Provided that the F1-score serves as the main performance metric used for the experiments conducted, it is crucial to ensure that these metrics are robust and unbiased, especially for multi-classification purposes. When taking the average F1-score for an unbalanced dataset, the amount of data per level may affect and skew the results for the mean F1-score in favour of the classes with the most amount of data. Therefore, we consider three different ways of calculating the mean F1-score, including the Weighted, Micro, and Macro F1-scores, in order to take into account for the dataset imbalance [69].

4. Results

4.1. Feature Extraction Results

Comparison of Cepstral, Spectral, and Spectro-Temporal Features

Per-level and average comparisons using MFCC and Log-Mel spectrogram features against the proposed CWTFT scalograms method are seen in Table 3, which is an average of three training trials. As observed, F1-score averaging is done using three different methods: Micro, Macro, and Weighted, in order to take into account the biasing that may be caused by the data imbalance. Further, the table also entails the comparison of the system performance between imbalanced and balanced data. To achieve a balanced data, the size of the dataset is reduced to match the lowest numbered category in both training and testing sets. As per Table 2, for each category, this turns out to be 1565 files for training, based on the "Slam" category, and 260 files for testing, based on the "Alarm" category. This adds up to a total of 17,215 training files and 2860 testing files.

The following results are achieved using the traditional AlexNet network, provided that this gives us the highest results as per our previous works [45,46]. Training for the imbalanced data is achieved at 10 epochs with 1016 iterations per epoch. However, it is important to note that the number of epochs for the balanced data is 75, as it has less iterations per epoch due to the lower amount of data per category. Hence, it requires more epochs in order to reach stability.

Table 3. Per-level comparison between imbalanced and balanced data between different types of features, with an average of three training trials.

| | CWTFT Scalograms | | | | | | | |
| | Imbalanced Data | | | | Balanced Data | | | |
Category	Accuracy	Precision	Recall	F1-Score	Accuracy	Precision	Recall	F1-Score
Silence	100.0%	99.3%	100.0%	99.7%	100.0%	98.4%	100.0%	99.2%
Alarm	65.4%	63.4%	65.4%	64.4%	75.2%	83.8%	75.2%	78.7%
Cat	97.2%	82.3%	97.2%	89.1%	94.8%	77.9%	94.8%	86.1%
Dog	74.8%	74.3%	74.8%	74.5%	84.7%	89.2%	84.7%	85.8%
Kitchen	82.3%	82.4%	82.3%	82.4%	76.5%	59.3%	76.5%	67.2%
Scream	83.7%	82.4%	83.7%	83.1%	85.9%	85.2%	85.9%	86.1%
Shatter	78.2%	72.2%	78.2%	75.1%	75.4%	89.8%	75.4%	83.2%
Shaver	71.5%	83.0%	71.5%	76.8%	66.7%	75.2%	66.7%	69.5%
Slam	65.4%	70.6%	65.4%	67.9%	71.5%	82.1%	71.5%	77.6%
Speech	100.0%	97.8%	100.0%	98.9%	100.0%	92.1%	100.0%	96.7%
Water	74.2%	85.8%	74.2%	79.6%	75.2%	82.2%	75.2%	78.1%
Micro	**86.0%**	**86.0%**	**86.0%**	**86.0%**	**82.4%**	**83.2%**	**82.4%**	**82.6%**
Weight	**86.0%**	**86.0%**	**86.0%**	**85.9%**	**82.4%**	**83.2%**	**82.4%**	**82.6%**
Macro	**81.2%**	**81.2%**	**81.2%**	**81.0%**	**82.4%**	**83.2%**	**82.4%**	**82.6%**

| | MFCCs | | | | | | | |
| | Imbalanced Data | | | | Balanced Data | | | |
Category	Accuracy	Precision	Recall	F1-Score	Accuracy	Precision	Recall	F1-Score
Absence	100.0%	98.6%	100.0%	99.3%	100.0%	98.7%	100.0%	99.3%
Alarm	53.2%	69.7%	53.2%	60.4%	52.3%	79.4%	52.3%	62.3%
Cat	75.6%	62.6%	75.6%	68.5%	74.1%	65.3%	74.1%	72.0%
Dog	74.8%	69.7%	74.8%	72.1%	76.9%	79.4%	76.9%	78.1%
Kitchen	64.1%	71.6%	64.1%	67.7%	51.8%	48.3%	51.8%	49.2%
Scream	75.8%	71.9%	75.8%	73.8%	76.4%	74.1%	76.4%	74.3%
Shatter	69.1%	53.8%	69.1%	60.5%	72.5%	70.2%	72.5%	73.1%
Shaver	53.8%	69.6%	53.8%	60.7%	48.6%	43.8%	48.6%	45.6%
Slam	37.9%	53.0%	37.9%	44.2%	50.1%	70.6%	50.1%	57.8%
Speech	99.1%	97.0%	99.1%	98.0%	99.1%	86.3%	99.1%	94.0%
Water	48.7%	55.5%	48.7%	51.9%	50.2%	50.9%	50.2%	50.1%
Micro	**77.6%**	**77.6%**	**77.6%**	**77.6%**	**68.4%**	**69.7%**	**68.4%**	**68.7%**
Weight	**77.6%**	**77.3%**	**77.6%**	**77.3%**	**68.4%**	**69.7%**	**68.4%**	**68.7%**
Macro	**68.4%**	**70.3%**	**68.4%**	**68.8%**	**68.4%**	**69.7%**	**68.4%**	**68.7%**

| | Log-Mel Spectrograms | | | | | | | |
| | Imbalanced Data | | | | Balanced Data | | | |
Category	Accuracy	Precision	Recall	F1-Score	Accuracy	Precision	Recall	F1-Score
Absence	100.0%	98.4%	100.0%	99.2%	100.0%	100.0%	100.0%	100.0%
Alarm	62.5%	61.2%	62.5%	61.8%	70.2%	62.2%	70.2%	65.6%
Cat	73.4%	65.0%	73.4%	68.9%	55.9%	60.3%	55.9%	61.2%
Dog	52.2%	51.4%	52.2%	51.8%	49.8%	54.9%	49.8%	51.9%
Kitchen	51.8%	42.6%	51.8%	46.7%	32.3%	31.6%	32.3%	32.6%
Scream	43.9%	47.4%	43.9%	45.6%	54.4%	53.6%	54.4%	54.3%
Shatter	58.2%	62.2%	58.2%	60.1%	66.8%	64.2%	66.8%	65.8%
Shaver	43.1%	41.2%	43.1%	42.1%	41.9%	31.4%	41.9%	38.1%
Slam	20.2%	36.3%	20.2%	26.0%	37.2%	56.1%	37.2%	44.4%
Speech	99.1%	92.8%	99.1%	95.9%	98.1%	82.9%	98.1%	89.5%
Water	32.2%	38.7%	32.2%	35.1%	35.2%	40.7%	35.2%	37.1%
Micro	**65.0%**	**65.0%**	**65.0%**	**65.0%**	**58.3%**	**58.0%**	**58.3%**	**58.2%**
Weight	**65.0%**	**63.9%**	**65.0%**	**64.2%**	**58.3%**	**58.0%**	**58.3%**	**58.2%**
Macro	**57.9%**	**57.9%**	**57.9%**	**57.6%**	**58.3%**	**58.0%**	**58.3%**	**58.2%**

As observed, the CWTFT scalograms have consistently achieved the highest F1-score across all categories, exceeding the performance of the MFCC features by over 10%. As mentioned earlier, this can be explained by the spectro-temporal properties of wavelets, which allows excellent time and frequency localization. The Log-Mel spectrograms gather the least F1-score out of the three features. In terms of the data imbalance, it is observed that once data is even across all categories, it improves the performance of the smaller categories. Nonetheless, the trade-off is that it reduces the F1-score for the categories with more data initially. It is also evident that performances associated with classes referring to acoustic scenes are higher than those associated to sound events. This is because sound events occur sporadically and at different instances throughout the 5-s intervals, whereas sound scenes are continuously present throughout the duration. Overall, the imbalanced dataset returns higher performance. Figure 5 accordingly shows the relevant confusion matrices for imbalanced and balanced datasets.

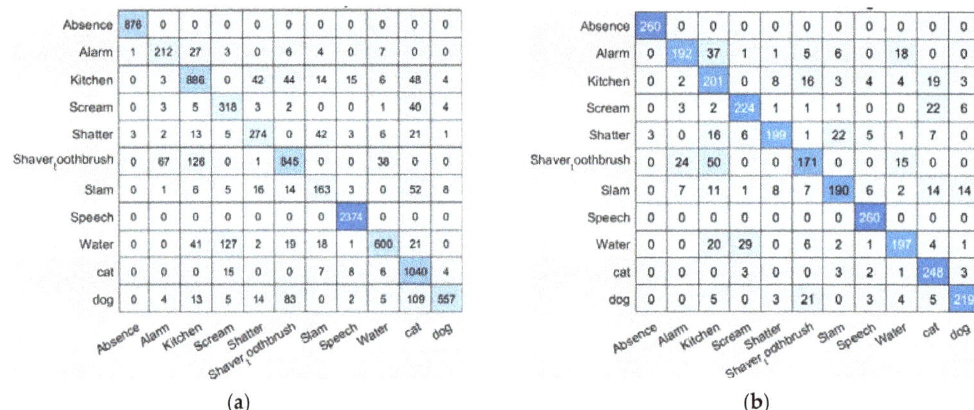

(a) (b)

Figure 5. Confusion matrices for the top performing algorithm—CWTFT scalograms for: (**a**) Imbalanced dataset using the full synthetic database; (**b**) balanced dataset with 1565 files for training, and 260 files for testing.

In our previous works, we examined the response of the system performance by concatenating the cepstra from individual channels [45,46]. This yielded a slightly better performance than using a single cepstrum after averaging the four time-aligned channels for the case of cepstral coefficients. Extracting cepstral coefficients for each channel allows a thorough consideration of all distinctive properties of the signal, which minimizes the loss of information. However, per-channel feature extraction did not cause improvement with Scalogram features, yielding a result of 90.72% as opposed to 92.33% for averaging the channels, as audio sources are already separated within its wavelet computation process.

Aside from the accuracy, execution time for the inference and resource requirements is another important consideration that must be made when selecting features. Table 4 details the execution time information for the three features compared, in terms of extracting the relevant features and translating them into a 227 × 227 image. Recording the execution time was achieved through a machine with Intel Core i7-9850H CPU @ 2.60 GHz processor, operated in single core. The reported execution times are in seconds and are an average of 100 different readings. As observed, scalograms also returned the shortest overall time duration across all three features compared. The numerous processes involved with the MFCC and Log-mel features justify the longer extraction time.

Table 4. Average execution time for inference (in s).

Parameter	Scalograms	MFCC	Log-Mel
Feature Extraction Execution Time	0.1981	1.0076	1.0640

CWTFT coefficients are derived through taking the product between the DFT of the signal and the analyzing function through FFT, and inverting this in order to extract the wavelet coefficients. On the other hand, both MFCC and Log-Mel are based on the Mel-scale filter bank. This is based on the short-term analysis, from where vectors are computed per frame. Further, windowing is performed to remove discontinuities, prior to utilizing the DFT to generate the Mel filter bank. Further processes, such as the use of triangular filters and warping, are also necessary prior to the application of the IDFT and transformation.

It is important to note that in terms of memory usage, there are negligible differences between the three features compared. This is because the features are being resized and translated into a 227×227 image through bi-cubic interpolation, in order to fit the classifier. Nonetheless, each image translation occupies between 4–12 KB of memory, depending on the sound class.

4.2. Architecture of Modified AlexNet-33 (MAlexNet-33)

This section discusses the results achieved through the detailed study of the effects of modifying the traditional AlexNet architecture. The AlexNet model was found to result in the highest F1-scores based on our previous work experiments [45,46]. In this work, we aim to improve this network by decreasing the overall network size while maintaining its performance. To begin with, the original layer structure of the AlexNet network is presented in Figure 6. As observed, it contains 25 layers, with 2 regular convolution layers, 3 group convolution layers, and 3 fully connected layers.

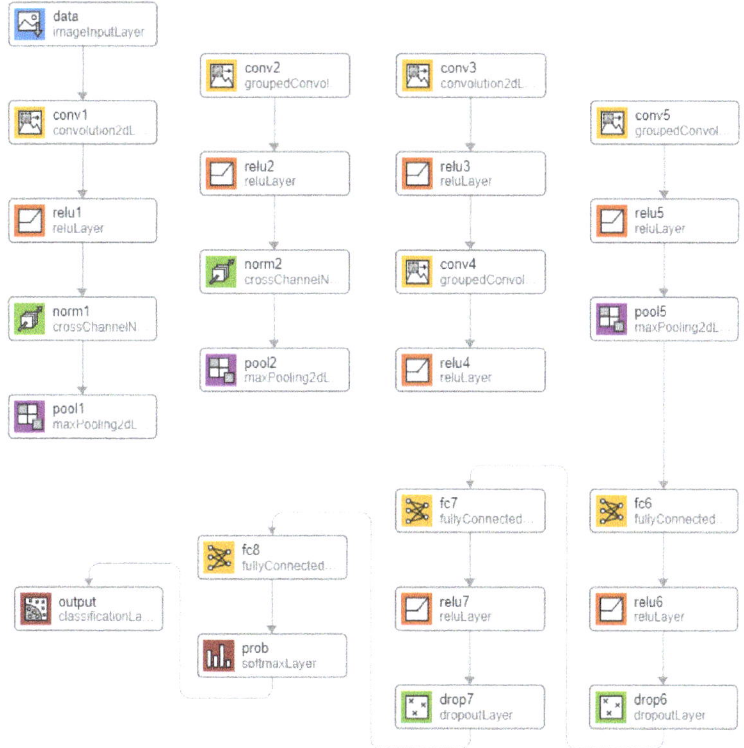

Figure 6. AlexNet Network Layer Structure: This is a 25-layer series architecture imported via Matlab Deep Network Designer. The CNN model accepts 227×227 image inputs and is trained to classify between 1000 image classes via ImageNet.

4.2.1. Exploring Variations of the Rectified Linear Unit and the Number of Layers

For this experiment, the response of the system to reducing the number of layers is investigated. Further, different variations of the ReLU activation function are also examined. Table 5 displays the different combinations tested for this experiment with regards to decreasing the number of layers and changing the activation function, presented as an average between 11 classes. Hence, throughout the results, it is apparent that the micro averaging results between the four measures are the same and there are close similarities between some of the measures. This is due to the total number of false negatives and false positives being the same. More distinct differences between the classes can be seen in the per-level comparison, such as that of Table 3.

Table 5. Performance Measures of Different Networks using Variations of the F1-score.

Network	Type	Accuracy	Precision	Recall	F1-Score
	Micro	86.36%	86.36%	86.36%	86.36%
AlexNet	Weighted	86.36%	86.72%	86.36%	86.24%
	Macro	81.03%	82.99%	81.03%	81.69%
	Micro	85.19%	85.19%	85.19%	85.19%
AlexNet-20	Weighted	85.19%	86.01%	85.19%	85.02%
	Macro	79.68%	82.42%	79.68%	80.44%
AlexNet-20	Micro	84.30%	84.30%	84.30%	84.30%
with eLU (1)	Weighted	84.30%	84.80%	84.30%	84.18%
	Macro	78.08%	81.20%	78.08%	79.22%
AlexNet-20	Micro	85.70%	85.70%	85.70%	85.70%
with Leaky	Weighted	85.70%	86.37%	85.70%	85.58%
ReLU (0.01)	Macro	79.45%	83.62%	79.45%	80.99%
AlexNet-20	Micro	84.10%	84.10%	84.10%	84.10%
with Clipped	Weighted	84.10%	84.25%	84.10%	84.04%
ReLU (6)	Macro	78.38%	78.55%	78.38%	78.26%
AlexNet-17	Micro	81.89%	81.89%	81.89%	81.89%
with Leaky	Weighted	81.89%	82.67%	81.89%	81.74%
ReLU (0.01)	Macro	75.04%	76.39%	75.04%	75.13%

From Table 5, AlexNet-20 was achieved by removing one grouped convolutional, two ReLU, one fully connected, and one 50% dropout layer from the original network. It is observed that removing convolutional and fully connected layers from the network reduces its performance as well.

However, it is also apparent that using other activation functions improves the performance. For instance, using a Leaky ReLU with a 0.01 parameter in place of the ReLU activation function increased the weighted F1-score to 85.58%, having less than 1% difference from the original network's performance. Such improvement is reportedly due to the Leaky ReLU's added parameter to solve the dying ReLU problem. Due to having less layers in the system, a reduction of about 30% from the original size was also achieved. MAlexNet-20 with a Leaky ReLU activation function has a network size of about 150 MB, compared against AlexNet's 220 MB network size.

Subsequent to this, the concept of a successive activation function was also looked at. For this, two activation function layers were placed successively throughout the network. However, as per Table 6, it is implied that using two successive activation functions does not necessarily improve the overall system performance. However, it is also apparent that using more than one activation function does not affect the overall size of the network.

Table 6. Successive Activation Function Combination Summary.

Activation Function 1	Activation Function 2	Accuracy	Network Size
ReLU	ReLU	83.27%	157.92 MB
Leaky ReLU (0.01)	Leaky ReLU (0.01)	84.32%	157.92 MB
ReLU	Leaky ReLU (0.01)	85.38%	157.92 MB
Tanh	Leaky ReLU (0.01)	73.49%	157.92 MB

4.2.2. Parameter Modification

The AlexNet contains three fully connected layers with parameter values of 9216, 4096, and 4096 for the inputs, and 4096, 4096, and 1000 for the outputs. In this experiment, we reduce the output parameters across the first two fully connected layers within the network through scaling. The results achieved from this experiment are reported in Table 7.

Table 7. Parameter Modification Results.

Activation Function	Input to FC6	FC6	FC7	Num. of Layers	Scale	Epochs	Network Size	Weighted F1
ReLU	9216	4096	4096	25 (orig *)	None	10	221.4 MB	86.24%
ReLU	4608	574	574	25 (equ *)	Equ.	30	31.90 MB	85.76%
ReLU	4608	576	256	25 (div 16)	16	30	31.23 MB	85.15%
Leaky ReLU (0.01)	4608	576	256	25 (div 16)	16	30	31.23 MB	85.48%
Leaky ReLU (0.01)	4608	384	172	25 (div 24)	24	30	23.82 MB	86.82%
ReLU	4608	384	192	25 [64]	None	30	23.85 MB	85.76%

* orig—refers to the original AlexNet layer; equ—refers to using equal fully connected layer parameters.

In here, FC6 refers to the output of the first fully connected layer, and FC7 refers to the output of the second fully connected layer. It is important to note that the output of the last fully connected layer corresponds to the number of classes the system aims to identify and is not determined by parameter scaling.

As observed from Table 7, a notable improvement is observed through scaling the output parameters of the fully connected layers through a division of 24 (from the input parameter and fully connected sizes of the original network), which provided slightly higher F1-score compared to the original AlexNet. Further, this results in an almost 90% reduction in size of the network compared to the original (23.82 MB as opposed to 221.4 MB). Uniform scaling also returns better performance compared to keeping an equal number of parameters across all fully connected layers. Further, it also achieved a higher weighted F1-score than the combination used by previous recent studies, for which the exact parameters used are represented by the last entry on Table 7 [66]. It is important to note that the input size for FC6 is automatically calculated for the modified networks. After the convolution stages, this is found to be 4608 parameters. Quantitatively, it is implied that the output parameters of all fully connected layers subsequent to the last fully connected layer can be scaled down extensively, depending on the number of classes that the model is designed to predict, keeping in mind that the fully connected output parameters are higher than the number of possible predictions.

The number of epochs required is determined through the training accuracy and losses graph. Generally, a lower number of output parameters slows down the training, requiring more epochs in order to reach a well-learned network. Figure 7 displays the difference between a traditional AlexNet and a version with lower numbers of output parameters in the fully connected layers. The comparison was done for 10 epochs.

4.2.3. The Combination of Layer and Parameter Modification

Provided that uniformly scaling the fully connected layer parameters has proven beneficial, in this section, we combine this technique with the advantages of modifying the number of layers. This is done in two ways, the results for which are presented in Table 8:

- Decreasing the number of layers: Similar to the experiment conducted in Section 3.2.1, this reduces the number of convolutional and fully connected layers within the network. For example, MAlexNet-23 refers to the removal of conv4 and relu4, maintaining all fully connected layers. On the other hand, MAlexNet-20 is the same network structure examined in Section 3.2.1.
- Increasing the number of layers: For this experiment, another grouped convolutional layer/s with the relevant activation function was added to the network structure. From the original AlexNet model, the grouped convolutions carry bias learnable weights of $1 \times 1 \times 192 \times 2$ and $1 \times 1 \times 128 \times 2$, respectively. For this work, additional grouped convolution functions were added, such that it has a bias learnable weight of $1 \times 1 \times 64 \times 2$ for MAlexNet-27, and $1 \times 1 \times 64 \times 2$ and $1 \times 1 \times 32 \times 2$ for MAlexNet-33. Accordingly, Leaky ReLu (0.01) activation functions were utilized for all grouped convolutional layers.

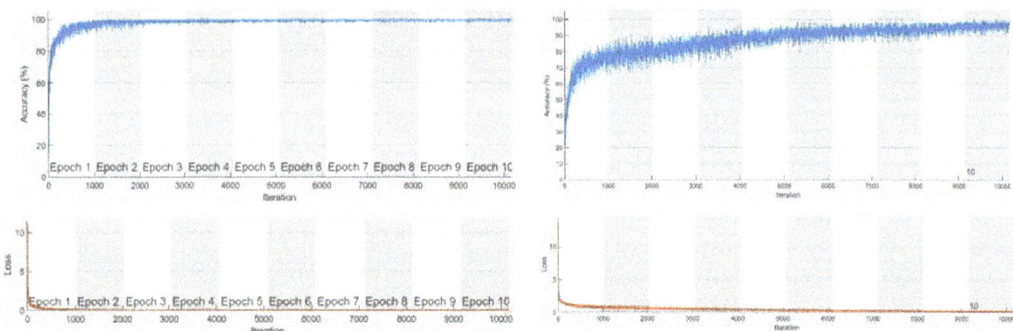

Figure 7. Training accuracy and losses graph (**Left**) AlexNet; (**Right**) Modified AlexNet with less parameters.

Table 8. Results for the combination of layer and parameter modifications.

Activation Function	FC6	FC7	Num. of Layers	Scale	Epochs	Network Size	Weighted F1
ReLU	384	192	23 (n.conv4)	None	30	21.19 MB	84.63%
Leaky ReLU (0.01)	384	192	23 (n.conv4)	None	30	21.19 MB	83.05%
Leaky ReLU (0.01)	576	-	20 (n.conv4)	None	30	27.98 MB	83.80%
Leaky ReLU (0.01)	1064	-	20 (n.conv4)	Equ.	30	45.99 MB	84.66%
ReLU	1064	-	20 (n.conv4)	Equ.	30	45.99 MB	82.54%
ReLU	576	-	20 (n.conv4)	Equ.	30	27.99 MB	83.63%
Leaky ReLU (0.01)	576	-	20 (n.conv4)	Equ.	30	27.99 MB	83.71%
Leaky ReLU (0.01)	576	-	22 (w.conv4)	Equ.	30	30.64 MB	85.00%
Leaky ReLU (0.01)	384	-	22 (w.conv4)	24	30	23.56 MB	85.76%
Leaky ReLU (0.01)	384	172	23 (n.conv4)	24	30	21.16 MB	84.76%
Leaky ReLU (0.01)	384	172	27 (gconv64)	24	30	17.34 MB	86.89%
Leaky ReLU (0.01)	192	86	27 (gconv64)	48	30	13.59 MB	85.61%
Leaky ReLU (0.01)	384	172	33 (gconv32)	24	30	14.33 MB	87.92%

As per Table 8, it is observed that the top performing algorithm is the MAlexNet-33, which is designed as a combination of both fully connected parameter scaling, as well as the addition of two new grouped convolutional layers with bias learnable weights of $1 \times 1 \times 64 \times 2$ and $1 \times 1 \times 32 \times 2$, and relevant activation layers. This provided a weighted F1-score of 87.96%, exceeding the performance of the AlexNet, with a network size of 14.33 MB. This suggests an over 95% decrease in the size of the resource requirements when compared to the original model. When compared to [66], this also improved both the performance and the network size, exceeding the performance by around 2.16%

and decreasing the network size by over 40%. Aside from the improvement in resource requirements, decreasing the network size also returned a notable improvement in the inference execution time, provided that they are factors linearly related to one another.

4.2.4. Comparison with Other Compact Networks

In this section, a comparison of the proposed architecture to currently existing compact networks is presented. For this work, several compact pre-trained models including SqueezeNet [42], MobileNet-v2 [70], NasNet Mobile [71], and ShuffleNet [72], are considered. A summary of the comparison is seen in Table 9, in terms of the total number of layers, depth, type, network size in MB, the activation function used, the weighted F1-score, the training time for 30 epochs, the network loading time, and the execution inference time average. The network loading time is an average of five trials, while the execution time is measured in 100 trials.

Table 9. Detailed Comparison with other Compact Neural Networks.

	MAlexNet-33	SqueezeNet	MobileNet-v2	NasNet Mobile	ShuffleNet
Number of Layers	33	68	155	913	173
Depth	8	18	53	N/A	50
Type	Series Network	DAG	DAG	DAG	DAG
Network Size	14.33 MB	3.07 MB	9.52 MB	19.44 MB	3.97 MB
Activation Function	Leaky ReLU (0.01)	Fire ReLU	Clipped ReLU (C: 6)	ReLU	ReLU
Weighted F1-score	87.92%	84.48%	86.85%	83.38%	86.91%
Training time	178 min	273 min	599 min	1668 min	792 min
Epochs	30	30	30	30	30
Loading time average	1.10 s	1.04 s	1.32 s	2.59 s	1.62 s
Execution time average	0.0148 s	0.0159 s	0.0338 s	0.1345 s	0.0348 s

Throughout the comparison, it is important to note that, while MAlexNet-33 is a series network, all other compact networks are DAG networks, which have a complex architecture and a significantly larger number of layers.

As observed, our proposed network consistently provided the highest weighted F1-score in comparison to the other compact networks. Despite having a 14.33 MB network size, this provided negligible time differences (about 0.08-s against SqueezeNet) in terms of loading the network. Further, it also possesses the least training and execution time compared to the other networks.

It is also apparent that other compact networks possess a higher loading time despite the smaller network size, which is caused by the DAG network configuration, and the multiple layers within the architecture. Provided that the MAlexNet-33 has the least number of layers, it creates a highly customizable network architecture. Adding more layers of neurons increases the complexity of the neural networks. Although hidden layers are crucial for extracting the relevant features, having too many hidden layers may cause overfitting. In this case, the network would be limited in terms of its generalization abilities. In order to avoid this effect, this work focuses on designing a smaller network with fewer neurons and weights than a traditional compact neural network.

5. Discussion

Interpreting the presented results, we conclude that the use of CWTFT scalograms returns the best results for audio scene and event classification applications. This is supported by our previous experiments, which were performed using the SINS database [45,46] and the experiments conducted in this work. This can be justified by the fact that scalograms possess excellent time and frequency localization. Furthermore, another advantage is that it also separates audio sources upon the wavelet computation process. Using an FFT-based wavelet transform also returns favourable time duration requirements, which exceeded that of cepstral and spectral features.

There are three main discoveries found in this study:

Hypothesis 1: *The Leaky ReLU activation function returned higher performance for multi-level classification as opposed to the traditional ReLU in the majority of cases.*

Verification of Hypothesis 1: This is true on a case-by-case scenario. This can be explained by the presence of the dying ReLU problem in feature sets, which is ameliorated through the small parameter added through the Leaky ReLU. However, it is important to note that the presence of the dying ReLU problem could depend on several factors, including the nature of the data being trained. In cases where this does not occur, replacing the activation function to a Leaky ReLU may not return any advantages.

Hypothesis 2: *Decreasing the number of fully connected and convolutional layers throughout the network also slightly decreases the performance.*

Verification of Hypothesis 2: Generally, convolutional layers represent high level features within the network. Accordingly, fully connected layers flatten and combine these features. Hence, reducing the number of these layers negatively affects the performance of the network.

Hypothesis 3: *Decreasing parameters, weight factors, and biases within the fully connected and convolutional layers helps decrease the size of the network more, compared to when these layers are removed completely.*

Verification of Hypothesis 3: Both convolutional and fully connected layers contribute to the high and low-level features from which the network learns, and are therefore essential. However, since pre-trained models are originally trained on very large data, large parameters, weight, and bias factors are often not necessary for the smaller dataset by which transfer learning is being implemented for. This explains the maintenance of the system performance despite decreasing the parameters for these layers accordingly. Based on our experiments, scaling the parameters uniformly across fully connected layers returns the best performance.

6. Conclusions

This study started with a per-level performance comparison against top-performing feature extraction methodologies, which demonstrated the robustness of the proposed CWTFT features. Further, an extensive study on pre-trained neural network modification was also presented, aiming to reduce the size of the AlexNet model whilst maintaining the accuracy. The top performing methodology involved the use of FFT-based CWT Scalogram features, with a modified AlexNet model with 33 layers (MAlexNet-33). This model uses the Leaky ReLU as its main activation function, combining strategies of both including additional convolutional layers and uniformly scaling the parameters of convolutional and fully connected layers in order to create the optimum network. The best performance resulted in an 87.92% weighted F1-score at a network size of 14.33 MB. This suggests a good improvement when compared with using the original AlexNet network with the same features, which resulted in an F1-score of 86.24%, at a size of 221.4 MB.

Author Contributions: Conceptualization, A.C., C.R., N.A. and S.F.; methodology, A.C., C.R., N.A. and S.F., software and validation, A.C.; formal analysis and investigation, A.C., data curation, A.C.; writing—original draft preparation, A.C.; writing—review and editing, C.R., N.A. and S.F., supervision, C.R., N.A. and S.F., funding acquisition, C.R. All authors have read and agreed to the published version of the manuscript.

Funding: This research received no external funding.

Institutional Review Board Statement: Not applicable.

Informed Consent Statement: Not applicable.

Data Availability Statement: The dataset used in this work has been released publicly in order to be utilized for future research, and can be downloaded from the following Kaggle link: www.kaggle. com/dataset/9e2e3c726425eb38c5b65349e5622964cc4bb454cfff46a76db3ecf0291bcc57 (accessed on 1 May 2021).

Conflicts of Interest: The authors declare no conflict of interest.

References

1. Almaadeed, N.; Asim, M.; Al-ma'adeed, S.; Bouridane, A.; Beghdadi, A. Automatic Detection and Classification of Audio Events for Road Surveillance Applications. *Sensors* **2018**, *18*, 1858. [CrossRef]
2. Lozano, H.; Hernaez, I.; Picon, A.; Camarena, J.; Navas, E. Audio Classification Techniques in Home Environments for Elderly/Dependant People. In Proceedings of the ICCHP 2010, Vienna, Austria, 14–16 July 2010; pp. 320–323.
3. Lecouteux, B.; Vacher, M.; Portet, F. Distant Speech Recognition in a Smart Home: Comparison of Several Multisource ASRs in Realistic Conditions. In Proceedings of the INTERSPEECH 2011, Florence, Italy, 27–31 August 2011.
4. Mitilineos, S.A.; Potirakis, S.M.; Tatlas, N.A.; Rangoussi, M. A Two-level Sound Classification Platform for Environmental Monitoring. *Hindawi J. Sens.* **2018**, *2018*, 2–13. [CrossRef]
5. Imoto, K. Introduction to acoustic event and scene analysis. *Acoust. Sci. Technol.* **2018**, *39*, 182–188. [CrossRef]
6. Adavanne, S.; Parascandolo, G.; Pertila, P.; Heittola, T.; Virtanen, T. Sound Event Detection in Multichannel Audio Using Spatial and Harmonic Features. In Proceedings of the DCASE 2016, Budapest, Hungary, 3 September 2016.
7. Dekkers, G.; Vuegen, L.; van Waterschoot, T.; Vanrumste, B.; Karsmakers, P. DCASE 2018—Task 5: Monitoring of domestic activities based on multi-channel acoustics. *arXiv* **2018**, arXiv:1807.11246.
8. Serizel, R.; Bisot, V.; Essid, S.; Richard, G. Acoustic Features for Environmental Sound Analysis. In *Computational Analysis of Sound Scenes and Events*; Springer: Cham, Switzerland, 2017; pp. 71–101.
9. Valenti, M.; Squartini, S.; Diment, A.; Parascandolo, G.; Virtanen, T. A Convolutional Neural Network Approach for Acoustic Scene Classification. In Proceedings of the DCASE2016 Challenge, Budapest, Hungary, 8 February–7 September 2016.
10. Chen, H.; Zhang, P.; Bai, H.; Yuan, Q.; Bao, X.; Yan, Y. Deep Convolutional Neural Network with Scalogram for Audio Scene Modeling. In Proceedings of the INTERSPEECH 2018, Hyderabad, India, 2–6 September 2018.
11. Lee, M.; Lee, Y.K.; Lim, M.T.; Kang, T.K. Emotion Recognition using Convolutional Neural Network with Selected Statistical Photolethysmogram Features. *Appl. Sci.* **2020**, *10*, 3501. [CrossRef]
12. Srinivasu, P.N.; SivaSai, J.G.; Ijaz, M.F.; Bhoi, A.K.; Kim, W.; Kang, J.J. Classification of Skin Disease Using Deep Learning Neural Networks with MobileNet V2 and LSTM. *Sensors* **2021**, *21*, 2852. [CrossRef] [PubMed]
13. Ristea, N.C.; Radoi, A. Complex Neural Networks for Estimating Epicentral Distance, Depth, and Magnitude of Seismic Waves. *IEEE Geosci. Remote. Sens. Lett.* **2021**, 1–5. [CrossRef]
14. Peeters, G. A large set of audio features for sound description (similarity and classification) in the CUIDADO project. In Proceedings of the IRCAM, Paris, France, 23–24 June 2004.
15. Sahidullah, M.; Saha, G. Design, analysis and experimental evaluation of block based transformation in MFCC computation for speaker recognition. *Speech Commun.* **2012**, *54*, 543–565. [CrossRef]
16. Zheng, F.; Zhang, G.; Song, Z. Comparison of different implementations of MFCC. *J. Comput. Sci. Technol.* **2001**, *16*, 582–589. [CrossRef]
17. Ravindran, S.; Demirogulu, C.; Anderson, D. Speech Recognition using filter-bank features. In Proceedings of the 37th Asi-lomar Conference on Signals, Systems, and Computers, Pacific Grove, CA, USA, 9–12 November 2003.
18. Le Roux, J.; Vincent, E.; Mizuno, Y.; Kameoka, H.; Ono, N.; Sagayama, S. Consistent Wiener Filtering: Generalized Time-Frequency Masking Respecting Spectrogram Consistency. In *Latent Variable Analysis and Signal Separation. LVA/ICA 2010. Lecture Notes in Computer Science*; Vigneron, V., Zarzoso, V., Moreau, E., Gribonval, R., Vincent, E., Eds.; Springer: Berlin/Heidelberg, Germany, 2010; Volume 6365.
19. Choi, W.; Kim, M.; Chung, J.; Lee, D.; Jung, S. Investigating Deep Neural Transformations for Spectrogram-based Musical Source Separation. In Proceedings of the International Society for Music Information Retrieval, Montreal, QC, Canada, 11–16 October 2020.
20. Gerkmann, T.; Krawezyk-Becker, M.; Roux, J. Phase processing for single-channel speech enhancement: History and recent advances. *IEEE Signal Process. Mag.* **2015**, *32*, 55–66. [CrossRef]
21. Zheng, W.; Mo, Z.; Xing, X.; Zhao, G. CNNs-based Acoustic Scene Classification using Multi-Spectrogram Fusion and Label Expansions. *arXiv* **2018**, arXiv:1809.01543.
22. Chu, S.; Kuo, C.; Narayanan, S.; Mataric, M. Where am I? Scene Recognition for Mobile Robots using Audio Features. In Proceedings of the 2006 IEEE International Conference on Multimedia and EXPO, Toronto, ON, USA, 9–12 July 2006.
23. Inou, T.; Vinayavekhin, P.; Wang, S.; Wood, D.; Greco, N.; Tachibana, R. Domestic Activities Classification based on CNN using Shuffling and Mixing Data Augmentation. In Proceedings of the DCASE2018, Surrey, UK, 19–20 November 2018.

24. Tanabe, R.; Endo, T.; Nikaido, Y.; Ichige, T.; Nguyen, P.; Kawaguchi, Y.; Hamada, K. Multichannel Acoustic Scene Classification by Blind Dereverberation, Blind Source Separation, Data Augmentation, and Model Ensembling. In Proceedings of the DCASE2018, Surrey, UK, 19–20 November 2018.

25. Mesaros, A.; Heittola, T.; Virtanen, T. Acoustic Scene Classification: An Overview of DCASE 2017 Challenge Entries. In Proceedings of the 2018 16th International Workshop on Acoustic Signal Enhancement (IWAENC), Tokyo, Japan, 17–20 September 2018.

26. Schroder, J.; Goetze, S.; Anemuller, J. Spectro-Temporal Gabor Filterbank Features for Acoustic Event Detection. *IEEE/ACM Trans. Audio Speech Lang. Process.* **2015**, *23*, 2198–2208. [CrossRef]

27. Cotton, C.V.; Ellis, D.P.W. Spectral vs. spectro-temporal features for acoustic event detection. In Proceedings of the 2011 IEEE Workshop on Applications of Signal Processing to Audio and Acoustics (WASPAA), New Paltz, NY, USA, 20–23 October 2011; pp. 69–72.

28. Wolf, G.; Mallat, S.; Shamma, S. Audio source separation with time-frequency velocities. In Proceedings of the 2014 IEEE International Workshop on Machine Learning for Signal Processing (MLSP), Reims, France, 21–24 September 2014; pp. 1–6.

29. Sejdic, E.; Djurovic, I.; Stankovic, L. Quantitative Performance Analysis of Scalogram as Instantaneous Frequency Estimator. *IEEE Trans. Signal Process.* **2008**, *56*, 3837–3845. [CrossRef]

30. Komorowski, D.; Pietraszek, S. The Use of Continuous Wavelet Transform Based on the Fast Fourier Transform in the Analysis of Multi-channel Electrogastrography Recordings. *J. Med Syst.* **2016**, *40*, 1–15. [CrossRef] [PubMed]

31. Zhou, Y.; Hu, W.; Liu, X.; Zhou, Q.; Yu, H.; Pu, Q. Coherency feature extraction based on DFT-based continuous wave-let transform. In Proceedings of the IEEE PES Asia-Pacific Power and Energy Engineering Conference (APPEEC), Brisbane, Australia, 15–18 November 2015.

32. Phan, H.; Hertel, L.; Maass, M.; Koch, P.; Mazur, R.; Mertins, A. Improved Audio Scene Classification Based on Label-Tree Embeddings and Convolutional Neural Networks. *IEEE/ACM Trans. Audio Speech Lang. Process.* **2017**, *25*, 1278–1290. [CrossRef]

33. Dang, A.; Vu, T.; Wang, J. Acoustic Scene Classification using Convolutional Neural Network and Multi-scale Multi-Feature Extraction. In Proceedings of the IEEE International Conference on Consumer Electronics, Las Vegas, NV, USA, 12–14 January 2018.

34. Krishna, S.; Kalluri, H. Deep Learning and Transfer Learning Approaches for Image Classification. *Int. J. Recent Technol. Eng.* **2019**, *7* (Suppl. 4), S427–S432.

35. Curry, B. *An Introduction to Transfer Learning in Machine Learning*; Medium: San Francisco, CA, USA, 2018.

36. Zabir, M.; Fazira, N.; Ibrahim, Z.; Sabri, N. Evaluation of Pre-Trained Convolutional Neural Network Models for Object Recognition. *Int. J. Eng. Technol.* **2018**, *7*, 95. [CrossRef]

37. Krizhevsky, A.; Sutskever, I.; Hinton, G.E. Imagenet classification with deep convolutional neural networks. *Commun. ACM* **2012**, *60*, 1097–1105. [CrossRef]

38. Szegedy, C.; Liu, W.; Jia, Y.; Sermanet, P.; Reed, S.; Anguelov, D.; Erhan, D.; Vanhoucke, V.; Rabinovich, A. Going Deeper with Convolutions. In Proceedings of the 2015 IEEE Conference on Computer Vision and Pattern Recognition (CVPR), Boston, MA, USA, 7–12 June 2015; pp. 1–9.

39. He, K.; Zhang, X.; Ren, S.; Sun, J. Deep residual learning for image recognition. In Proceedings of the IEEE Conference on Computer Vision and Pattern Recognition, Las Vegas, NV, USA, 27–30 June 2016.

40. Szegedy, C.; Ioffe, S.; Vanhoucke, V.; Alemi, A. Inception-v4, Inception-ResNet and the Impact of Residual Connections on Learning. In Proceedings of the Thirty-First AAAI Conference on Artificial Intelligence (AAAI-17), San Francisco, CA, USA, 4–9 February 2017.

41. Chollet, F. Xception: Deep Learning with Depthwise Separable Convolutions. In Proceedings of the IEEE Conference on Computer Vision and Pattern Recognition (CVPR), Honolulu, HI, USA, 21–26 July 2017.

42. Iandola, F.; Han, S.; Moskewicz, M.; Ashraf, K.; Dally, W.; Keutzer, K. SqueezeNet: AlexNet-Level Accuracy with $50\times$ Fewer Parameters and <0.5MB model size. In Proceedings of the ICLR 2017, Toulon, France, 24–26 April 2017.

43. Simonyan, K.; Zisserman, A. Very deep convolutional networks for large-scale image recognition. *arXiv* **2014**, arXiv:1409.1556.

44. LeCun, Y.; Bottou, L.; Bengio, Y.; Haffner, P. Gradient-based learning applied to document recognition. *Proc. IEEE* **1998**, *86*, 2278–2324. [CrossRef]

45. Copiaco, A.; Ritz, C.; Fasciani, S.; Abdulaziz, N. Scalogram Neural Network Activations with Machine Learning for Domestic Multi-channel Audio Classification. In Proceedings of the 2019 IEEE International Symposium on Signal Processing and Information Technology (ISSPIT), Ajman, United Arab Emirates, 10–12 December 2019; pp. 1–6.

46. Copiaco, A.; Ritz, C.; Abdulaziz, N.; Fasciani, S. Identifying Optimal Features for Multi-channel Acoustic Scene Classification. In Proceedings of the ICSPIS Conference, Dubai, United Arab Emirates, 18–19 December 2019; pp. 1–4.

47. Hebrew SeniorLife. Available online: https://www.hebrewseniorlife.org/newbridge/types-residences/independent-living/independent-living-apartments (accessed on 27 January 2021).

48. Fonseca, E.; Plakal, M.; Font, F.; Ellis, D.P.; Serra, X. Audio Tagging with Noisy Labels and Minimal Supervision. In Proceedings of the Detection and Classification of Acoustic Scenes and Events 2019 Workshop (DCASE2019), New York, NY, USA, 25–26 October 2019.

49. Takahashi, N.; Gygli, M.; Pfister, B.; Van Gool, L. Deep Convolutional Neural Networks and Data Augmentation for Acoustic Event Recognition. In Proceedings of the INTERSPEECH 2016, San Francisco, CA, USA, 8–12 September 2016.

50. Turpault, N.; Serizel, R.; Salamon, J.; Shah, A.P. Sound Event Detection in Domestic Environments with Weakly Labeled Data and Soundscape Synthesis. In Proceedings of the Detection and Classification of Acoustic Scenes and Events 2019 Workshop (DCASE2019), New York, NY, USA, 25–26 October 2019.
51. He, F.; Chu, S.H.; Kjartansson, O.; Rivera, C.; Katanova, A.; Gutkin, A.; Demirsahin, I.; Johny, C.; Jansche, M.; Sain, S.; et al. Open-source Multi-speaker Speech Corpora for Building Gujarati, Kannada, Malayalam, Marathi, Tamil and Telugu Speech Synthesis Systems. In Proceedings of the 12th LREC Conference, Marseille, France, 11–16 May 2020.
52. Hafezi, S.; Moore, A.H.; Naylor, P.A. Room Impulse Response for Directional source generator (RIRDgen). 2015. Available online: http://www.commsp.ee.ic.ac.uk/~{}ssh12/RIRD.htm (accessed on 31 March 2021).
53. MATLAB Documentation, Continuous Wavelet Transform and Scale-Based Analysis. 2019. Available online: https://www.mathworks.com/help/wavelet/gs/continuous-wavelet-transform-and-scale-based-analysis.html (accessed on 31 March 2021).
54. Tiwari, R.; Agrawal, K.K. Normalized Cepstral Coefficients based Isolated Word Recognition for Oral-tradition Tribal Languages using Scaled Conjugate Gradient Method. *J. Crit. Rev.* **2020**, *7*, 2097–2107.
55. Dinkar Apte, S. *Random Signal Processing*; CRC Press: Boca Raton, FL, USA, 2018.
56. Han, D. Comparison of Commonly Used Image Interpolation Methods. In Proceedings of the 2nd International Conference on Computer Science and Electronics Engineering (ICCSEE 2013), Hangzhou, China, 22–23 March 2013.
57. Hirvonin, T. Classification of Spatial Audio Location and Content Using Convolutional Neural Networks. In Proceedings of the Audio Engineering Society 138th Convention, Warsaw, Poland, 7–10 May 2015.
58. Wang, Y.; Li, Y.; Song, Y.; Rong, X. The Influence of the Activation Function in a Convolution Neural Network Model of Facial Expression Recognition. *Appl. Sci.* **2020**, *10*, 1897. [CrossRef]
59. Weir, M. A method for self-determination of adaptive learning rates in back propagation. *Neural Netw.* **1991**, *4*, 371–379. [CrossRef]
60. Shi, S.; Chu, X. Speeding up Convolutional Neural Networks by Exploiting the Sparsity of Rectifier Units. *arXiv* **2017**, arXiv:1704.07724.
61. Hu, W.; Wang, M.; Liu, B.; Ji, F.; Ma, J.; Zhao, D. Transformation of Dense and Sparse Text Representations. In Proceedings of the 28th International Conference on Computational Linguistics, Barcelona, Spain, 8–13 December 2020; pp. 3257–3267.
62. Lu, L.; Shin, Y.; Su, Y.; Karniadakis, G. Dying ReLU and Initialization: Theory and Numerical Examples. *Commun. Comput. Phys.* **2020**, *28*, 1671–1706. [CrossRef]
63. Doshi, C. *Why Relu? Tips for Using Relu. Comparison between Relu, Leaky Relu, and Relu-6*; Medium: San Francisco, CA, USA, 2019.
64. Maas, A.; Hanuun, A.; Ng, A. Rectifier nonlinearities improve neural network acoustic models. In Proceedings of the ICML, Atlanta, GA, USA, 16–21 June 2013.
65. Djork-Arne, C.; Unterthiner, T.; Hochreiter, S. Fast and accurate deep network learning by exponential linear units (ELUs). In Proceedings of the ICLR, San Juan, Puerto Rico, 2–4 May 2016.
66. Romanuke, V. An Efficient Technique for Size Reduction of Convolutional Neural Networks after Transfer Learning for Scene Recognition Tasks. *Appl. Comput. Syst.* **2018**, *23*, 141–149. [CrossRef]
67. Mathworks. DAG Network, Matlab Documentation. 2017. Available online: https://www.mathworks.com/help/deeplearning/ref/dagnetwork.html (accessed on 31 March 2021).
68. Phung, S.L.; Bouzerdoum, A.; Nguyen, G.H. Learning pattern classification tasks with imbalanced data sets. In *Pattern Recognition*; Yin, P., Ed.; In-Tech: Vukovar, Croatia, 2009; pp. 193–208.
69. Shmueli, B. Multi-Class Metrics Made Simple, Part II: The F1-Score, towards Data Science. 2019. Available online: https://towardsdatascience.com/multi-class-metrics-made-simple-part-ii-the-f1-score-ebe8b2c2ca1 (accessed on 31 March 2021).
70. Sandler, M.; Howard, A.; Zhu, M.; Zhmoginov, A.; Chen, L. Mobilenetv2: Inverted Residuals and Linear Bottlenecks. In Proceedings of the 2018 IEEE Conference on Computer Vision and Pattern Recognition (CVPR), Salt Lake City, UT, USA, 18–23 June 2018.
71. Zoph, B.; Vasudevan, V.; Shlens, J.; Le, Q.V. Learning Transferable Architectures for Scalable Image Recognition. In Proceedings of the 2018 IEEE/CVF Conference on Computer Vision and Pattern Recognition, Salt Lake City, UT, USA, 18–23 June 2018; pp. 8697–8710.
72. Zhang, X.; Zhou, X.; Lin, M.; Sun, J. ShuffleNet: An Extremely Efficient Convolutional Neural Network for Mobile Devices. In Proceedings of the 2018 IEEE/CVF Conference on Computer Vision and Pattern Recognition, Salt Lake City, UT, USA, 18–23 June 2018.

Article

Empirical Evaluation on Utilizing CNN-Features for Seismic Patch Classification

Chunxia Zhang [1], Xiaoli Wei [1] and Sang-Woon Kim [2,*]

[1] School of Mathematics and Statistics, Xi'an Jiaotong University, Xi'an 710049, China; cxzhang@mail.xjtu.edu.cn (C.Z.); wxl18847162006@stu.xjtu.edu.cn (X.W.)

[2] Department of Computer Engineering, Myongji University, Yongin 17058, Korea

* Correspondence: kimsw@mju.ac.kr

Featured Application: Automatic seismic fault detection, exploration of underground resources, and other areas of image recognition where large-scale artificial data is possible but real-world data is extremely limited.

Abstract: This paper empirically evaluates two kinds of features, which are extracted, respectively, with traditional statistical methods and convolutional neural networks (CNNs), in order to improve the performance of seismic patch image classification. In the latter case, feature vectors, named "CNN-features", were extracted from one trained CNN model, and were then used to learn existing classifiers, such as support vector machines. In this case, to learn the CNN model, a technique of transfer learning using synthetic seismic patch data in the source domain, and real-world patch data in the target domain, was applied. The experimental results show that CNN-features lead to some improvements in the classification performance. By analyzing the data complexity measures, the CNN-features are found to have the strongest discriminant capabilities. Furthermore, the transfer learning technique alleviates the problems of long processing times and the lack of learning data.

Keywords: seismic patch classification; CNN-features; transfer learning; data complexity

Citation: Zhang, C.; Wei, X.; Kim, S.-W. Empirical Evaluation on Utilizing CNN-Features for Seismic Patch Classification. *Appl. Sci.* **2022**, *12*, 197. https://doi.org/10.3390/app12010197

Academic Editor: Juan J. Rodríguez

Received: 2 December 2021
Accepted: 23 December 2021
Published: 25 December 2021

Publisher's Note: MDPI stays neutral with regard to jurisdictional claims in published maps and institutional affiliations.

1. Introduction

Seismic faults are important subsurface structures that have significant geologic implications for hydrocarbon accumulation and migration in a petroleum reservoir. On the basis of these characteristics, it is very important to detect faults with advanced techniques. Recently, seismic fault detection using deep-learning techniques has been actively studied [1–5]. In this approach, seismic images are first divided into patches of a certain size. The fault detection problem then becomes a two-class classification problem that classifies fault and nonfault (normal) patches. The fault detection problem can be solved by identifying the location of the patches classified as abnormal patches in the fault line. This paper focuses on the classification of patch data. First, feature vectors were extracted from seismic patch data and were then classified as fault and nonfault patches using existing classifiers to find fault lines.

Convolutional neural networks (CNNs) are now state-of-the-art approaches for a lot of applications. On the basis of their good performance, CNNs have recently been used to detect seismic faults [4]. However, two constraints can be found in this approach: one is the need to provide a huge amount of interpreted data (e.g., fault and nonfault patches); the other is the significant amount of time required to process them. To address the first, a synthetic dataset, having simple fault geometries, was built. Therein, the input to the CNN was the seismic amplitude only; that is, the approach did not require the calculation of the other seismic attributes, but the second constraint remains without any solution.

As is commonly known, CNNs take a tremendous amount of time to learn when allowing "sufficient" training data. Recently, it has been observed that the convolutional

(C) and fully connected (FC) layers take the most time to run [6]. In particular, since the latter is responsible for the multiplication of large-scale matrices, it consumes up to almost 60% of the computation time. From the above review results, as well as from the findings in [6–11], we may consider replacing the FC layers responsible for classifying seismic patch data in a CNN with an existing classifier, such as support vector machines (SVMs). The role of the C layer in this framework corresponds to, for example, the role of the principal component analysis (PCA) in traditional statistical-based classification.

The use of SVMs instead of FC layers is known to improve the classification accuracy [7,12], but no analysis has been made on why. Rather than embarking on a general analysis, in this paper, we will consider a comparative study that will be taken as the basis for the above improvements. In addition, the measures of the data complexity can be used to estimate the difficulty in separating the sample points into their expected classes. Especially, it has been reported that the measurements can be performed to figure out a variety of characteristics related to data classification [13]. From this point of view, to derive an intuitive comparison and to answer the above question, we will consider the complexity measurements.

In this paper, we use CNN-features in seismic patch classification. Using this feature vector, we can avoid the problems of a long learning time and a lack of training data, without deteriorating the classification performance. We also analyzed the data complexity to compare the discriminating powers of the features. A preliminary short version of this paper was published as a conference paper [14]. In particular, the current version has been expanded to include: First, additional techniques to improve the classification performance of CNNs; second, a quadratic pooling strategy that was reviewed and tested in order to further increase the discriminative capabilities of the CNN-features [15]; and third, new experiments were conducted and analyzed to consider how improving CNNs in classification systems that classify CNN-features using conventional classifiers changes the system performance.

In this paper, we focus on the following research issues regarding the use of CNN-feature vectors. First, we attempt to hybridize deep-learning techniques and pattern recognition (PR) methods for seismic patch classification. Traditional PR proceeds in two stages: data representation and feature classification, while CNN is implemented in two stages: feature extraction and classification. It is well known that the representation phase of PR is sensitive to the domain properties, while the classification phase of a CNN has a long training time. The main issue of this paper is the hybridization of the above two stages to address these shortcomings. On the basis of this hybrid strategy, we can combine the feature extraction capabilities of CNNs with the various classifiers that have been developed so far and can utilize the appropriate classifiers for applications.

Second, we attempt to adopt transfer learning [16] in model learning for seismic patch classification. In general, while artificially generating experimental data is not expensive, there is a limit to presenting the detailed state of the real world. On the other hand, the real-world data can reflect the true state well, but it is difficult to obtain a sufficient amount. The second research point of the paper is the use of the transfer learning technique in the learning of models for the extraction of CNN-features. That is, we first learned CNN models using well-prepared artificial data, and then leveraged them as pre-trained models to learn classification models on real-world data.

The third issue of the paper is the application of data complexity in an effort to find reasons why this hybrid approach leads to improved classification results. To achieve our purpose, we measured the data complexity of the feature vectors extracted with traditional methods, including PCA, and compared them with that of the CNN-features.

Finally, the fourth concern of the paper is to represent the seismic patches in CNN-features. To avoid the curse of dimensionality in image classification, various approaches have been used so far. The CNN-features can be utilized regardless of their sensitivity to the cardinality and dimensions of the dataset since they are extracted from CNNs.

The remainder of the paper is organized as follows: in Section 2, a brief introduction to the latest findings on seismic fault detection, using CNN-features and data complexities, is provided; in Section 3, a classification method related to this empirical study and the structure of seismic patch image data are presented, in turn; in Section 4, experimental studies, such as experimental data and methods, and classification accuracy rates and data complexity measurements, are described in detail; in Section 5, conclusive arguments and limitations that are worthy of further research are summarized.

2. Related Work

This section briefly reviews some of the latest results for detecting seismic faults, using CNN-features and data complexities for measuring the discrimination power of the features.

2.1. Seismic Fault Detection

In geophysics, current fault detection methods can be broadly categorized into two classes: traditional methods [17–19], and machine-learning-based algorithms [1–5,20–23]. In general, traditional methods work by detecting the local discontinuity in seismic images on the basis of some of the seismic features, such as semblance, variance, etc. For example, Wang and Alregib [19] propose a combination of the Hough transform and the tracking vector to extract faults from coherence maps. Although these methods can produce satisfactory results in early studies, they are time-consuming, and they also require professional background knowledge to calculate the informative attributes.

With the great success of various machine-learning and deep-learning techniques in many fields, scholars have developed many automatic fault detection procedures to alleviate the shortcomings of traditional ones. For instance, SVMs [20] and the multilayer perceptron technique [21] are separately applied to analyze the attributes (including the detected edges, and the geometric and texture features) extracted from different seismic images, and, finally, to classify whether there are faults or not. In [5], Wang et al. provide a review of the use of image-processing and machine-learning methods to identify faults. Considering that the performance of machine-learning algorithms strongly relies on the calculated attributes, it is popular, in recent years, to employ deep-learning approaches to achieve automatic fault detection. Among them, the CNN is the most commonly used technique because of its powerful ability to extract useful features. Recently, it has been combined with transfer learning [1–3] in order to further enhance its performance while alleviating the difficulty in obtaining sufficient labeled data.

2.2. CNN-Features

CNN-features are composed of values taken from the activation unit of the first FC layer of the CNN architecture [8]. Various studies have been conducted using CNN-features in many applications. In [24], it was evaluated whether the CNN weights obtained from large-scale source tasks could be transferred to a target task with a limited amount of training material. Along with this study, many other studies related to the extraction and utilization of CNN-features have been conducted [6,9,10].

In [25], it was reported that reusing a previously trained CNN as a generic feature extractor leads to a state-of-the-art result, meaning that CNNs are able to learn generic feature extractors that can be used for different tasks. Thus, some studies have recently been reported in the industry on the techniques of extracting, and classifying feature vectors with this approach [7,11]. In [7], top-performing hand-crafted descriptors, including the LBP (local binary pattern) and the HOG (histogram of oriented gradients), were compared with CNN-based models using variants of AlexNet [26]. Experiments on three datasets reported that CNN-based models were superior to other models. However, it was also pointed out that, to extract meaningful features from raw data, the approach requires huge amounts of training data. In case the generation of data is expensive, it might not be appropriate, as in [27].

In the meantime, CNNs have improved performances while building deeper and wider networks [28]. In a different way, there are studies to improve the performance by simply using a pooling based on quadratic statistics, such as covariance, rather than sum (mean) or max pooling. A few examples are: second-order pooling (O2P) [15]; bilinear pooling (BiP) [12]; compact BiP [29]; deep architecture for O2P [30]; improved BiP [31]; matrix power normalized covariance (MPN-COV) [32]; and the iterative matrix square root normalization of covariance pooling (iSQRT-COV) [33].

In [12], an effective architecture, called a bilinear CNN (B-CNN), was developed for visual classification. B-CNNs represent an image as a pooled outer product of features, derived from two CNNs, which capture the localized feature interactions. The work in [31] shows that feature normalization and domain-specific fine-tuning provide additional benefits, improving the accuracy using identical networks. In subsequent studies, various methods were proposed, in turn, to overcome the disadvantages of the B-CNN, such as the high dimensionality of the feature vectors, or GPU-unfriendly algorithms.

2.3. Data Complexity

An attempt to identify the relationship between the data and the classifier that classifies it was specifically initiated in [34]. Since then, it has been studied a lot. A few examples, but not all, can be found in [13,35,36]. Referring to these findings, we selected the data complexity measures to be used in our experiments as the following four indicators: the Fisher's discriminant ratio ($F1$); the directional-vector Fisher's discriminant ratio ($F1v$, or simply, Fv); the distance of erroneous instances to a linear classifier, or its training error rate ($L2$); and the volume of the local neighborhood ($D2$).

Each measurement of complexity mentioned above is made as follows. First, $F1$ is used to measure the separation capacity of a single feature between two classes. One way to obtain the value for multiple features is:

$$F1 = \sum_{i=1}^{2} p_i (m - m_i) C^{-1} (m - m_i), \tag{1}$$

where m_i (and m) is the mean of each class (and c_i the entire class); C is the pooled covariance matrix derived from averaging the covariance of each class; and p_i is the proportion of examples in the class, c_i. As a result, the higher the value, the less redundancy and the easier it is to distinguish between the two classes.

Second, the Fv value, developed as a complement to $F1$, is calculated as in [29]:

$$Fv = \frac{1}{1 + d^T B d / d^T W d}, \tag{2}$$

where B is the between-class scatter matrix; W is the within-class scatter matrix; and $d = W^{-1}(m_1 - m_2)$. That is, d is a directional vector in which the data are projected to maximize class separation, and the value of Fv approaches zero by maximizing $W^{-1}B$, which becomes a simpler classification. Therefore, the lower the measurement, the simpler the classification problem is.

Third, $D2$ is related to the density measurement, defined as the average number of samples per unit volume in the space where all samples are distributed [36]. The value of $D2$ in n training samples is determined by measuring the average volume occupied by the k nearest neighbors in each sample, x_i, referred to as $N_k(x_i)$. The value is counted as:

$$D2 = \frac{1}{n} \sum_{i=1}^{n} \prod_{h=1}^{d} (max(f_h, N_k(x_i)) - min(f_h, N_k(x_i))), \tag{3}$$

where $max(f_h, N_k(x_i))$ and $min(f_h, N_k(x_i))$ denote the maximum and minimum values of the feature, f_h, among the k-NN of x_i, respectively.

Finally, the *L*2 value is decided by referring to the error rate of the linear SVM classifiers. Therefore, the higher the value, the more errors, and the more difficult it is to classify linearly, which increases the complexity. Here, each measurement is briefly described to the minimum required. A detailed description of each of these measurements, and other measures that were not selected here, but that were closely related to the selected ones, can be found in the relevant literature.

3. Methods and Data

This section first introduces the classification methods associated with the current empirical research, and then presents the seismic wave image data used in this paper.

3.1. Classification Methods

The classification system to be developed in this paper (named HYB) is a method of **hybri**dizing an existing linear (or nonlinear) classifier (e.g., an SVM) with a CNN. HYB is a classification framework that performs feature extraction on CNNs, and then trains SVMs using the extracted features. In this approach, when learning the CNN in the target domain, the problem of insufficient learning data can be avoided through transfer learning, using a model learned in advance in the source domain.

The CNN architecture extracting the feature vector consists of an input layer, a hidden layer, and an FC layer. The hidden layer includes layers that perform convolution and subsampling. In this structure, the feature vectors we are trying to extract are made from input neurons directly connected to the FC layer. The extracted vector is a midlevel representation, just before the input image passes through the hidden layers and is transferred to the FC layer. Therefore, the dimension of the feature vector is equal to the number of neurons that make up the first FC layer. As a result, we can adjust the number of FC input neurons to find the optimal dimension for the given application.

To extract CNN-features as described above, the weights of the CNN that make up the C layers should be fixed in advance. To this end, transfer learning can be used [16]. In transfer learning, the cardinality of the target task's dataset is usually less than that of the source task's training dataset. Because of this, the proposed CNN-feature extraction method can avoid the general difficulty of CNN learning, which requires a large amount of data. Moreover, CNN training in the transfer learning mode is made up of fine-tuning. Therefore, the learning time can be significantly reduced with a small amount of fine-tuning.

The CNN model is learned by repeatedly performing forward propagation (FP) and backward propagation (BP), in an end-to-end manner. At this time, one of sum-, max-, or O2P-pooling is used to reduce the amount of processed data. When using O2P-pooling, the learning process is performed as follows: First, for the FP, it is assumed that a feature tensor, $X \in \mathbb{R}^{h \times w \times d}$, of the height, h, width, w, and channel, d, is generated from the last C layer, and that this tensor is reshaped into a matrix, $X \in \mathbb{R}^{n \times d}$, consisting of $n = hw$ features of the d-dimension. Then the second-order pooling, $\Sigma \in \mathbb{R}^{d \times d}$, for X is computed as:

$$\Sigma = X \bar{I} X^{\mathrm{T}}, \tag{4}$$

where $\bar{I} = \frac{1}{n}\left(I - \frac{1}{n}\mathbf{1}\right)$; I and $\mathbf{1}$ are the $n \times n$ identity matrix and the matrix of all the ones, respectively. To improve the representation power, Σ is transformed into Z by performing matrix power normalization using SVD (singular value decomposition) and is then sent to the FC layer.

Next, BP proceeds in the opposite direction. Given the gradient of a loss function, l w.r.t. Z, propagated from the FC layer, i.e., $\partial l / \partial Z$, the gradient of l w.r.t. Σ is estimated sequentially. First, we have prepared all the relevant derivations (see Equation (2) in [31]), and then we can calculate the $\partial l / \partial \Sigma$. Using this derivative, the gradient of l w.r.t. X, which is used to update the weights of the network, is determined as:

$$\frac{\partial l}{\partial X} = \bar{I}X\left(\frac{\partial l}{\partial \Sigma} + \left(\frac{\partial l}{\partial \Sigma}\right)^T\right). \tag{5}$$

Here, it should be noted that the BP contains GPU-nonfriendly computations, such as SVD, which leads to costly training. To solve this problem, fast end-to-end methods were developed, such as MPN-COV [32] and iSQRT-COV [33], which are suitable for parallel implementation.

3.2. Seismic Data

A total of 500 synthetic seismic images were prepared, and each of them contained one fault line with different slopes and positions. The corresponding fault position in each image was indicated in white by generating binary masks, referring to the seismic amplitude information. Figure 1 presents an example of a seismic wave image (a), and a fault line (b) to be extracted from it. The dataset of the synthetic seismic wave images (501 × 501 pixels), reproduced through open-source code, IPF [37], is an artificial implementation of sequential rock deformation over time.

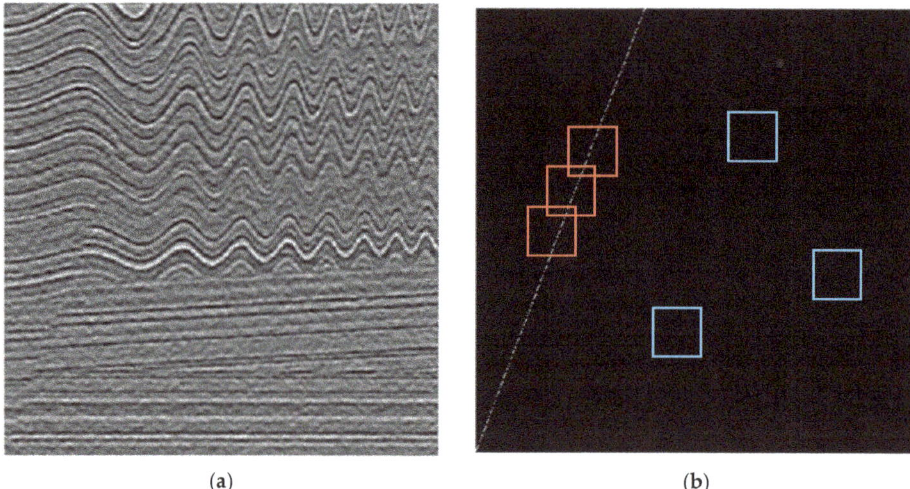

(a)   (b)

Figure 1. Plots presenting synthetic seismic wave data: (**a**) a synthetic seismic image; (**b**) a fault line. Here, two blocks, marked in the red color and in blue, indicate the fault and nonfault patches, respectively.

From the dataset composed of image pairs, shown in Figure 1 (where the two images on the left and right sides include seismic waves and fault lines, respectively), the fault and nonfault patches were extracted. One patch is a (45 × 45)-dimensional matrix with one candidate pixel in the center, and 2024 pixels adjacent to it (this is the smallest patch size generating satisfactory results [4]). This patch can be classified as a fault patch or a nonfault patch, according to the following rule: If the candidate pixel is a pixel forming a fault line, then it becomes a fault patch; otherwise, it becomes a nonfault patch. For example, referring to the fault line matrix shown in Figure 1b, all possible fault patches were first extracted, and then the same number of nonfault patches were randomly extracted from the seismic wave image, shown in Figure 1a.

4. Experimental Study

In this section, we present the evaluation results for the classification performance and data complexity of the CNN-features. These were compared to those of the feature vectors, extracted in three ways: PCA; kPCA using "Gaussian" kernel [38]; and discriminative auto-encoder (AE) [39]. kPCA and AE were selected as the nonlinear and the network-based version of the PCA, respectively, and the implementations of the original authors were used without modification. Regarding CNNs, two versions were implemented [40]. The first is a typical CNN with sum-pooling, and the second is an improved CNN using O2P-pooling. Then, the SVMs were realized from the LIBSVM [41].

4.1. Synthetic and Real Experimental Data

As mentioned above, we first generated synthetic seismic waves and fault images. From the total of 500 seismic wave images, two patch sets of 'Test1' and 'Train1' were constructed as follows: Test1 is a set of 76,038 patches extracted from the first 100 images; and Train1 is a set of 284,850 patches extracted from 400 seismic images from the remaining 101 to 500 images. Train1 (and Test1) was used to pretrain the CNN models needed to extract the CNN-features.

For real-world seismic images, we considered the real seismic wave image data, cited from the Project Netherlands Offshore F3 Block—Complete [42]. We first asked some experienced experts to manually mark the fault lines according to the structural information (the work of assigning fault lines to real seismic wave images should be done by human labelers using appropriate tools, as was performed in ImageNet [28]). Next, for a simple comparative analysis, only ten seismic images along with fault lines were selected. Then, a total of 52,026 (26,013 for each class) fault and nonfault patches were extracted from the ten seismic images. When extracting the patches, the number of patches on the larger side was adjusted to the smaller side by randomly selecting them in order to prevent an imbalance between the classes. Hereafter, this real-world patch data is referred to as "RealPatch".

4.2. CNN Models Implemented

Figure 2 and Table 1 show the details of the CNN architecture designed for the experiment. This CNN consists of two convolutional (and subsampling: SUB) layers, and one FC layer, which is one of the smallest required scales for our goal. The parameters involved in the CNN are listed in Table 1.

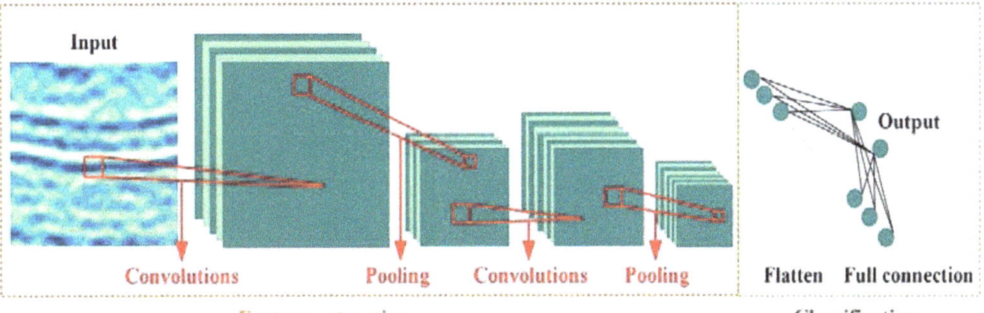

Figure 2. The architecture of the CNN model.

Table 1. The details of the CNN architecture.

Layer Descriptions	Input Size	Output Size	Feature Maps	Kernel Size
1st convolution and subsampling	45×45	40×40	4 (8)	6×6
2nd convolution and subsampling	20×20	16×16	8 (16)	5×5
Fully connected	512×1	2×1		

In Table 1, the architecture is implemented with two different models, a conventional sum-pooling CNN and an improved O2P-pooling CNN, which are referred to as "CNN-sum" and "CNN-O2P", respectively. Details related to the models are as follows: In CNN-sum, the number of output maps of the two C layers was set to 4 and 8, respectively, but, in CNN-O2P, the number was adjusted to 8 and 16, as indicated by the parentheses in the fourth column of the table.

Second, in the two SUB layers, sampling was performed by sum-pooling, which reduces the dimension of each axis by half. At this time, the CNN-sum directly connected the output (which was reshaped to a vector) of the second SUB layer to the FC layer, but, in CNN-O2P, O2P-pooling was additionally implemented, and the obtained result was connected to the FC layer. In all these operations, pooling was performed with a stride of 2.

Third, the last component at the bottom of both models is the FC layer, and the number of input neurons of this layer is defined as the number of weights output from the previous layer. The CNN-features we are trying to extract consist of these input neurons that connect directly to the FC layer. Therefore, the dimension of the feature vector can be determined by adjusting the number of these neurons.

Fourth, each model deals with the two-class problem of classifying the seismic image patches as normal and abnormal. Therefore, it is necessary to adjust the number of neuron units in the output layer when expanding to multiclass problems. In addition, a soft-max function is required to obtain the normalized probabilities for each class.

Finally, the parameters for learning the CNN models were experimentally set to a learning rate of 1, a momentum of 0.5, a batch size of 10, and the number of epochs was set to 100. The number of epochs was optimized by referring to the learning curve to prevent overfitting.

4.3. Classification Accuracy Rates

Our CNN-features were extracted as follows: We first prepared a CNN model using source training data, Data1 (# of epochs: 100). Next, we fine-tuned this pretrained model using target training data, Data2 (# of epochs: 200). Here, as for Data1, Train1 was used as it was, and Data2, of 10,000 patches (5000 per class), was randomly selected from RealPatch.

In classification work performed by a feature extractor and a classifier, in general, there is no optimal feature extractor for all classifiers, and vice versa. Thus, to validate the performance of our CNN-based extractor, we conducted a simple classification experiment using only two types of classifiers: the k-nearest neighbor rules (kNNs: k = 1, 3), and SVMs (of the polynomial and RBF kernels). (Here, each of these two classifiers was chosen as the easiest to implement and the most widely used classifiers. Although kNNs have a long execution time, the classifiers were chosen because the discriminative characteristics of the input vectors can be directly reflected in the output determination. Therefore, thorough evaluation using other classifiers, such as AdaBoost and the decision tree, is a future challenge).

Table 2 presents a numerical comparison of the classification accuracy rates (%) of the features extracted by the five methods, i.e., PCA, kPCA, AE, CNN-sum, and CNN-O2P. Here, the values were averaged after repeating 10 times. Each time, 10,000 (and 10,000) patches were randomly selected as the training (and test) data from RealPatch, and the five different types of feature vectors were extracted. For a fair comparison, the dimensions of all the feature vectors were adjusted to 256. The highest (second) accuracy of each row is emphasized in bold (underlined).

Table 2. Classification accuracy rates (mean) (%).

Classifiers (Types)	Feature Extraction Methods				
	PCA	kPCA	AE	CNN-Sum	CNN-O2P
kNN (k = 1)	92.47	93.10	86.33	<u>95.67</u>	**99.03**
kNN (k = 3)	84.52	86.56	74.59	<u>93.19</u>	**98.47**
SVM (polyn.)	80.20	<u>86.69</u>	72.69	85.88	**96.21**
SVM (RBF)	91.16	93.32	83.58	<u>95.91</u>	**97.99**

From the comparison shown in Table 2, it can be observed that the use of the CNN-based extractor, in conjunction with existing classifiers, can significantly improve the classification accuracy compared to the other extractors included in the comparison. In particular, it is noteworthy that the CNN-based extractor uses not only training data, but also pretrained CNN models that could not be used for other extractions. Under this condition, a direct comparison of the five feature extractors may not be fair. However, from the perspective of machine learning, such as transfer learning, the results of this experiment suggest one possibility related to the "new" extractor. From this consideration, the performance improvements observed in the last column of the table may have resulted from the discriminating ability of the pretrained models.

Finally, in Table 2, CNN-sum and CNN-O2P were included as feature extractors, but they can also serve as classifiers with the FC layers. The classification performances of the CNN-sum and CNN-O2P trained for the experiments in the table were 96.13 (%) and 96.61 (%), respectively.

4.4. Data Complexity Measures

To highlight the reason why the use of the CNN-features improves the classification performance, the complexity measures were also measured and compared. Figure 3 shows a comparison of the complexity measurements obtained by applying the four extractors of PCA, kPCA, AE, and a CNN (CNN-O2P or CNN-sum) to RealPatch, prepared for this experiment.

In Figure 3, the CNN (CNN-O2P or CNN-sum) bar has the highest height for the $F1$ complexity, while the PCA (or kPCA) bar has the highest for the Fv, $D2$, and $L2$ complexities. Through this comparison, we can consider as follows: First, the result of comparing the data complexity shown in Figure 3 is consistent with the result of comparing the classification accuracy shown in Table 2. That is, compared to the accuracy of PCA and CNN-features in Table 2, CNN-based is superior to PCA. In Figure 3, when comparing the $F1$ values for these two extractors, the CNN bar height is higher than that of PCA.

Second, a comparison of the Fv, $D2$, and $L2$ complexities is the opposite of $F1$: the CNN-based bar height is lower than the PCA height. Specifically, the fact that the CNN-based accuracy rate in Table 2 is greater than that of the PCA is consistent with the fact that the CNN-based $F1$ bar in Figure 3 is higher than that of the PCA. However, the Fv value is small when the accuracy rate is large. As reported in the relevant literature [13], the increasing value of $F1$ reduces the overlapping feature space and, thus, allows for a better separation of the feature space of the two classes. Unlike $F1$, the smaller the value measured in Fv, the simpler the classification problem.

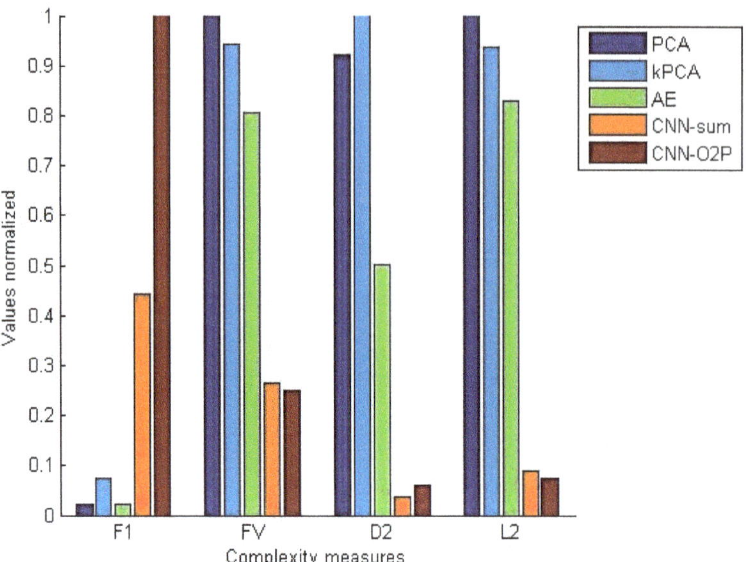

Figure 3. Plot comparing the complexity measures obtained in PCA, kPCA, AE, CNN-sum, and CNN-O2P from RealPatch. For ease of comparison, each of the complexity values measured was adjusted so that the maximum value was 1. In addition, CNN-sum and CNN-O2P were clearly compared only in F1. Thus, to clarify the focus of the paper, the two were analyzed only with a CNN.

Similar comparative analyses can be applied for the other complexity measures, $D2$ and $L2$. In general, for $D2$ and $L2$, the larger the bar height, the greater the overlap between classes, which makes classification difficult. The results of Figure 3 are very consistent with the above fact. From the observations made above, it can be argued that the CNN-based extractor is a good feature extractor when compared to the AE-based extractor, as well as when compared to the PCA (and kPCA). In particular, the extractor can be applied to situations where existing extractors do not work well because of the nature of the data. In various practices dealing with high-dimensional data, PCAs, which rely on covariance matrices, are known to be rarely used because they are not efficient.

4.5. Time Complexity

Finally, to make the comparison complete, the time complexity of the extraction methods was explored. Rather than embarking on another analysis of the computational complexities, however, the time consumption levels for the datasets were simply measured and compared. In the interest of brevity, the processing CPU times for each dataset is the time consumed by repeating the feature extraction several times and then averaging it. Table 3 presents a numerical comparison of the processing CPU times. Here, the times recorded are the required CPU times on a laptop computer with a CPU of 2.60 GHz and a RAM of 16.0 GB, and that is operating on a Window 64-bit platform.

Table 3. The processing CPU times (in seconds).

Data	PCA	kPCA	AE	CNN
Train1	8.3	7775.0	9560.8	35.5

In Table 3, it can be observed that the extractor of the CNN-features requires a much shorter processing CPU time than the traditional nonlinear algorithms for the datasets used. Particularly, this demonstrates that the extractor can dramatically reduce the processing time compared to other kernel- and network-based methods (e.g., kPCA and AE). However, it should be noted that extracting the CNN-features requires a pretrained model. In Table 3, Train1 was used for the seismic data to obtain the pretrained model (# of epochs: 100). The training time for the model was excluded when counting the processing time in the table.

4.6. Summary and Future Challenges

As a result, from the above experiment, we can observe that the classification of seismic patch data can be improved by means of hybridizing CNN-features with existing classifiers. The summary of the experimental results and future tasks are as follows.

In this approach, CNN models were first trained to extract CNN-features. To this end, we first prepared a CNN model using synthetic data and then used it as a pretrained model when learning models for the feature extraction of real-life data. Therefore, it is necessary to analyze the extent to which such transfer learning has affected the CNN-feature extraction process.

More specifically, the task of analyzing the impact of the learning results of the source domain (i.e., CNN models learned from the synthetic data) on the classification performance of the target domain is a future challenge. We also experimented with a very limited number of real-world data here. Therefore, the task of investigating the optimal cardinality of the training set required for the target domain is also an open problem.

It is also interesting to see what has happened when using enhanced CNNs to extract CNN-features with the hybrid method. Here, however, only CNN-sum and CNN-O2P were implemented in one simple CNN and its weakly enhanced CNN, respectively, and were compared to each other. Therefore, experiments on strongly enhanced CNNs, including LesNet [43] and VGG-Net [44], remain open.

In order to find out why this hybrid approach leads to improved results, we measured the data complexity of the CNN-features and compared it with that of the existing extractors. However, complexity measurements were limited here, using only a limited number of measurements. In addition, the feature-extraction performance was compared using only a few traditional extractors. Therefore, introducing more diverse complexity metrics and including handcrafted extractors in comparison, such as the LBP, the HOG, etc., is a future challenge.

5. Conclusions

In this paper, the CNN-feature was first extracted using one CNN model, and was then used to learn an existing classifier, such as an SVM. Here, the model was first pre-trained using synthetic seismic data and was then fine-tuned using real-world data. By measuring the data complexity, the answer to the question of why this approach works effectively is presented.

However, the simulation was simply performed, by referring to the smallest structure that can express basic algorithms. Therefore, a comprehensive evaluation, using various architectures, such as ResNet and VGG-Net, and a comparison with the latest results in this domain remain a challenge in the future. In addition, detecting the occurrence of negative transfer that may occur in this approach is an important topic for future work. In addition to these limitations, the problem of theoretically investigating the CNN-based extractor remains unresolved.

Author Contributions: Software, methodology, conceptualization, formal analysis, funding acquisition, C.Z.; software, methodology, investigation, formal analysis, X.W.; software, methodology, formal analysis, Writing—original draft, S.-W.K. All authors have read and agreed to the published version of the manuscript.

Funding: This work was supported, in part, by the National Key Research and Development Program of China (No. 2018AAA0102201).

Informed Consent Statement: Written informed consent has been obtained from the patient(s) to publish this paper.

Data Availability Statement: Data associated with this research are available from the corresponding author.

Acknowledgments: The work of the third author was conducted while visiting Xi'an Jiaotong University, China. The authors appreciate the providers of the software and the related data used in this study, and, especially, Dave Hale for making the IPF codes available online. We also appreciate the hard work of the anonymous referees. Their detailed comments greatly improved this paper.

Glossary

Abbreviation	Meaning
AE	auto-encoder
B-CNN	bilinear convolutional neural network
BP	backward propagation
BiP	bilinear pooling
C	convolutional
CNN	convolutional neural network
FC	fully connected layer
FP	forward propagation
GPU	graphics computing units
HOG	histogram of oriented gradients
HYB	hybrid method of combining linear (nonlinear) classifier with CNN
iSQRT-COV	iterative matrix square root normalization of covariance pooling
LBP	local binary pattern
MPN-COV	matrix power normalized covariance pooling
O2P	second-order pooling
PCA	principal component analysis
PR	pattern recognition
SVM	support vector machine

References

1. Cunha, A.; Pochet, A.; Lopes, H.; Gattass, M. Seismic fault detection in real data using transfer learning from a convolutional neural network pre-trained with synthetic seismic data. *Comput. Geosci.* **2020**, *135*, 104344. [CrossRef]
2. Di, H.; Wang, Z.; AlRegib, G. Seismic fault detection from post-stack amplitude by convolutional neural networks. In Proceedings of the 80th EAGE Conference & Exhibition, Copenhagen, Denmark, 11–14 June 2018; pp. 1–5. [CrossRef]
3. Hung, L.; Dong, X.; Clee, T. A scalable deep learning platform for identifying geologic features from seismic attributes. *Lead. Edge* **2017**, *36*, 249–256. [CrossRef]
4. Pochet, A.; Diniz, P.H.B.; Lopes, H.; Gattass, H. Seismic fault detection using convolutional neural networks trained on synthetic poststacked amplitude maps. *IEEE Geosci. Remote Sens. Lett.* **2019**, *16*, 352–356. [CrossRef]
5. Wang, Z.; Di, H.; Shafiq, M.A.; Alaudah, Y.; AlRegib, G. Successful leveraging of image processing and machine learning in seismic structural interpretation: A review. *Lead. Edge* **2018**, *37*, 451–461. [CrossRef]
6. Donahue, J.; Jia, Y.; Vinyals, O.; Hoffman, J.; Zhang, N.; Tzeng, E.; Darrell, T. DeCAF: A deep convolutional activation feature for generic visual recognition. In Proceedings of the 31st International Conference on Machine Learning (ICML), Beijing, China, 21–26 June 2014; pp. 647–655.
7. Alshazly, H.; Linse, C.; Barth, E.; Martinetz, T. Handcrafted versus CNN features for ear recognition. *Symmetry* **2019**, *11*, 1493. [CrossRef]
8. Athiwaratkun, B.; Kang, K. Feature representation in convolutional neural networks. *arXiv* **2015**, arXiv:1507.02313.
9. Girshick, R.; Donahue, J.; Darrell, T.; Malik, J. Rich feature hierarchies for accurate object detection and semantic segmentation. In Proceedings of the 2014 IEEE Conference on Computer Vision and Pattern Recognition (CVPR), Columbus, OH, USA, 23–28 June 2014; pp. 580–587. [CrossRef]
10. Razavian, L.; Azizpour, H.; Sullivan, J.; Carlsson, S. CNN features off-the-shelf: An astounding baseline for recognition. In Proceedings of the IEEE Conference on Computer Vision and Pattern Recognition Workshops, Columbus, OH, USA, 23–28 June 2014; pp. 806–813. [CrossRef]
11. Weimer, D.; Scholz-Reiter, B.; Shpitalni, M. Design of deep convolutional neural network architectures for automated feature extraction in industrial inspection. *CIRP Ann. Manuf. Technol.* **2016**, *65*, 117–120. [CrossRef]

12. Lin, T.-Y.; RoyChowdhury, A.; Maji, S. Bilinear CNN models for fine-grained visual recognition. In Proceedings of the 2015 IEEE International Conference on Computer Vision (ICCV), Santiago, Chile, 7–13 December 2015; pp. 1449–1457. [CrossRef]
13. Lorena, A.C.; Garcia, L.P.F.; Lehmann, J.; Souto, M.C.P.; Ho, T.K. How complex is your classification problem? A survey on measuring classification complexity. *ACM Comput. Surv.* **2018**, *52*, 1–34. [CrossRef]
14. Zhang, C.-X.; Wei, X.-L.; Kim, S.-W. Empirical evaluation on utilizing CNN-features for seismic patch classification. In Proceedings of the 10th International Conference on Pattern Recognition Applications and Methods (ICPRAM), Online, 4–6 February 2021; pp. 166–173. [CrossRef]
15. Carreira, J.; Caseiro, R.; Batista, J.; Sminchisescu, C. Semantic segmentation with second-order pooling. In Proceedings of the 12th European Conference on Computer Vision (ECCV), Florence, Italy, 7–13 October 2012; pp. 430–443. [CrossRef]
16. Weiss, K.; Khoshgoftaar, T.M.; Wang, D. A survey of transfer learning. *J. Big Data* **2016**, *3*, 9. [CrossRef]
17. Marfurt, K.; Kirlin, R.; Farmer, S.; Bahorich, M. 3-D seismic attributes using a semblance-based coherency algorithm. *Geophysics* **1998**, *63*, 1150–1165. [CrossRef]
18. Marfurt, K.J.; Sudhaker, V.; Gersztenkorn ACrawford, K.D.; Nissen, S.E. Coherency calculations in the presence of structural dip. *Geophysics* **1999**, *64*, 104–111. [CrossRef]
19. Wang, Z.; Alregib, G. Interactive fault extraction in 3-D seismic data using the Hough Transform and tracking vectors. *IEEE Trans. Comput. Imaging* **2017**, *3*, 99–109. [CrossRef]
20. Di, H.; Amir Shaq, M.; AlRegib, G. Seismic fault detection based on multi-attribute support vector machine analysis. In *SEG Technical Program Expanded Abstracts*; SEG: Houston, TX, USA, 2017; pp. 2039–2244. [CrossRef]
21. Di, H.; Shaq, M.; AlRegib, G. Patch-level MLP classification for improved fault detection. In *SEG Technical Program Expanded Abstracts*; SEG: Anaheim, CA, USA, 2018; pp. 2211–2215. [CrossRef]
22. An, Y.; Guo, J.; Ye, Q.; Childs, C.; Walsh, J.; Dong, R. Deep convolutional neural network for automatic fault recognition from 3D seismic datasets. *Comput. Geosci.* **2021**, *153*, 104776. [CrossRef]
23. Wei, X.; Zhang, C.; Kim, S.; Jing, K.; Wang, Y.; Xu, S.; Xie, Z. Seismic fault detection using convolutional neural networks with focal loss. *Comput. Geosci.* **2022**, *158*, 104968. [CrossRef]
24. Oquab, M.; Bottou, L.; Laptev, I.; Sivic, J. Learning and transferring mid-level image representations using convolutional neural networks. In Proceedings of the 2014 IEEE Conference on Computer Vision and Pattern Recognition (CVPR), Columbus, OH, USA, 6–7 September 2014; pp. 1717–1724. [CrossRef]
25. Hertel, L.; Barth, E.; Käster, T.; Martinetz, T. Deep convolutional neural networks as generic feature extractors. In Proceedings of the 2015 International Joint Conference on Neural Networks (IJCNN), Killarney, Ireland, 12–17 July 2015; pp. 1–4. [CrossRef]
26. Krizhevsky, A.; Sutskever, I.; Hinton, G.E. ImageNet classification with deep convolutional neural networks, Advances. In Proceedings of the 26th Annual Conference on Neural Information Processing Systems (NIPS), Lake Tahoe, NV, USA, 3–6 December 2012; pp. 1097–1105.
27. Sun, C.; Shrivastava, A.; Singh, S.; Gupta, A. Revisiting unreasonable effectiveness of data in deep learning era. In Proceedings of the 16th IEEE International Conference on Computer Vision (ICCV), Venice, Italy, 22–29 October 2017; pp. 843–852. [CrossRef]
28. Tan, M.; Le, Q.V. EfficientNet: Rethinking model scaling for convolutional neural networks. In Proceedings of the International Conference on Machine Learning Research (PMLR), Vancouver, CA, USA, 9–15 June 2019; Volume 97, pp. 6105–6114.
29. Gao, Y.; Beijbom, O.; Zhang, N.; Darrell, T. Compact bilinear pooling. In Proceedings of the 2016 IEEE Conference on Computer Vision and Pattern Recognition (CVPR), Las Vegas, NV, USA, 27–30 June 2016; pp. 317–326. [CrossRef]
30. Ionescu, C.; Vantzos, O.; Sminchisescu, C. Training deep networks with structured layers by matrix backpropagation. *arXiv* **2016**, arXiv:1509.07838v4. [CrossRef]
31. Lin, T.-Y.; Maji, S. Improved bilinear pooling with CNNs. In Proceedings of the 31st British Machine Vision Conference (BMVC), London, UK, 7–10 September 2017. [CrossRef]
32. Li, P.; Xie, J.; Wang, Q.; Zuo, W. Is second-order information helpful for large-scale visual recognition? In Proceedings of the 2017 IEEE International Conference on Computer Vision (ICCV), Venice, Italy, 22–29 October 2017. [CrossRef]
33. Li, P.; Xie, J.; Wang, Q.; Zuo, W. Towards faster training of global covariance pooling networks by iterative matrix square root normalization. In Proceedings of the 2018 IEEE/CVF Conference on Computer Vision and Pattern Recognition, Salt Lake City, UT, USA, 18–23 June 2018; pp. 947–955. [CrossRef]
34. Ho, T.K.; Basu, M. Complexity measures of supervised classification problems. *IEEE Trans. Pattern Anal. Mach. Intell.* **2002**, *24*, 289–300. [CrossRef]
35. Cano, J.-R. Analysis of data complexity measures for classification. *Expert Syst. Appl.* **2013**, *40*, 4820–4831. [CrossRef]
36. Sotoca, J.M.; Mollineda, R.A.; Sánchez, J.S. A meta-learning framework for pattern classification by means of data complexity measures. *Revista Iberoamericana de Inteligencia Artificial* **2006**, *10*, 31–38. [CrossRef]
37. Hale, D. Seismic Image Processing for Geologic Faults. 2014. Available online: https://github.com/dhale/ipf (accessed on 28 August 2020).
38. Wang, Q. Kernel principal component analysis and its applications in face recognition and active shape models. *arXiv* **2012**, arXiv:1207.3538.
39. Gogna, A.; Majumdar, A. Discriminative autoencoder for feature extraction: Application to character recognition. *Neural Process. Lett.* **2019**, *49*, 1723–1735. [CrossRef]

40. Palm, R.B. Prediction as a Candidate for Learning Deep Hierarchical Models of Data. 2012. Available online: https://github.com/rasmusbergpalm/DeepLearnToolbox (accessed on 15 October 2020).
41. Chang, C.-C.; Lin, C.-J. LIBSVM: A library for support vector machines. *ACM Trans. Intell. Syst. Technol.* **2011**, *2*, 1–27. Available online: https://www.csie.ntu.edu.tw/~jlin/libsvm (accessed on 8 February 2021). [CrossRef]
42. Netherland Offshore F3 Block Complete. Available online: https://www.opendtect.org/osr/Main/NetherlandsOffshoreF3BlockComplete4GB (accessed on 15 September 2021).
43. Lecun, Y.; Bottou, L.; Bengio, Y.; Haffner, P. Gradient-based learning applied to document recognition. *Proc. IEEE* **1998**, *86*, 2278–2324. [CrossRef]
44. Simonyan, K.; Zisserman, A. Very deep convolutional networks for large-scale image recognition. In Proceedings of the International Conference on Learning Representations (ICLR), Banff, AB, Canada, 14–16 April 2014; pp. 1–14.

Article

Foreground Objects Detection by U-Net with Multiple Difference Images

Jae-Yeul Kim [1] and Jong-Eun Ha [2,*]

1   Graduate School of Automotive Engineering, Seoul National University of Science and Technology, Seoul 01811, Korea; jaeyorkim@naver.com
2   Department of Mechanical and Automotive Engineering, Seoul National University of Science and Technology, Seoul 01811, Korea
*   Correspondence: jeha@seoultech.ac.kr

Abstract: In video surveillance, robust detection of foreground objects is usually done by subtracting a background model from the current image. Most traditional approaches use a statistical method to model the background image. Recently, deep learning has also been widely used to detect foreground objects in video surveillance. It shows dramatic improvement compared to the traditional approaches. It is trained through supervised learning, which requires training samples with pixel-level assignment. It requires a huge amount of time and is high cost, while traditional algorithms operate unsupervised and do not require training samples. Additionally, deep learning-based algorithms lack generalization power. They operate well on scenes that are similar to the training conditions, but they do not operate well on scenes that deviate from the training conditions. In this paper, we present a new method to detect foreground objects in video surveillance using multiple difference images as the input of convolutional neural networks, which guarantees improved generalization power compared to current deep learning-based methods. First, we adjust U-Net to use multiple difference images as input. Second, we show that training using all scenes in the CDnet 2014 dataset can improve the generalization power. Hyper-parameters such as the number of difference images and the interval between images in difference image computation are chosen by analyzing experimental results. We demonstrate that the proposed algorithm achieves improved performance in scenes that are not used in training compared to state-of-the-art deep learning and traditional unsupervised algorithms. Diverse experiments using various open datasets and real images show the feasibility of the proposed method.

Keywords: visual surveillance; deep learning; object detection

Citation: Kim, J.-Y.; Ha, J.-E. Foreground Objects Detection by U-Net with Multiple Difference Images. *Appl. Sci.* **2021**, *11*, 1807. https://doi.org/10.3390/app11041807

Academic Editor: Cheonshik Kim
Received: 29 December 2020
Accepted: 10 February 2021
Published: 18 February 2021

1. Introduction

In video surveillance, the main aim is to detect foreground objects, such as pedestrians, vehicles, animals, and other moving objects. This can be used for object tracking or behavior analysis by further processing. Foreground detection in video surveillance is usually done by comparing a background model image and the current image. Traditional approaches to video surveillance require many steps, including initialization, representation, maintenance of a background model, and foreground detection operation [1–3]. Illumination changes, camera jitter, camouflage, ghost object motion, and hard shadows make the robust detection of foreground objects difficult in video surveillance. Many approaches have been proposed to cope with these problems. Since the introduction of deep learning, it has also been adopted in video surveillance. Most algorithms are supervised, while most traditional algorithms are unsupervised. Methods based on deep learning have led to a huge improvement in video surveillance like other domains of image classification, detection, and recognition. However, the use of deep learning in video surveillance has two disadvantages. One is that they have little generalization power. Deep learning achieves improved results compared to the traditional machine learning algorithm, but it still requires improvement in the generalization power. Domain transfer algorithms shows some improvement in this problem. It is well known that, as more training data are used, more accurate results can be obtained through deep learning. The other disadvantage is that deep learning requires a lot of labeled data. In video surveillance, it requires pixel-level labeled data, which are more expensive one than those of image classification and detection. Recently, various datasets that satisfy this requirement have been established with diverse scenarios for video surveillance. In this study, we used the CDnet 2014 dataset [4]. It consists of 53 scenes that cover diverse situations in video surveillance. Typical deep learning algorithms for video surveillance train a new model for each scene using some portion of the data and apply it to the remaining images.

Our goal is to achieve improved generalization power in comparison to recent deep learning-based algorithms in video surveillance. The main contribution of the proposed method is summarized as follows. We present a deep learning-based approach which shows better generalization power than the traditional non-deep learning-based state-of-the-art approach. Deep learning-based approaches achieve better performance than the non-deep learning-based traditional state-of-the-art approach on scenes that are similar to learning environments. However, it requires foreground label images, which require designation per pixel. Therefore, the preparation of training data requires a huge amount of time, while traditional non-deep learning-based algorithms do not require training images. When they are applied to scenes that are different from the training environment without new training on that scene, they show even worse performance than the traditional approach. We present a deep learning-based algorithm that achieves better generalization performance than the traditional non-deep learning-based state-of-the-art approach, and at the same time, it does not require training images. This is possible due to two factors. One is to use multiple difference images as the input of U-Net. The other is to train networks using all training samples from publicly open datasets in visual surveillance. We show the feasibility of the presented method through diverse experimental results.

The rest of the paper is organized as follows. Section 2 gives related works, Section 3 shows the proposed algorithm. The experimental results are shown in Section 4 and, finally, Section 5 gives conclusions.

2. Related Works

Background subtraction and foreground detection in video surveillance have been studied widely. Good surveys of this research are available [5–8]. We divide them into two groups, namely, approaches that do not use deep learning and those that are based on deep learning.

2.1. Earlier Approaches

Stauffer and Grimson [9] proposed a method called mixture of Gaussian (MOG) that represents the brightness value of each pixel as the combination of Gaussian distributions. They suggested a method to determine the number of the Gaussian mixture and each parameter of the Gaussian distribution using the expectation and maximization algorithm [10]. No special initialization is required because it adapts their parameters as a sequence goes on. Pixels are considered as background when their brightness values belong to the Gaussian mixture model, otherwise, they are considered as foreground. Elgammal et al. [11] proposed a probabilistic non-parametric method using kernel density estimation. Barnich et al. [1] introduced a sample-based method in background modeling. Samples from previous predefined frames are used in background modeling. If there is a predefined group of samples that is close to the current pixel, then it is considered as background, otherwise, it is considered as foreground. Kim et al. [12] proposed a method that uses a codebook. At the initial stage, codewords from intensity, color, and temporal features are constructed. They build up a codebook for later segmentation. The current frame's pixel values of intensity and color are compared to those of the codewords in the code book. Finally, a foreground or background label is assigned to each pixel by comparing the distance with codewords in the codebook. In the case of a background pixel, the matching codeword is updated. Oliver et al. [13] proposed a method based on principal component analysis, which is called the eigenbackground. The mean and the covariance matrix are computed using a predefined number of images. Here, N eigenvectors are chosen corresponding to the N largest eigenvalues, and they are used as the background model. Incoming images are projected into those eigenvectors, and their distance in those spaces is used to identify the foreground and background.

Wang et al. [14] proposed a method that uses a Gaussian mixture model for the background and uses single Gaussian for the foreground. They employed a flux tensor [15] that can explain variations of optic flow within a local 3D spatio-temporal volume, and it is used in detecting blob motion. With information from blob motion, foreground and background models are integrated to find moving and static foreground objects. Additionally, edge matching [16] is used to classify static foreground objects as ghosts or intermediate motions. Varadarajan et al. [3] proposed a method that applies a region-based mixture of Gaussians for foreground object segmentation to cope with the sensitivity of the dynamic background. Additionally, Chen et al. [17] proposed an algorithm that uses a mixture of Gaussians in a local region. At each pixel level, the foreground and background are modeled using a mixture of Gaussians. Each pixel is determined to be foreground or background by finding the highest probability of the center pixel around an N × N region.

Sajid and Cheung [18] proposed an algorithm to cope with sudden illumination changes by using multiple background models through single Gaussians and different color representations. K-means clustering is used to classify the pixels of input images. For each pixel, K models are compared, and the group that shows the highest normalized cross-correlation is chosen. An RGB and YCbCr color frame is used, and segmentation is done for each color, which yields six segmentation masks. Finally, background segmentation is performed by integrating all available segmentation masks.

Hofmann et al. [19] proposed an algorithm that improves Barnich et al. [1]. They replace the global threshold R with an adaptive threshold R(x) that depends on the pixel location and a metric of the background model which is called background dynamics. The threshold R(x) and the model update rate are determined by a feedback loop using the additional information from the background dynamics. They showed that it can cope with a dynamic background and highly structured scenes. Tiefenbacher et al. [20] proposed an algorithm that improves the algorithm introduced by Hofmann et al. [19] by controlling the updates of the pixel-wise thresholds using a PID controller. St-Charles et al. [2] also proposed an improved algorithm by using local binary similarity patterns [21] as additional features of pixel intensities and slight modification of the update mechanism of the thresholds and the background model.

2.2. Deep Learning-Based Approaches

Braham and Droogenbroeck [20] proposed the first scene-specific convolutional neural network (CNN)-based algorithm for background subtraction. A fixed background model is generated by a temporal median operation over the first 150 video frames. Then, image patches centered on each pixel are extracted from both the current and background images. The combined patches are used as the input of the trained CNN, and it outputs the probability of foreground. They evaluated their algorithm on the 2014 ChangeDetection.net dataset (CDnet 2014) [22]. The CNN requires training for each sequence in CDnet 2014. It requires a long computation time because patches from each pixel are required to pass the CNN, and it is similar to the sliding window approach in object detection. Babaee et al. [23] proposed a method that uses a CNN to perform the segmentation of foreground objects, and they use a background model that is generated using the SuBSENSE [2] and Flux Tensor [14] algorithms. Spatial median filtering is used for the post-processing of the network outputs. Wang et al. [24] proposed multi-scale convolutional neural networks with cascade structure for background subtraction. Additionally, they trained a network for each video in the CDnet 2014 dataset. More recently, Lim et al. [25] proposed an encoder–decoder-type neural network for foreground segmentation called FgSegNet. It uses a pretrained convolutional network of VGG-16 [26] as the encoding part with a triplet network structure. In the decoding part, a transposed convolutional network is used. Their network is trained by randomly selecting some training samples for each video in CDnet 2014.

Zeng et al. [27] proposed a multi-scale fully convolutional network architecture that takes advantage of various layer features for background subtraction. Zheng et al. [28] proposed an algorithm that combines traditional background subtraction and semantic segmentation [29]. The output of semantic segmentation is used to update the background model through feedback. Their result shows that it achieves the best performance among unsupervised algorithms in CDnet 2014. Sakkos et al. [30] presented a robust model that consists of a triple multi-task generative adversarial network (GAN) that can detect foreground even in exceptionally dark or bright scenes and in continuously varying illumination. They generate low- and high-brightness image pairs using the gamma function from a single image and use them in training by simultaneously minimizing GAN loss and segmentation loss. Patil et al. [31] proposed a motion saliency foreground network (MSFgNet) to estimate the background and to find the foreground in video frames. Original video frames are divided into a number of small video streams, and the background is estimated for each divided video stream. The saliency map is computed using the current video frame and the estimated background. Finally, an encoder–decoder network is used to extract the foreground from the estimated saliency maps. Varghese et al. [32] investigated visual change, aiming to accurately identify variations between a reference image and a new test image. They proposed a parallel deep convolutional neural network for localizing and identifying the changes between image pairs.

Akilan et al. [33] proposed a 3D convolutional neural network with long short-term memory (LSTM) to include temporal information in a deep learning framework for background subtraction. This is similar to our approach in terms of using temporal information. We use multiple difference images as the input of networks, while they extracted temporal information by LSTM. Yang et al. [34] proposed a method to apply multiple images to fully convolutional networks (FCNs). When selecting multiple input images, images close to the current are selected more. The studies in [33,34] belong to the method of using multiple input images in the same way as the proposed method. In the case of [33,34], multiple original images are used, whereas the proposed method is different in using multiple difference images.

3. Proposed Method

Unlike general object segmentation, proper acquisition of temporal information as well as spatial information is essential for robust foreground object detection in video surveillance. If we rely only on spatial information in the foreground object detection process, it may be difficult to determine whether the vehicle is moving or not. However, this problem can be solved if temporal information from past images is used.

Figure 1 shows the difference images between the current image and a number of past images. Using only spatial information existing in the current image has a limitation in distinguishing between the driving vehicle in the red box in Figure 1 and the parked vehicle in the blue box. On the other hand, when the difference image is used as input data for a deep learning model, it is possible to distinguish between a moving object and a stationary object. However, as can be seen from the difference images in Figure 1, there is a problem in that both the location where the foreground object existed in the past and the location that existed in the present view are displayed in the difference image between the current image and the past image. In addition, elements such as snow and rain and dynamic background objects such as moving bushes in bad weather conditions show high difference values even though they are background objects. In order to solve these problems, the proposed method uses many difference images, not a single difference image, as input data.

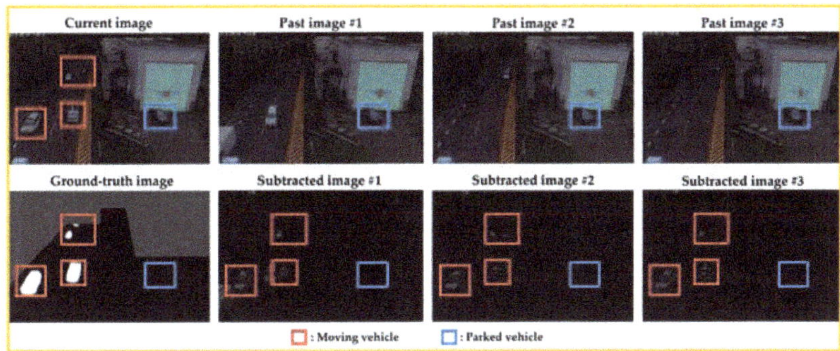

Figure 1. Use of temporal information by multiple difference images.

We adopt U-Net [35], which uses a gray or color image as the input of the network. We adjust it to use multiple difference images. Figure 2 shows the overall structure of the proposed algorithm. A network structure that uses multiple difference images as the input of U-Net [35] is shown in Figure 3. Difference images are obtained by subtracting each past image from the current image. The total number of difference images and the frame interval in subtraction are the hyper-parameters. We choose them through the analysis of experimental results. We choose 10 difference images as the number of inputs of the network through experiments. The size of the input is changed to 320(W) $\times$ 240(H) $\times$ 10(C), while the original U-Net uses input images of 572(W) $\times$ 572(H) $\times$ 1(C). U-Net [35] does not use padding in the convolutional layer and uses "copy and crop" in the layer connection process, so it outputs an image of 388x388 in size, which is different from the input image size of 572 $\times$ 572. In visual surveillance, all areas of the image need to be classified into foreground or background. Therefore, we prevented the size reduction of the output according to the convolutional layer by using padding in all layers of U-Net, and layers were connected using concatenation without cropping to make the size of the input image and the output image the same. The size of the input image was 320 $\times$ 240, which is an image size mainly used in the visual surveillance.

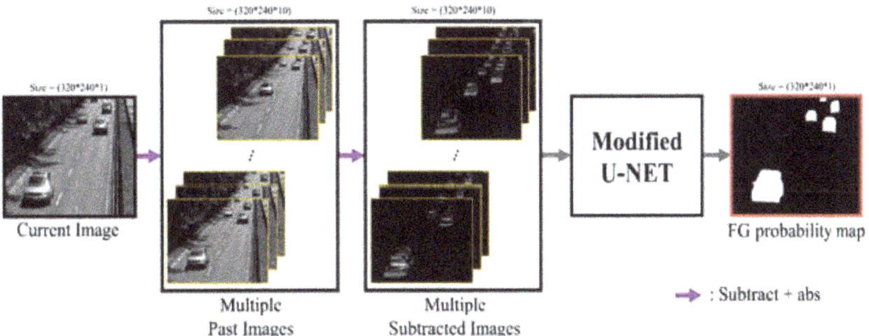

Figure 2. Foreground object detection by U-Net with multiple difference images as input.

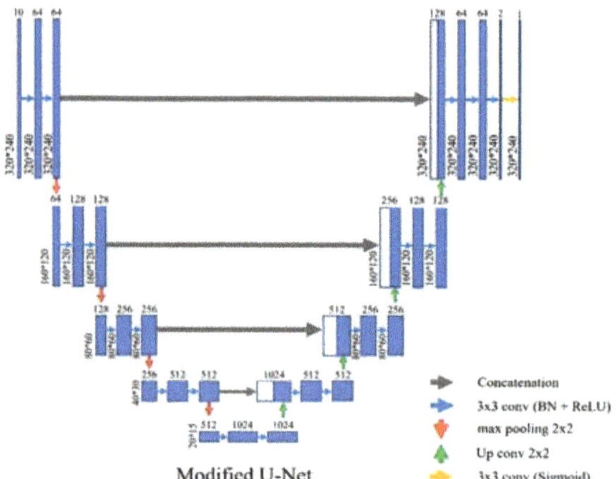

Figure 3. The structure of modified U-Net.

Batch normalization is used between each convolutional layer and the nonlinear function. Here, 2×2 max polling is used and the filter size of all convolution layers is 3×3. A rectified linear unit (ReLU) is used as the activation function in all layers except the last layer where a sigmoid function is used. We use the sigmoid function on the final layer to make the foreground and background map have a value between 0 and 1. The output of the final convolution layer gives the segmentation map by the sigmoid function. Finally, a segmentation map of $320(W) \times 240(H) \times 1(C)$ is obtained. The total number of parameters of the proposed structure is 31,064,261, and the number of learnable parameters is 31,050,565.

Binary cross-entropy is used as the loss, which is defined as

$$L = \frac{1}{N} \sum_{i=0}^{N} -(y_i log \hat{y}_i + (1 - y_i) \log(1 - \hat{y}_i)) \tag{1}$$

where y_i is the ground truth label of the i-th pixel is, $\hat{y}_i$ is the label estimated by networks, and N is the total number of pixels in the image. We train the proposed structure using the CDnet 2014 dataset, and 24 scenes are selected from the total of 53 scenes. We select 200 images for each scene and randomly divide them into 160 training images and 40 validation images. When 24 scenes consisting of 4800 images are used in the training, 3840 images and 960 images are used for the training and validation, respectively.

The Keras framework [36] with TensorFlow as a backend is used in implementation. The initial values of parameters in the networks are initialized using the He normal initializer [37]. We do not use the pretrained weights of VGG-16 [26] for our model because we use multiple difference images as the input, while VGG deals with raw input images. We train our network using the Adam optimizer [38] with an initial learning rate as 0.001, β_1 as 0.9, β_2 as 0.999, and ε as 10^{-8}. If the validation loss does not decrease in five successive epochs, the learning rate is reduced by half. The learning process is stopped if the validation loss does not decrease in 10 successive epochs within the maximum of 100 epochs. The CDnet 2014 dataset provides four labels of static, hard shadow, outside region of interest, and unknown motion as the ground truth of the segmentation map. Preprocessing is performed to divide the pixel value of the ground truth images by 255. We set static as 0 and motion as 1 in the computation of loss. The outside region of interest area and the unknown motion are not used in the computation of loss.

4. Experimental Results

In the experiment, the proposed algorithm is compared with the traditional algorithms of SuBSENSE [2], CwisarD [39], Spectral360 [40], GMM [9], and PAWCS [41] and deep learning-based algorithms of FgSegNet-v2 [42] and modified FgSegNet-v2. The original FgSegNet-v2 [42] algorithm uses one RGB image as the input of a network. We modify it to use multiple difference images as the input of a network, like the proposed algorithm, and we denote it as modified FgSegNet-v2. Data for training consisted of a training set and validation set, and the performance of each algorithm was evaluated using a test set that was not used for training.

The following experiment was performed to show the performance of the proposed algorithm.

(1) Comparison when using multiple original images and multiple difference images as the input of a network; we show the superiority of the proposed algorithm through this.

(2) Comparison between learning using data obtained in one environment and learning using all data obtained in various environments; we show that the proposed algorithm gives improved results in unknown scenes.

Experiments were done using the CDnet 2014 dataset [4]. They consisted of 53 scenes from 11 categories, as shown in Table 1, and they dealt with diverse situations that could occur during visual surveillance. We evaluated the foreground object detection algorithms using a variety of metrics that are widely used in visual surveillance, namely, recall, precision, F-measure (FM), percentage of wrong classification (PWC), false positive rate (FPR), false negative rate (FNR), and specificity (SP):

$$\text{Precision} = \frac{TP}{TP + FP}$$

$$\text{Recall} = \frac{TP}{TP + FN}$$

$$\text{FM} = 2 \times \frac{Precision \times Recall}{Precision + Recall}$$

$$\text{PWC} = \frac{FP + FN}{TP + TN + FP + FN}$$

$$\text{FPR} = \frac{FP}{FP + TN}$$

$$\text{FNR} = \frac{FN}{TP + FN}$$

$$\text{SP} = \frac{TN}{TN + FP}$$

where TP, TN, FP, and FN mean true positive, true negative, false positive, and false negative, respectively.

Table 1. List of scenes in the CDnet 2014 dataset (bold indicates the scene used in training).

Categories (Total Number of Scenes/Number of Scenes Used for Training)	Scene Names
Baseline (4/2)	**Highway, Office**, Pedestrians, PETS2006
Camera Jitter (4/2)	**Badminton, Sidewalk**, Traffic, Boulevard
Bad Weather (4/2)	**Skating, Wet snow**, Blizzard, Snowfall
Dynamic Background (6/3)	**Boats, Canoe, Fountain1**, Fountain2, Fall, Overpass
Intermittent Object Motion (6/2)	**Abandoned box, Street light**, Parking, Sofa, Tram stop, Winter drive way
Low Framerate (4/2)	**Port_0_17 fps, Tram crossroad_1 fps**, Tunnel exit_0_35fps, Turnpike_0_5fps

Table 1. *Cont.*

Categories (Total Number of Scenes/Number of Scenes Used for Training)	Scene Names
Night Videos (6/3)	**Bridge entry, busy boulevard, fluid highway**, Street corner at night, Tram station, Winter street
PTZ (4/0)	Continuous pan, Intermittent pan, Two position ptz cam, Zoom in zoom out
Shadow (6/3)	**Back door, Copy machine, Bungalows**, Bus station, Cubicle, People in shade
Thermal (5/3)	**Corridor, Library, Lakeside**, Dining room, Park
Turbulence (4/2)	**Turbulence0, Turbulence1**, Turbulence2, Turbulence3

In Table 1, bold letters represent the 24 scenes used in the training. We used 200 images for each scene in training. Test statistics are obtained by using scenes not used in the training in Table 1. Ten difference images under a five-frame interval were used as the input of networks. Subtracting by mean was used for the preprocessing of the input data. In both cases, four scenes of the pan–tilt–zoom (PTZ) category were not used in the training. The proposed method has a weakness for the category of PTZ where images are obtained through panning of the camera. In video surveillance, cameras are usually fixed at a predefined location. Scenes in the PTZ category are not common situations in video surveillance. Therefore, experiments are done using 49 scenes from 10 categories, excluding the PTZ category. Computation was done using one Intel i7-7820X CPU and an NVIDIA RTX 2080Ti GPU. The computation time for each input was 30 ms, which was obtained by averaging the processing time of 100 trials. In the case of FgSegNet-v2 [42], a deep learning-based method, it took 9 ms to process one image in the same PC environment. This is a faster processing speed than the proposed method, but the proposed method shows much better generalization ability in the real environment. Additionally, the proposed method can be computed over 30 fps. Therefore, it was judged that the proposed method has an appropriate level of model size and computational cost to use in a real environment.

First, we present the experimental results of training using only one scene in the CDnet 2014 dataset. After training using one scene, we applied it to other scenes to assess the generalization ability. Table 2 shows a comparison of the proposed method and FgSegNet-v2 [42], which produces state-of-the-art results on the CDnet 2014 dataset. FgSegNet-v2 was used to train a separate network for each scene in the CDnet 2014 dataset, and test statistics were obtained for each scene. The proposed method and FgSegNet-v2 were trained using only a highway scene in the CDnet 2014 dataset. FgSegNet-v2 uses one RGB image as input, while the proposed algorithm uses 10 difference images as input. FgSegNet-v2 achieved a better result than the proposed algorithm in a scene that was used in the training. Though the proposed method showed dramatic improvement in comparison to FgSegNet-v2 for other scenes that were not used in the training, the overall performance of the proposed method still requires improvement because it achieves much lower performance than SuBSENSE [2]. We can conclude that training using only a highway scene does not guarantee generalization power for other scenes. Therefore, we trained the proposed algorithm using all the scenes except the PTZ category in the CDnet 2014 dataset to improve the generalization ability.

Table 2. Comparison of results obtained by training using one scene (Highway) in the CDnet 2014 dataset.

	Highway Scene		All Scenes		Scenes Not Used in Training	
	FM	PWC	FM	PWC	FM	PWC
Proposed	0.99	0.08	0.47	3.61	0.46	3.68
FgSegNet-v2 [42]	1.00	0.02	0.25	11.1	0.23	11.4

Tables 3 and 4 show the result of the proposed algorithm, which was trained using 24 scenes in CDnet 2014 dataset. The proposed method shows superior performance compared to other algorithms, with FM scores of 0.927 and 0.895, respectively, even in "Bad Weather" and "Dynamic Background" categories, where a large amount of noise is included in the difference image.

Table 3. Results of proposed method which is trained using 24 scenes in CDnet 2014 dataset.

Categories	FM	PWC	Recall	Precision	FPR	FNR	SP
Baseline	0.9535	0.1301	0.9481	0.9597	0.0006	0.0519	0.9994
Camera Jitter	0.7759	3.5410	0.7563	0.8461	0.0084	0.2437	0.9916
Bad Weather	0.9266	0.1741	0.9628	0.9007	0.0012	0.0372	0.9988
Dynamic BG	0.8952	0.2805	0.8892	0.9033	0.0012	0.1108	0.9988
Int. Obj. Motion	0.7509	3.0798	0.9169	0.6792	0.0306	0.0831	0.9694
Low Framerate	0.7854	0.9016	0.9256	0.7204	0.0088	0.0744	0.9912
Night Videos	0.8553	0.5602	0.8717	0.8437	0.0035	0.1283	0.9965
Shadow	0.9108	0.6420	0.9251	0.9037	0.0042	0.0749	0.9958
Thermal	0.9319	0.6305	0.9688	0.9006	0.0059	0.0312	0.9941
Turbulence	0.8536	0.2404	0.9766	0.7881	0.0023	0.0235	0.9977
Average	0.8635	1.0301	0.9130	0.8437	0.0072	0.0870	0.9928
Scenes used in training	0.9649	0.1580	0.9788	0.9529	0.0013	0.0212	0.9987
Scenes not used in training	0.7662	1.8674	0.8499	0.7389	0.0128	0.1501	0.9872

Table 4. Comparison result of FM score by proposed method and other methods on the CDnet 2014 dataset.

Scenes	Proposed	Modified FgSegNet-v2	FgSegNet-v2 [42]	SuBSENSE [2]	CwisarD [39]	Spectral-360 [40]	GMM [9]
Baseline	**0.954**	0.940	0.814	0.950	0.908	0.933	0.825
Camera Jitter	0.776	0.769	0.613	**0.815**	0.781	0.716	0.597
Bad Weather	0.927	0.919	0.876	0.862	0.684	0.757	0.738
Dynamic Background	**0.895**	0.883	0.619	0.818	0.809	0.787	0.633
Int. Obj. Motion	**0.751**	0.719	0.584	0.657	0.567	0.566	0.633
Low Framerate	**0.785**	0.750	0.742	0.645	0.641	0.644	0.537
Night Videos	**0.855**	0.831	0.703	0.560	0.374	0.483	0.410
Shadow	**0.911**	0.893	0.734	0.899	0.841	0.884	0.737
Thermal	**0.932**	0.929	0.799	0.817	0.762	0.776	0.662
Turbulence	0.854	**0.896**	0.521	0.779	0.723	0.543	0.466
Average	**0.864**	0.850	0.697	0.777	0.706	0.706	0.619

Table 4 shows a comparison of the results obtained by the proposed algorithm and other algorithms. The proposed method, FgSegNet-v2 [42], and modified FgSegNet-v2 were trained using the same 24 scenes in the CDnet 2014 dataset. The proposed method, FgSegNet-v2, and modified FgSegNet-v2 are deep learning-based algorithms that require training samples. SuBSENSE [2], CwisarD [39], Spectral-360 [40], and GMM [9] are traditional algorithms that do not require training samples, and their experimental statistics shown in Table 4 are those reported in the literature. The proposed algorithm achieved the best performance, except for camera jitter and turbulence categories in the CDnet 2014 dataset. Training the original FgSegNet-v2 using 24 scenes in the CDnet 2014 dataset produced an even worse performance than the traditional SuBSENSE algorithm [2]. Simply training using multiple scenes without changing the network cannot guarantee generalization power. The proposed algorithm, which uses multiple difference images as input, achieves a meaningful improvement. We can conclude that the proposed algorithm provides greater generalization ability than other algorithms.

The original FgSegNet-v2 has no generalization ability in other scenes that are not used in training. Modifying its input to be multiple difference images, like in the proposed method, leads to dramatic improvement. Therefore, we can conclude that using multiple difference images as the input of the network could increase its generalization ability.

4.1. Multiple Difference Images vs. Multiple Original Images

In this section, we show experimental results according to the types of input images. We compare two cases of using multiple original images and multiple difference images as the input of networks. FgSegNet-v2 [42] predicts foreground objects using only the current image as the input of the networks. We modify it to use multiple original images or multiple difference images. In both cases, subtracting with mean is used as preprocessing. Training is done using 24 scenes in Table 1. Two hundred images from each scene are used, so 4800 images in total are used in training.

Table 5 shows the performance of the trained network on CDnet 2014 dataset according to the input of original images and multiple difference images. The numbers of original images and difference images are varied according to the interval between frames, as shown in Table 5, where 50 frames are considered for the input of the network. We show performance in two different aspects. One is applying a trained network on scenes used in training. The other is applying a trained network on scenes that are not used in the training. Using multiple original images gives a slightly better result than using multiple difference images in scenes used in training. However, using multiple difference images shows a distinctly better performance than for scenes which are not used in the training. At 10-frame intervals, we could reach a 27.5% reduction in false detection by using five difference images compared to using six original images, and we could reach a 28.6% reduction in false detection by using 10 difference images compared to using 11 original images. We can conclude that using multiple difference images as the input of networks gives improved accuracy and generalization power.

Table 5. Comparison results of using multiple original images and multiple difference images as the input of networks within a 50-frame range.

Number of Original or Difference Images	Overall		Scenes Used for Training		Scenes Not Used for Training	
	FM	PWC	FM	PWC	FM	PWC
6 (org)	0.84	1.43	0.96	0.12	0.72	2.69
5 (diff)	0.86	1.06	0.97	0.13	0.76	1.95
11 (org)	0.84	1.36	0.98	0.06	0.71	2.62
10 (diff)	0.86	1.03	0.96	0.16	0.77	1.87

4.2. Frame Intervals in Multiple Difference Images

We show the experimental results by varying the number of difference images and the interval between frames in difference image computation. Training is done using 24 scenes in Table 1. Two hundred images are used for each scene, so 4800 images in total are used in training. Table 6 shows the experimental results by varying the number of difference images at the fixed interval of five frames. Table 7 shows the experimental results by varying intervals between frames in computing difference images at the fixed range of 50 frames. Evaluation is done using CDnet 2014 datasets except the PTZ category. Three experimental statistics of performance using all scenes, scenes used for training, and scenes not used for training are presented in Tables 6 and 7. The number of difference images and the frame interval between successive images are closely related to the speed of moving foreground objects. We think that there are different optimal numbers of difference images and intervals according to the speed of moving foreground objects. Small differences in performance appear according to the variation of the number of difference images and frame intervals in the Cdnet 2014 dataset. Finally, we set the number of difference images as 10 and the interval between frames as five based on these experimental results which show better performance in scenes not used in training.

Table 6. A comparison result by changing the number of difference images under a five-frame interval.

Number of Difference Images	All Scenes		Scenes Used for Training		Scenes Not Used for Training	
	FM	PWC	FM	PWC	FM	PWC
5	0.84	1.10	0.95	0.30	0.74	1.86
10	0.86	1.03	0.96	0.16	0.77	1.87
15	0.85	1.06	0.96	0.26	0.75	1.84
20	0.84	1.09	0.93	0.30	0.75	1.86

Table 7. A comparison result by changing frame intervals to within 50 frames.

Number of Difference Images	All Scenes		Scenes Used for Training		Scenes Not Used for Training	
	FM	PWC	FM	PWC	FM	PWC
2	0.85	1.24	0.96	0.18	0.75	2.26
5	0.86	1.06	0.97	0.13	0.76	1.95
10	0.86	1.03	0.96	0.16	0.77	1.87
50	0.84	1.22	0.96	0.18	0.72	2.21

4.3. Generalization Ability Test Using Scenes Not Used in Training

Having a good generalization power is one of the main goals of machine learning. Though deep learning has shown a big jump in performance in various areas, it still requires an improvement in the generalization power. We show the improved generalization power of the proposed method by applying it on the scenes that are not used in the training. The proposed algorithm is compared to three algorithms, SuBSENSE [2], modified FgSegNet-v2, and FgSegNet-v2 [42]. We adjust FgSegNet-v2 to use multiple difference images as input, like the proposed method, and we denote it as modified FgSegNet-v2. Experiments were done by training the proposed method, modified FgSegNet-v2, and FgSegNet-v2 using the same 24 scenes in the CDnet 2014 dataset, which are shown in Table 1.

First, we evaluate the generalization power on the CDnet 2014 dataset. We investigated the generalization ability by applying the trained networks to the other 29 scenes that were not used in training in the CDnet 2014 dataset. Second, we present the results obtained by applying the trained networks to scenes in the SBI2015 dataset [43] and scenes that we acquired ourselves. The SBI2015 dataset and scenes that we acquired were not used in training. Figure 4 shows the results obtained on scenes used for training in the CDnet 2014 dataset. Figure 4a shows the original image and the corresponding frame number of the scene. Figure 4b shows the ground truth segmentation map. The results of SuBSENSE [2], the proposed method, modified FgSegNet-v2, and FgSegNet-v2 [42] are presented in Figure 4b–e. The deep learning-based methods of the proposed method, modified FgSegNet-v2, and FgSegNet-v2 give better results than the traditional approach of SuBSENSE [2]. Through this, we can ascertain that deep learning-based algorithms give superior results compared to traditional a BGS algorithm in scenes used in the training.

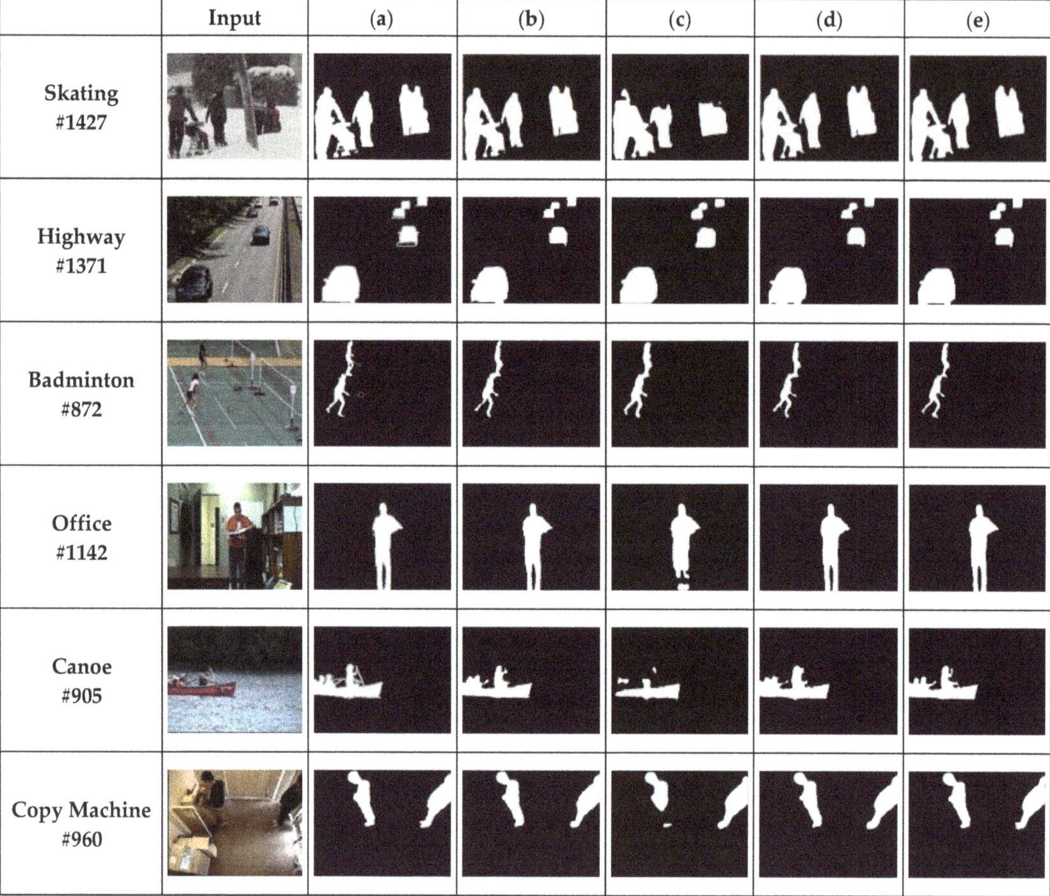

Figure 4. Test results on scenes used for the training in the CDnet 2014 dataset: (**a**) ground truth foreground maps, (**b**) proposed method, (**c**) SuBSENSE [2], (**d**) modified FgSegNet-v2, (**e**) FgSegNet-v2 [42].

Figure 5 shows the results obtained on scenes that were not used for training in the CDnet 2014 dataset. We can notice a clearly different tendency in Figure 4. FgSegNet-v2 [42] produces the worst results among the four algorithms. It produces even worse results than the non-deep learning-based method of SuBSENSE [2]. We can conclude that the original FgSegNet-v2 is efficient in scenes that were used for training, and it has little generalization ability. This can also be noticed quantitatively in Table 4. The proposed method and modified FgSegNet-v2 achieve better results than SuBSENSE even in scenes not used in training. Through this, it can be confirmed that the generalization ability is improved considerably by simply changing the input structure without changing the structure of the deep learning model.

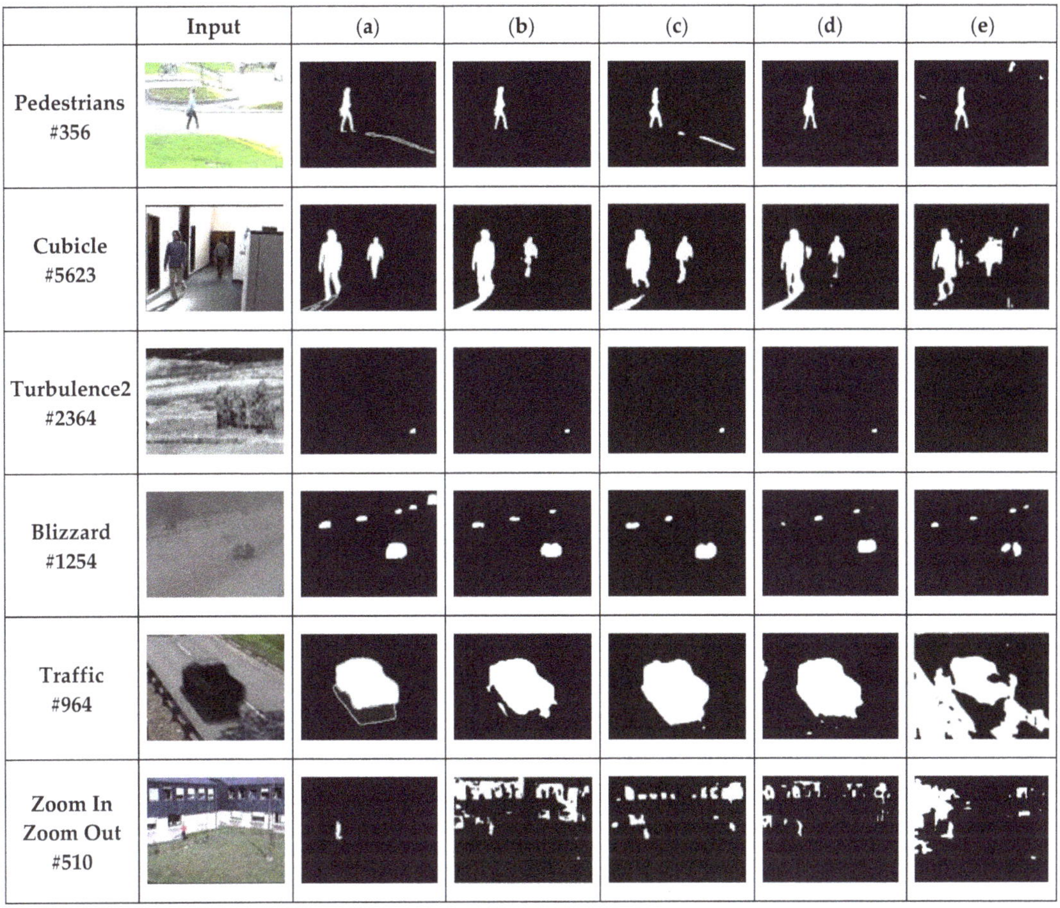

Figure 5. Test results obtained on scenes not used for training in the CDnet 2014 dataset: (**a**) ground truth foreground maps, (**b**) proposed method, (**c**) SuBSENSE [2], (**d**) modified FgSegNet-v2, (**e**) FgSegNet-v2 [42].

We present quantitative results obtained using the SBI2015 dataset [43] to show the generalization ability of proposed method. SBI2015 provides 14 scenes in total. We do not use the Toscana scene because it consists of six images that are not continuous. In addition, "Snellen" and "Foliage" scenes treated moving leaves as foreground labels. This classification differs from the foreground concept used in video surveillance. Moving leaves are generally classified as dynamic background, and we think that they should be treated as background labels. Therefore, in experiments, "Snellen" and "Foliage" scenes were also excluded from the evaluation. The proposed method, FgSegNet-v2 [42], PAWCS [41], and the SuBSENSE [2] algorithm were compared, and the results are shown in Table 8. The proposed method achieved a better performance than other algorithms. The proposed method shows low FM scores in the "Candela" and "People&oliage" scenes. Since the proposed method receives images in a range of 50 frames, it shows insufficient performance in the "Candela" scene where there is a foreground object that has been stopped for a long time. Additionally, in the "People&Foliage" scene, both moving people and bushes are classified as foreground objects. In visual surveillance, moving bushes should be classified as dynamic background, but in the scene they are classified as foreground, so most methods, including the proposed method, show very low performance. Furthermore, FgSegNet-v2 achieved much lower performance than the PAWCS and SuBSENSE algorithms, as seen in Figure 5.

Table 8. Comparison of FM score by the proposed method and other algorithms on the SBI2015 dataset.

Scene	Ours	Modified FgSegNet-v2	PAWCS [41]	SuBSENSE [2]	FgSegNet-v2 [42]
Board	**0.8114**	0.8086	0.7798	0.5777	0.5816
CAVIAR1	**0.9566**	0.9342	0.8589	0.9144	0.9115
CAVIAR2	0.8094	**0.8192**	0.6772	0.8714	0.0306
CaVignal	0.8634	**0.9102**	0.3697	0.3980	0.7704
Candela	0.6402	0.6646	**0.8725**	0.5356	0.4144
Hall&Monitor	**0.9384**	0.8878	0.7411	0.7758	0.7365
Highway1	0.8465	**0.8619**	0.7015	0.5523	0.4263
Highway2	**0.9559**	0.9537	0.9031	0.8937	0.2277
HumanBody2	**0.9415**	0.9342	0.7013	0.8346	0.5978
IBMtest2	**0.9574**	0.9548	0.9386	0.9390	0.4197
People&Foliage	0.4474	0.3033	0.3162	0.2660	**0.4930**
Mean	**0.8335**	0.8211	0.7145	0.6871	0.5100

Figure 6 shows some representative results obtained using images in the SBI2015 dataset. We can notice that the proposed algorithm shows more improvement than traditional background model-based algorithms in the SBI dataset compared to the CDnet 2014 dataset. We think that this is caused by the differences in those datasets. The CDnet 2014 dataset provides a preparation section to generate background model images before a test, but the SBI dataset does not provide this. Therefore, background model-based algorithms have difficulties in the generation of good background model images in the first part of the SBI dataset.

Figure 7 shows the results obtained by using scenes that we acquired ourselves. We only show qualitative results because obtaining a ground truth segmentation map with pixel-wise resolution would requires a huge amount of time. Two sets of results obtained using the proposed method are presented in Figure 7. One was trained using 24 scenes in the CDnet 2014 dataset. The other was trained using 49 scenes in the CDnet 2014 dataset. In Figure 7, the SeoulTech #175 image was acquired with a small jitter of the camera and there are no foreground objects in the scene. Overall, the proposed method trained using 49 scenes achieved better results than when it was trained using 24 scenes. Through this, we can see that if additional datasets can be obtained, better performance can be expected. We can conclude that the proposed method can stably detect foreground objects even in new environments that are not shown in the CDnet dataset.

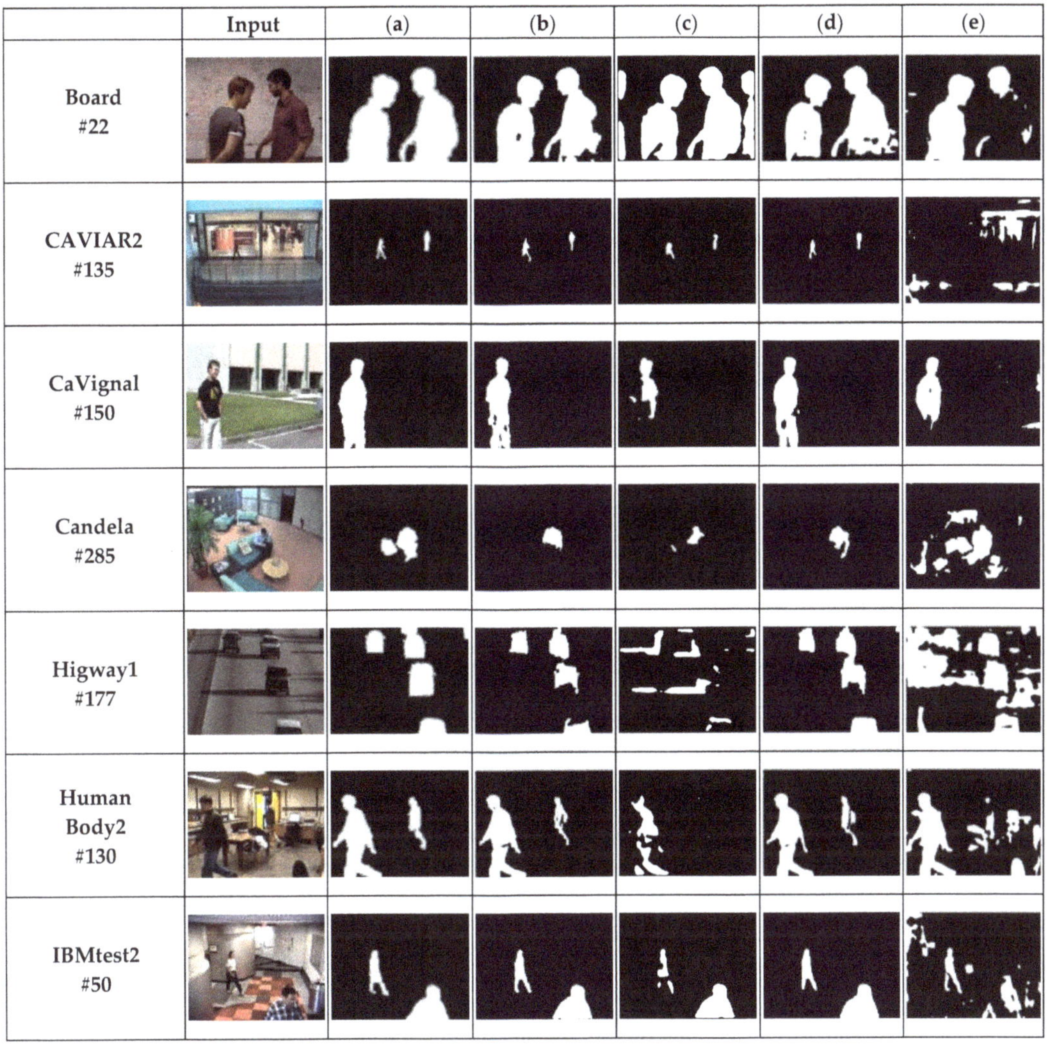

Figure 6. Test results on SBI2015 dataset: (**a**) ground truth foreground maps, (**b**) proposed method, (**c**) SuBSENSE [2], (**d**) modified FgSegNet-v2, (**e**) FgSegNet-v2 [42].

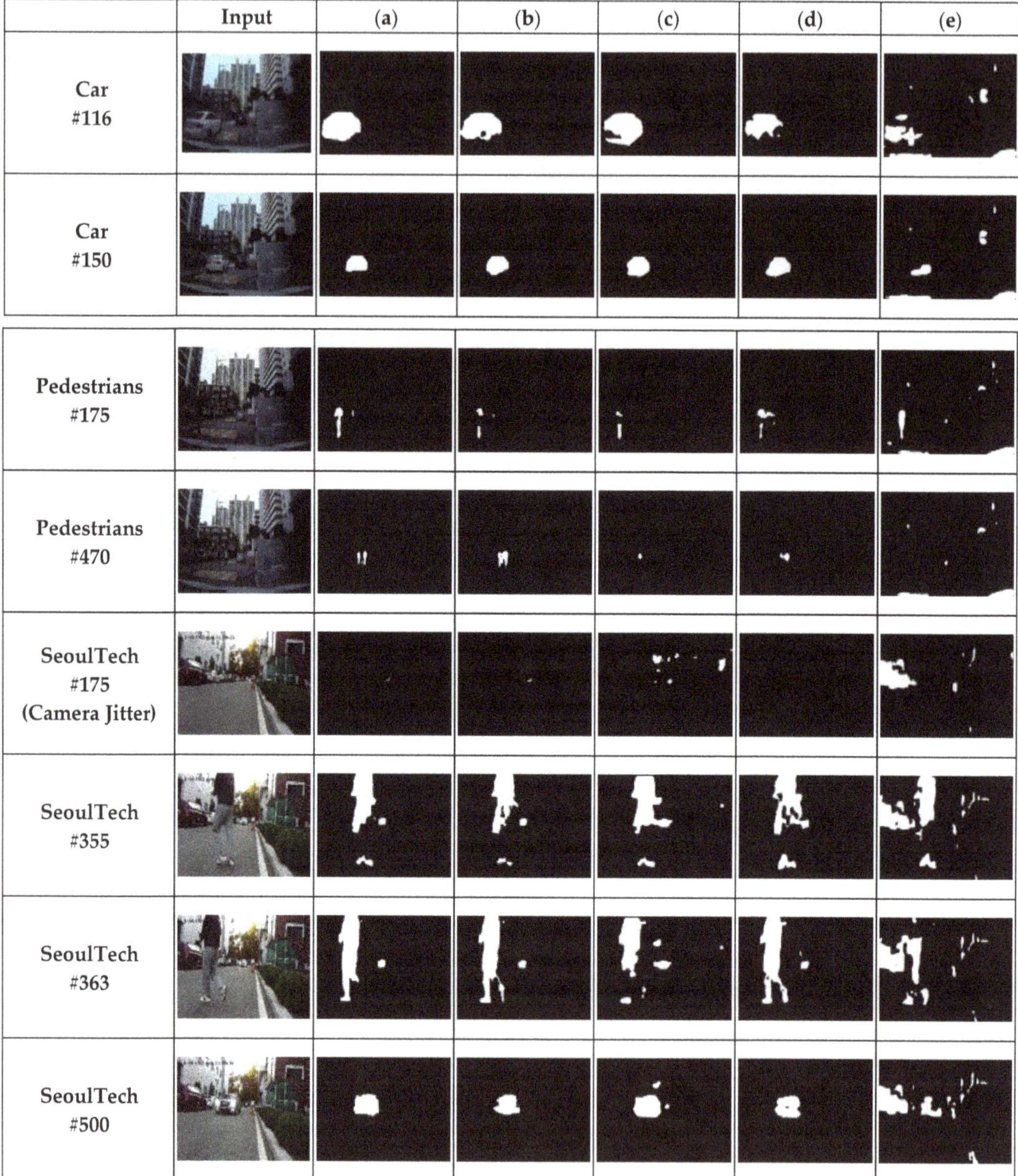

Figure 7. Qualitative results on real scenes that were not used in training: (**a**) proposed method trained using 49 scenes in the CDnet 2014 dataset, (**b**) proposed method trained using 24 scenes in the CDnet 2014 dataset, (**c**) SuBSENSE [2], (**d**) modified FgSegNet-v2, (**e**) FgSegNet-v2 [42].

Deep learning-based algorithms with supervised learning show the best performance in scenes that are similar to training scenes. Therefore, they require training before application to unknown scenes. However, they require a large set of training data. In particular, visual surveillance requires a ground truth segmentation map per pixel, which requires a large amount of time and is high cost. The best option would be a deep learning-based algorithm that does not require training samples of unknown scenes. We want to have an algorithm for visual surveillance that achieves a performance comparable to that of a deep learning-based algorithm, at the same time, requires little effort in preparing samples for training.

The proposed algorithm trained using many samples can achieve better performance than SuBSENSE [2] in situations where there are no training samples. We can conclude that the proposed method achieves better results on scenes that deviate from the training environment, in comparison to traditional and deep learning-based algorithms, from these experimental results. The proposed method is based on deep learning, and it does not require training samples before application to unknown scenes. Our goal is to have a foreground detection algorithm that achieves better performance than traditional unsupervised visual surveillance algorithms. The proposed algorithm meets this requirement by adjusting U-Net to use multiple difference images and training it using multiple scenes.

5. Conclusions

In this paper, we proposed an algorithm that has better generalization power than recent deep learning-based approaches and traditional unsupervised approaches in video surveillance. Using multiple difference images as the input of U-Net and training using all scenes in the CDnet 2014 dataset have made this possible. We demonstrated the improved generalization power of the proposed algorithm through diverse experiments using the CDnet 2014 dataset, the SBI 2015 dataset, and real scenes that we acquired ourselves. We have shown that the generalization ability can be improved by only using multiple difference images as input to other deep learning methods. However, since the frame range of the input data is limited, it is difficult to detect foreground objects that have been stopped for a long time. Additionally, because the proposed algorithm uses multiple difference images as input, it has a shortcoming for scenes acquired by a camera in motion. In further research, we are going to apply recurrent neural networks to cope with these problems. In addition, we plan to do research to cope with the problems that are caused by moving camera using a spatio-temporal network that properly considers spatial and temporal information.

Author Contributions: Conceptualization, J.-Y.K. and J.-E.H.; implementation, J.-Y.K.; analysis, J.-Y.K. and J.-E.H.; writing, original draft preparation; J.-Y.K.; draft modification, J.-E.H.; funding acquisition, J.-E.H. All authors have read and agreed to the published version of the manuscript.

Funding: This work was supported by a National Research Foundation of Korea (NRF) grant funded by the Korean government (MSIT) (2020R1A2C1013335).

Institutional Review Board Statement: Not applicable.

Informed Consent Statement: Not applicable.

Conflicts of Interest: The authors declare no conflict of interest.

References

1. Barnich, O.; Van Droogenbroeck, M. Vibe: A universal background subtraction algorithm for video sequences. *IEEE Trans. Image Process.* **2011**, *20*, 1709–1724. [CrossRef]
2. St-Charles, P.L.; Bilodeau, G.A.; Bergevin, R. Subsense: A universal change detection method with local adaptive sensitivity. *IEEE Trans. Image Process.* **2015**, *24*, 359–373. [CrossRef]
3. Varadarajan, S.; Miller, P.; Zhou, H. Region-based mixture of Gaussians modelling for foreground detection in dynamic scenes. *Pattern Recogn.* **2015**, *38*, 3488–3503. [CrossRef]
4. Wang, Y.; Jodoin, P.M.; Porikli, F.; Konrad, J.; Benezeth, Y.; Ishwar, P. Cdnet 2014: An expanded change detection benchmark dataset. In Proceedings of the IEEE Conference on Computer Vision and Pattern Recognition Workshops, Columbus, OH, USA, 23–28 June 2014; pp. 387–394.
5. Bouwmans, T. Traditional and recent approaches in background modeling for foreground detection: An overview. *Comput. Sci. Rev.* **2014**, *11*, 31–66. [CrossRef]
6. Sobral, A.; Vacavant, A. A comprehensive review of background subtraction algorithms evaluated with synthetic and real videos. *Comput. Vis. Image Underst.* **2014**, *122*, 4–21. [CrossRef]

7. Garcia-Garcia, A.; Orts-Escolano, S.; Oprea, S.; Villena-Martinez, V.; Garcia-Rodriguez, J. A review on deep learning techniques applied to semantic segmentation. *arXiv* **2017**, arXiv:1704.06857.
8. Bouwmans, T.; Javed, S.; Sultana, M.; Jung, S.K. Deep neural network concepts for background subtraction: A systematic review and comparative evaluation. *arXiv* **2018**, arXiv:1811.05255.
9. Stauffer, C.; Grimson, W.E.L. Adaptive background mixture models for real-time tracking. In Proceedings of the IEEE Computer Society Conference on Computer Vision and Pattern Recognition, Fort Collins, CO, USA, 23–25 June 1999; pp. 246–252.
10. Dempster, A.P.; Laird, N.M.; Rubin, D.B. Maximum likelihood from incomplete data via the EM algorithm. *J. R. Stat. Soc. Ser. B (Methodol.)* **1977**, *39*, 1–38.
11. Elgammal, A.; Duraiswami, R.; Harwood, D.; Davis, L.S. Background and foreground modeling using nonparametric kernel density estimation for visual surveillance. *Proc. IEEE* **2002**, *90*, 1151–1163. [CrossRef]
12. Kim, K.; Chalidabhongse, T.H.; Harwood, D.; Davis, L. Real-time foreground–back- ground segmentation using codebook model. *Real-Time Imaging* **2005**, *11*, 172–185. [CrossRef]
13. Oliver, N.M.; Rosario, B.; Pentland, A.P. A Bayesian computer vision system for modeling human interactions. *IEEE Trans. Pattern Anal. Mach. Intell.* **2000**, *22*, 831–843. [CrossRef]
14. Wang, R.; Bunyak, F.; Seetharaman, G.; Palaniappan, K. Static and moving object detection using flux tensor with split gaussian models. In Proceedings of the IEEE Conference on Computer Vision and Pattern Recognition Workshops, Columbus, OH, USA, 23–28 June 2014.
15. Bunyak, F.; Palaniappan, K.; Nath, S.K.; Seetharaman, G. Flux tensor constrained geodesic active contours with sensor fusion for persistent object tracking. *J. Multimed.* **2007**, *2*, 20–33. [CrossRef] [PubMed]
16. Evangelio, R.H.; Sikora, T. Complementary background models for the detection of static and moving objects in crowded environments. In Proceedings of the 8th IEEE International Conference on Advanced Video and Signal-Based Surveillance (AVSS), Klagenfurt, Austria, 30 August–2 September 2011; pp. 71–76.
17. Chen, Y.; Wang, J.; Lu, H. Learning sharable models for robust background subtraction. In Proceedings of the IEEE International Conference on Multimedia and Expo (ICME), Turin, Italy, 29 June–3 July 2015; pp. 1–6.
18. Sajid, H.; Cheung, S.C.S. Background subtraction for static & moving camera. In Proceedings of the 2015 IEEE International Conference on Image Processing (ICIP), Quebec City, QC, Canada, 27–30 September 2015; pp. 4530–4534.
19. Hofmann, M.; Tiefenbacher, P.; Rigoll, G. Background segmentation with feedback: The pixel-based adaptive segmenter. In Proceedings of the IEEE Computer Society Conference on Computer Vision and Pattern Recognition Workshops, Providence, RI, USA, 16–21 June 2012; pp. 38–43.
20. Braham, M.; Van Droogenbroeck, M. Deep background subtraction with scene-specific convolutional neural networks. In Proceedings of the International Conference on Systems, Signals and Image Processing, Bratislava, Slovakia, 23–25 May 2016.
21. Bilodeau, G.A.; Jodoin, J.P.; Saunier, N. Change detection in feature space using local binary similarity patterns. In Proceedings of the International Conference on Computer and Robot Vision (CRV), Regina, SK, Canada, 28–31 May 2013; pp. 106–112.
22. Goyette, N.; Jodoin, P.M.; Porikli, F.; Konrad, J.; Ishwar, P. A novel video dataset for change detection benchmarking. *IEEE Trans. Image Process.* **2014**, *23*, 4663–4679. [CrossRef] [PubMed]
23. Babaee, M.; Dinh, D.T.; Rigoll, G. A deep convolutional neural network for video sequence background subtraction. *Pattern Recognit.* **2018**, *76*, 635–649. [CrossRef]
24. Wang, Y.; Luo, Z.; Jodoin, P.M. Interactive deep learning method for segmenting moving objects. *Pattern Recognit. Lett.* **2017**, *96*, 66–75. [CrossRef]
25. Lim, L.A.; Keles, H.Y. Foreground segmentation using a triplet convolutional neural network for multiscale feature encoding. *arXiv* **2018**, arXiv:1801.02225.
26. Simonyan, K.; Zisserman, A. Very deep convolutional networks for large-scale image recognition. *arXiv* **2014**, arXiv:1409.1556.
27. Zeng, D.; Zhu, M. Background subtraction using multiscale fully convolutional network. *IEEE Access* **2018**, *6*, 16010–16021. [CrossRef]
28. Zeng, D.; Chen, X.; Zhu, M.; Geosele, M.; Kuijper, A. Background Subtraction with Real-Time Semantic Segmentation. *IEEE Access* **2019**, *7*, 153869–153884. [CrossRef]
29. Braham, M.; Pi'erard, S.; Van Droogenbroeck, M. Semantic background subtraction. In Proceedings of the IEEE International Conference on Image Processing (ICIP), Beijing, China, 17–20 September 2017; pp. 4552–4556.
30. Sakkos, D.; Ho, E.S.L.; Shum, H.P.H. Illumination-aware multi-task GANs for foreground segmentation. *IEEE Access* **2019**, *7*, 10976–10986. [CrossRef]
31. Patil, P.W.; Murala, S. MSFgNet: A novel compact end-to-end deep network for moving object detection. *IEEE Trans. Intell. Transp. Syst.* **2018**, *20*, 4066–4077. [CrossRef]
32. Varghese, A.; Gubbi, J.; Ramaswamy, A.; Balamuralidhar, P. ChangeNet: A deep learning architecture for visual change detection. In Proceedings of the European Conference on Computer Vision (ECCV), Munich, Germany, 8–14 September 2018.
33. Akilan, T.; Jonathan, Q.; Safaei, A.; Huo, J.; Yang, Y. A 3D CNN-LSTM-based image-to-image foreground segmentation. *IEEE Trans. Intell. Transp. Syst.* **2019**, *21*, 959–971. [CrossRef]
34. Yang, L.; Li, J.; Luo, Y.; Zhao, Y.; Cheng, H.; Li, J. Deep background modeling using fully convolutional network. *IEEE Trans. Intell. Transp. Syst.* **2018**, *19*, 254–262. [CrossRef]

35. Ronneberger, O.; Fischer, P.; Brox, T. U-Net: Convolutional networks for biomedical image segmentation. In *International Conference on Medical Image Computing and Computer-Assisted Intervention*; Springer: Cham, Switzerland, 2015; pp. 234–241.
36. Chollet, F. Keras. 2015. Available online: https://github.com/keras-team/kreas (accessed on 15 February 2021).
37. He, K. Delving Deep into rectifiers: Surpassing human-level performance on ImageNet. *arXiv* **2015**, arXiv:1502.01852.
38. Kingma, D.P.; Ba, J. Adam: A method for stochastic optimization. *arXiv* **2014**, arXiv:1412.6980.
39. De Gregorio, M.; Giordano, M. A WiSARD-based approach to CDnet. In Proceedings of theBRICS Congress on Computational Intelligence and 11th Brazilian Congress on Computational Intelligence, Ipojuca, Brazil, 8–11 September 2013; pp. 172–177.
40. Sedky, M.; Moniri, M.; Chibelushi, C.C. Spectral-360: A physics-based technique for change detection. In Proceedings of the IEEE Conference on Computer Vision and Pattern Recognition Workshop, Columbus, OH, USA, 23–28 June 2014; pp. 405–408.
41. St-Charles, P.L.; Bilodeau, G.A.; Bergevin, R. Universal background subtraction using word consensus models. *IEEE Trans. Image Process.* **2016**, *25*, 4768–4781. [CrossRef]
42. Lim, L.A.; Keles, H.Y. Learning multi-scale features for foreground segmentation. *arXiv* **2018**, arXiv:1808.01477. [CrossRef]
43. Maddalena, L.; Petrosino, A. Towards benchmarking scene background initialization. In *Proceedings of International Conference on Image Analysis and Processing*; Springer: Cham, Switzerland, 2015; pp. 469–476.

Article

A Pipeline Approach to Context-Aware Handwritten Text Recognition

Yee Fan Tan [1], Tee Connie [1,*], Michael Kah Ong Goh [1] and Andrew Beng Jin Teoh [2,*]

1   Faculty of Information Science and Technology, Multimedia University, Melaka 75450, Malaysia; 1171202077@student.mmu.edu.my (Y.F.T.); michael.goh@mmu.edu.my (M.K.O.G.)
2   School of Electrical and Electronic Engineering, College of Engineering, Yonsei University, Seoul 03722, Korea
*   Correspondence: tee.connie@mmu.edu.my (T.C.); bjteoh@yonsei.ac.kr (A.B.J.T.); Tel.: +60-62523592 (T.C.)

Featured Application: The proposed handwritten text recognition pipeline can be used for practical documents transcription and context recognition.

Abstract: Despite concerted efforts towards handwritten text recognition, the automatic location and transcription of handwritten text remain a challenging task. Text detection and segmentation methods are often prone to errors, affecting the accuracy of the subsequent recognition procedure. In this paper, a pipeline that locates texts on a page and recognizes the text types, as well as the context of the texts within the detected region, is proposed. Clinical receipts are used as the subject of study. The proposed model is comprised of an object detection neural network that extracts text sequences present on the page regardless of size, orientation, and type (handwritten text, printed text, or non-text). After that, the text sequences are fed to a Residual Network with a Transformer (ResNet-101T) model to perform transcription. Next, the transcribed text sequences are analyzed using a Named Entity Recognition (NER) model to classify the text sequences into their corresponding contexts (e.g., name, address, prescription, and bill amount). In the proposed pipeline, all the processes are implicitly learned from data. Experiments performed on 500 self-collected clinical receipts containing 15,297 text segments reported a character error rate (CER) and word error rate (WER) of 7.77% and 10.77%, respectively.

Keywords: handwritten text recognition; Residual Network; Transformer model; object detection; named entity recognition

Citation: Tan, Y.F.; Connie, T.; Goh, M.K.O.; Teoh, A.B.J. A Pipeline Approach to Context-Aware Handwritten Text Recognition. *Appl. Sci.* **2022**, *12*, 1870. https://doi.org/10.3390/app12041870

Academic Editors: Cheonshik Kim and Byung-Gyu Kim

Received: 17 December 2021
Accepted: 10 February 2022
Published: 11 February 2022

Publisher's Note: MDPI stays neutral with regard to jurisdictional claims in published maps and institutional affiliations.

1. Introduction

Handwritten text recognition (HTR) has gained enormous research interest due to the potential benefits that can be derived from accurate text transcription that cases at tempts to digitize handwritten content [1,2]. An HTR system is applicable to a myriad of scenarios, ranging from reading bank cheque amounts to transcribing medical records and notes [3]. Although highly desirable in practical applications, HTR is faced with a number of challenges.

The current HTR systems generally apply a Convolutional Neural Network (CNN) and Long Short-Term Memory (LSTM) model for text transcription [3]. However, a variety of text styles, such as printed texts, handwritten texts, scribble, and images, exists in real-life documents. Therefore, using a single text recognition model is insufficient for the text transcription task. Ingle et al. proposed a text style classification approach using an LSTM-based and fully feed-forward network model for line-level segmentation. An optimal model was determined when the calculated probability of a particular class was greater than a predefined threshold [3]. In addition, Singh and Karayev presented a study on HTR by decomposing an image into one or more regions as texts, mathematical equations, tables, and scratched texts using a Residual Network (ResNet) [4]. These studies have demonstrated a way for precise text transcription by ignoring unknown or unreadable regions, and the proposed approaches are applicable in a wide range of applications, covering simple to complex scenarios. Nevertheless, the line-based segmentation methods [3] sometimes fail to recognize the texts correctly due to difficulty in segmenting the image into lines accurately. Full page-based models that only transcribe a particular region of texts in the page while skipping others [4], on the other hand, can only work well on a balanced dataset, such as when the layout of the page is the same.

In the literature, the combination of CNN with Recurrent Neural Network (RNN) [5,6] and LSTM [7] are widely applicable for sequence modeling in HTR. However, the RNN variations face vanishing and exploding gradient problems, where the models fail to learn the long sequence information [8]. Recently, the Transformer model with an attention mechanism has been introduced, and it has demonstrated superior performance over the conventional RNN and LSTM models for long sequence information processing [4,9]. The Transformer model yields outstanding performance on public benchmark datasets, especially on the IAM dataset [10]. For example, a CNN and LSTM architecture integrated with a Transformer model was applied on the IAM dataset and achieved a character error rate (CER) of 8.50% [2]. Another study that applied the Transformer model to handwriting document recognition reported a CER of 6.30% on the IAM dataset at paragraph-level [4].

In general, an HTR system development pipeline comprises two stages: (1) text localization and (2) text recognition [11]. Real-life documents generally contain a combination of text types such as printed texts, handwritten texts, signatures, and others. How to correctly localize and recognize these text types has become pivotal to avoid bias and ensure the right data sample distribution. Many solutions have been proposed to better perform these tasks [3,4,9], but there is still room for improvement. Apart from accurately recognizing the handwritten texts, how to associate the meaning or context of the recognized texts is also crucial to enable the automatic documentation of the transcribed texts. Being able to meet the computational requirement of a real-life HTR system is also of paramount importance.

In this paper, a context-aware HTR pipeline is proposed to overcome the limitations of the existing HTR systems by considering the accuracy and efficiency of text type classification and localization, text recognition, and text context recognition on real-life documents as a whole. Clinical receipts are used as the subject of study as they contain a combination of printed items on the receipts, handwritten texts of the clinicians, as well as non-text elements, such as the logo of the clinics. Towards this end, a dataset containing 500 samples of clinical receipts has been collected. The documents are further segmented based on regions such as patient names, address, prescription, and bill amount, yielding a total of 15,297 text segments.

The proposed HTR pipeline consists of a You Only Look Once v5 (YOLOv5) model for text localization and type classification, followed by a Residual Network with Transformer (RESNET-101T) for text recognition, and a Named Entity Recognition (NER) model for text context recognition. An integrated model comprising ResNet-101 and Transformer, which is coined as ResNet-101T, is introduced in this paper. ResNet-101 acts as a feature extractor, whose output is fed into the Transformer model to perform text recognition. ResNet-101 is well-known for its ability to alleviate the effect of vanishing gradient and avoid performance degradation when the network's depth is increased. The Transformer model is selected due to its outstanding performance in handling sequential data, and it has a low inductive bias compared to conventional RNN architectures. Nevertheless, the Transformer model requires a huge dataset for training. Therefore, data augmentation is performed to ensure the model is properly trained with a sufficient amount of data. The proposed pipeline aims to study the applicability of data-driven DL models on a close-to-real-life dataset, where it contains much noisier and challenging data compared to the existing benchmark datasets. The contributions of this paper are highlighted as follows:

-   A context-aware HTR pipeline made up of a series of carefully chosen pre-processing, text recognition, and context interpretation funnels is presented to deal with a close-to-real-life handwritten text dataset. The pipeline is designed to locate texts on a page and is able to recognize the text types such as handwritten text, printed text, non-text, as well as the text context.

- A ResNet-101T model that has a better ability in handling sequence data compared to RNN variations is proposed for text recognition. The proposed model is compared with the state-of-the-art HTR methods, including CNN-LSTM and Vision Transformer.
- A NER model is proposed to complement the pipeline to recognize the context of the transcribed texts. Transcription of the document can be performed in a fully automatic way.

The remaining paper is organized as follows. Section 2 presents the literature review. Section 3 describes the data and theoretical backgrounds of the methods applied for the proposed pipeline. Then, Section 4 presents the experimental result, and Section 5 discusses the insights into the proposed pipeline. Finally, the conclusion and future works are drawn in Section 6.

2. Literature Review

The emergence of deep learning (DL) has brought significant advances to HTR. The DL models have progressively improved the performance of HTR transcription over the years. This section discusses the different ideas proposed to solve HTR and the performance achieved.

Bluche presented an approach for joint line segmentation for HTR transcription [11]. The dataset was taken from paragraph-level images from the RIMES and IAM datasets. The author proposed the integration of the attention mechanism with the Multi-dimensional Long Short-Term Memory Recurrent Neural Network (MDLSTM-RNN) for implicit text segmentation and transcription. The collapse layer of the model was modified with an attention mechanism to provide the weights in identifying the input positions on a paragraph image iteratively to enable a free segmentation method for text transcription. As a result, the model achieved a CER of 4.9 and 2.5 on the IAM and RIMES datasets, respectively, containing images with 300 dpi resolution.

Wigington et al. proposed a model of Start, Follow, and Read (SFR) for historical HTR [12]. The SFR model could identify the text position by using a Region Proposal Network and a CNN-LSTM model for text transcription. The proposed SFR model was composed of a Start-of-Line (SOL) finder, which identified the text line of a given image, a Line Follower (LF), which segmented the position identified by the SOL finder iteratively and, lastly, the HTR model. The 2017 ICDAR full-page competition dataset was applied in the study of German handwriting. The proposed model achieved outstanding performance, with a BLEU score of 72.3. The model was also evaluated on the RIMES and IAM datasets, achieving a CER and WER of 2.1 and 9.3, as well as 6.4 and 23.2, respectively.

There was another study that applied CNN to a Kannada handwritten document [13]. In the paper, the author proposed CNN for training the data. In the experiment setup, the Chars74K dataset containing over 657 classes was applied. Each class has 25 handwritten characters. Data augmentation techniques, including denoising, contrast normalization, segmentation, gray-scale conversion, and binarization, were performed to expand the dataset size. After a hundred epochs of training, the model achieved 99% accuracy on the Chars74k dataset and 96% on a self-collected handwritten document.

Ingle et al. conducted a study for a scalable HTR system [3]. The authors integrated the proposed HTR system into a larger-scale OCR system. In the study, the authors applied LSTM-based models and gated recurrent convolutional layers (GRCL) as a fully feed-forward network model for line-level text recognition and classification. Online handwritten data from the IAM offline and online databases and a self-collected online handwritten sample were used. The authors trained separate models for both printed and handwritten words. The dataset used for both printed and handwritten text consisted of 508 and 433 images. After hyperparameter tuning, the proposed GRCL achieved a character error rate (CER) and word error rate (WER) of 4.0 and 10.8, respectively.

Wu et al. presented a method to detect and recognize handwritten text and text-line [14]. The authors presented a method named Multi-Level Convolutions Convolutional and Recurrent Network (MLC-CRNN), which combined different deep learning techniques, including CNN, RNN, and Connectionist Temporal Classification (CTC) loss function. In the paper, the Connectionist Text Proposal Network (CTPN) was used in training a new model for a handwriting text-line detector. Following handwriting text recognition, the team applied a refined CRNN model. To make it a multi-layer convolution (MLC), two more branches were added linearly on the original convolutional layers. The datasets used for training were obtained from three hundred students. The participants were asked to write on a standard answer sheet. The training set contained 3883 images, and the testing set contained 297 images. The proposed MLC-CRNN model, when integrated with two MLC modules, achieved the best performance by obtaining an accuracy of 91.4%.

Singh and Karayev presented a study that applied full-page handwriting document recognition using the Transformer model [4]. The authors aimed to recognize handwritten texts in a full-page manner. The model consists of a CNN network to extract the features from the document, followed by Transformer as an image-to-sequence model, which learns to map an image to a sequence. The model was trained on various datasets, including IAM, WIKITEXT, FREE FORM ANSWER, ANSWERS2. The model was then evaluated on the FREE FORM ANSWERS dataset and obtained a CER of 7.6%. Despite the promising performance, the method suffers from a biased multi-task problem. For example, if the model is trained using datasets that only contain one transcribed text region per sample (like the Free Form dataset), the model will have a tendency to transcribe only one main text region while skipping the others due to its full-page recognition nature. This is a challenge for text recognition. It is important to have a robust text recognition system that can deal with different text types, including printed texts, handwritten texts, and non-texts. Table 1 summarizes the different papers discussed in this section.

Table 1. Summary of included studies.

Author	Subject of Study	Proposed Solution	Dataset	Experimental Results
Bluche (2016) [11]	Joint line segmentation and transcription	MDLSTM-RNN	RIMES and IAM database	CER of 4.9 and 2.5 on IAM and RIMES data, respectively
Wigington et al. (2018) [12]	Historical document processing	SFT	ICDAR 2017 competition dataset, IAM, RIMES	BLEU score of 72.3 on ICDAR dataset. CER and WER of 2.1 and 9.3, 6.4 and 23.2 on both IAM and RIMES datasets, respectively
Asha and Krishnappa (2018) [13]	Kannada Handwritten Document Recognition	CNN	Chars74K dataset	99% of accuracy
Ingle et al. (2019) [3]	Line-level text style classification and recognition	GRCL	IAM online and offline database, self-collected handwritten online samples	CER and WER of 4.0 and 10.8
Wu et al. (2020) [14]	Recognition of handwritten text and text-line	CTPN to detect text lines, MLC-CRNN for text recognition	3883 training images and 297 testing images	Accuracy of 91.4%
Sign and Karayev (2021) [4]	Full-page handwritten document recognition	Transformer	IAM, WIKITEXT, FREE FORM ANSWERS, ANSWERS2	CER of 7.6%

3. Proposed Method

3.1. Data Collection

In this study, a dataset consisting of 500 clinical receipts is collected. The dataset is composed of 10 variants of medical receipts with 50 samples each. The empty receipt templates are obtained from online resources, with a resolution of 300 dpi and above. During data collection, empty medical receipts were distributed to the participants to fill with their own handwriting. No restriction was imposed on how the participants should write on the receipts. The participants come from various backgrounds and professions, were aged between 12 and 50 at the time of the study, and are from Malaysia. Every participant is literate and can write independently, with no known disability. A blank printed copy of the receipt template was given to them to fill. After the form was filled, the filled form was collected and scanned. Figure 1 shows some samples of the collected handwritten receipts.

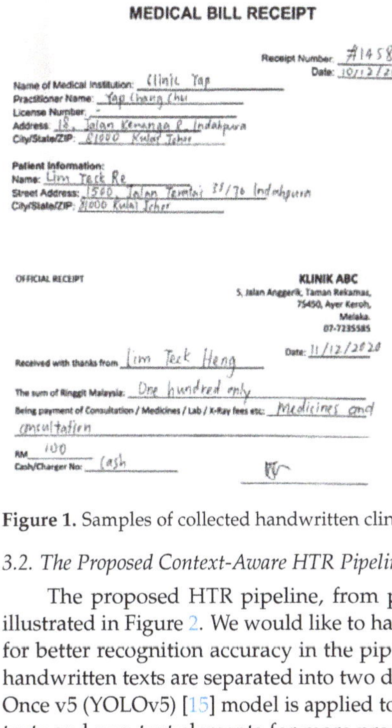

Figure 1. Samples of collected handwritten clinical receipts.

3.2. The Proposed Context-Aware HTR Pipeline

The proposed HTR pipeline, from pre-processed input to context recognition, is illustrated in Figure 2. We would like to have a model that is aware of the receipt's content for better recognition accuracy in the pipeline. Towards this end, the printed texts and handwritten texts are separated into two different processing funnels. The You Only Look Once v5 (YOLOv5) [15] model is applied to distinguish the handwritten texts from printed texts and non-text elements for more precise region of interest (ROI) localization. After that, an Optical Character Recognition (OCR) model is used to identify the printed texts, while the handwritten texts are recognized using a Residual Network (ResNet-101) and Transformer architecture (ResNet-101T). In this paper, Tesseract [16], a matured open-source OCR model, is applied for printed text recognition, and thus, no further evaluation is made for the model. Subsequently, both outputs from the text recognition models are processed by a Named Entity Recognition (NER) model to identify the context of the texts.

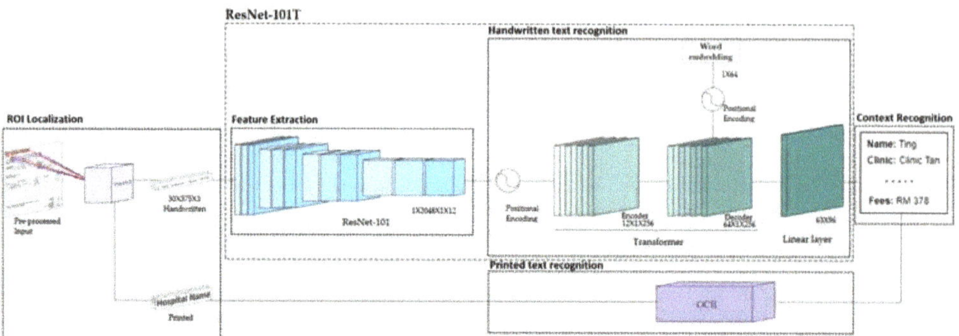

Figure 2. HTR pipeline process and architecture.

3.2.1. Pre-Processing

During the data pre-processing stage, various procedures were applied to ensure that the data were in an appropriate form for further model training. Considering different noises that might occur in a real-life document image, we took several aspects into consideration. We identified two problems: (a) the image is taken at a slanted or skewed angle; (b) there are lines in the image that could affect classification performance. Progressive Probabilistic Hough Line Transform (PPHLT) [17,18] was applied to rotate the image to the correct angle. The method first detects the image's edges by applying canny edge detection. The image is then rotated accordingly based on the computed angle from the PPHLT algorithm. Next, morphological operations and image inpainting [19] are applied to remove the lines from the image. After that, trimming is used to remove excessive white pixels, such as in the border regions in the image.

A YOLOv5 model has been trained to identify the text region of interest (ROI). A text ROI is categorized into three types: printed texts, handwritten texts, and non-text. More information about ROI localization and categorization is given in the next section. The segmented regions are then padded into the same size. For model training, the text ROIs are labeled manually as printed, handwritten, and others. There are a total of 5099 handwritten text segments, 7445 printed text segments, and 261 non-text segments in this study. Figure 3 shows a sample of the labeled image.

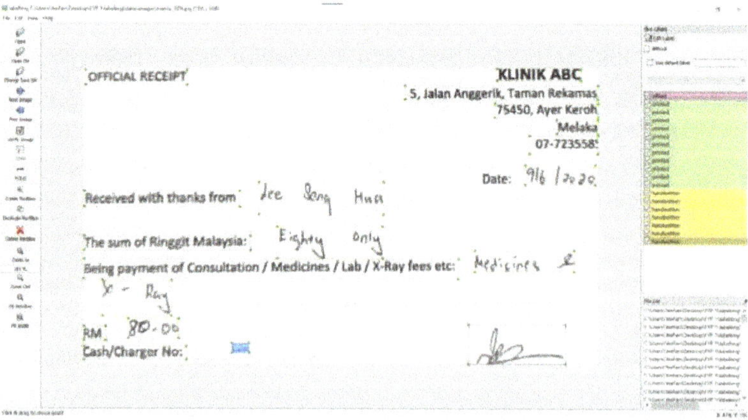

Figure 3. Sample labeled image.

The text segment is augmented randomly by reducing or adding line width, adding Gaussian noise, and blurring the images. The final dataset is composed of 15,297 segments of handwritten text images. Figure 4 presents the flow of the proposed pre-processing approach. Some samples of the text segment and the corresponding augmented images are shown in Figure 5.

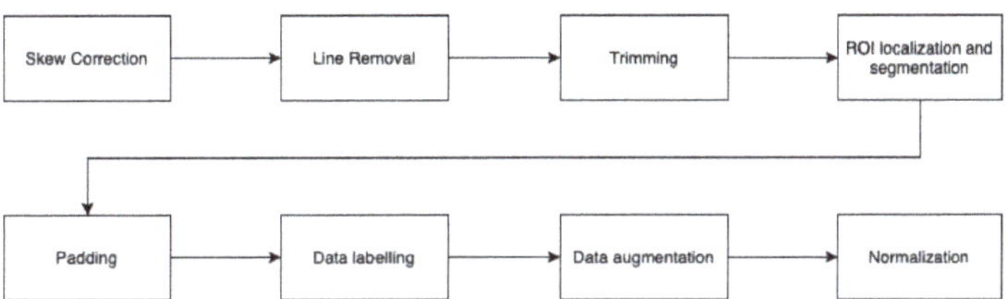

Figure 4. The flow of pre-processing approach.

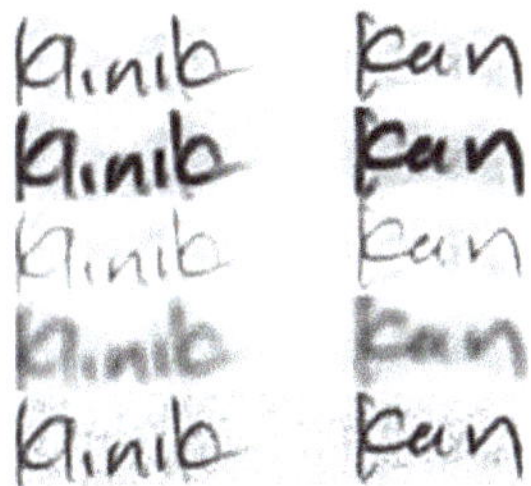

Figure 5. Samples of segmented and augmented images.

3.2.2. ROI Localization and Categorization

Text ROI location and categorization are crucial in making sure that the input fed to the HTR model is of good quality. Towards this end, YOLOv5, which is a successful object classification and detection method, is deployed. YOLOv5 works by dividing an image into a grid system, and the object will be detected within the cell of the grid. YOLOv5 is established and refined based on the YOLOv3 method presented by Joseph and Ali [20]. No academic publication for YOLOv5 is available; hence, the theoretical background of YOLOv3 is provided.

YOLOv3 predicts the coordinates of a bounding box, t_x, t_y, t_w, t_h, where x, y, w, h represent the x and y coordinates, width, and height. Sum squared error is used for training the model. Additionally, the model uses logistic regression to measure the objectness score, also known as the probability of being classified into a particular object. The feature extractor used consists of 53 layers (DarkNet-53) and is more efficient in utilizing the GPU for faster evaluation. Generally, YOLOv3 predicts the bounding boxes at three different scales, and features are extracted from those scales. The outcome is a 3d-tensor of the bounding box, objectness score, and classes prediction. The bounding boxes b_x, b_y, b_w, b_h are defined in Equation (1).

$$\begin{aligned} b_x &= \sigma(t_x) + c_x \\ b_y &= \sigma(t_y) + c_y \\ b_w &= p_w e^{t_w} \\ b_h &= p_h e^{t_h} \end{aligned} \tag{1}$$

where c_x, c_y are the offset from the top left corner of the image, and p_w, p_h are width and height of the bounding box prior. σ stands for the sigmoid function. Built upon the fundamentals of YOLOv3, different variants for YOLOv5 have been introduced [15]. Some examples include YOLO-v5n with the least parameters of 1.9 million, YOLO-v5s with 7.2 million parameters, YOLO-v5m with 21.2 million parameters, YOLO-v5l with 46.5 parameters, and YOLO-v5x with the most parameters of 86.7 million.

3.2.3. Residual Network with Transformer (ResNet-101T)

Resnet is proposed by He et al. [21] as a residual learning framework for better neural network training with a deeper architecture. The Residual network has shortcut connections inserted to the network to make a counterpart of the Residual version. There are ResNets with different lengths of layers, which can be implied from their names; for example, ResNet-101 stands for a ResNet architecture containing 101 layers.

In residual learning, the stacked layers are the building block for feature mapping. A building block can be defined as:

$$y = F(x, \{W_i\} + x)$$ (2)

where x is the input vector, y is the output vector, and the function, F stands for residual mapping. The operation for F can be performed by applying a shortcut connection or element-wise addition.

On the other hand, Transformer is a neural network proposed by Vaswani et al. [9], that applies an attention mechanism to connect the encoder and decoder. The encoder maps the input to a sequence of continuous representation. Given the mapped sequence, the decoder generates the output of one element at a time, where the model is autoregressive at each step as it uses the generated output at the previous step as additional input for output generation. Transformer with attention mechanism is introduced to tackle the problem of memory constraints in RNN variations by modeling the dependencies without considering the input and output sequence distances. Transformer has an advantage in that it has a relatively low inductive bias compared to RNN.

The encoder stack contains a multi-head attention mechanism, which is followed by a feed-forward neural network layer, and each layer is followed by a normalization layer. As opposed to the encoder, the decoder contains more layers. The output of the encoder is inserted into the third layer of the decoder. Additionally, a linear layer is appended to the end of the decoder for output prediction. Figure 6 shows the decoder–encoder structure of the Transformer model.

The attention in the Transformer model is known as scaled-dot attention. The attention function is defined as:

$$Attention(Q, K, V) = softmax\left(\frac{QK^T}{\sqrt{d_k}}\right)V$$ (3)

where Q is a matrix of a set of queries of the attention function, and K and V stand for keys and values. Dot-product attention is used as it is faster and has better space efficiency compared to additive attention. However, the attention is implemented as a multi-head mechanism. Thus, the outputs of the attention are concatenated as:

$$Multiheaad\ attention(Q, K, V) = Concatenate(head_1, \ldots, head_i)W^O$$ (4)

where $head_i$ stands for $Attention\left(QW_i^Q, KW_i^K, VW_i^V\right)$. The multi-head attention is applied in the encoder–decoder of the Transformer model to allow the decoder to look through every position of the input sequences, coming from the queries of previous layers and the memory keys and values. Additionally, the self-attention layers in the encoder and decoder allow the output of previous layers to be attended for its position.

The Transformer model uses positional encoding, PE, to make use of the sequence order by providing information of the token of the sequences,

$$PE_{(pos,2i)} = \sin\left(\frac{pos}{10000^{\frac{2i}{d_{model}}}}\right), \quad PE_{(pos,2i+1)} = \cos\left(\frac{pos}{10000^{\frac{2i}{d_{model}}}}\right)$$ (5)

where pos and i are the position and dimension of the input, and d_{model} is the hidden size of the Transformer model.

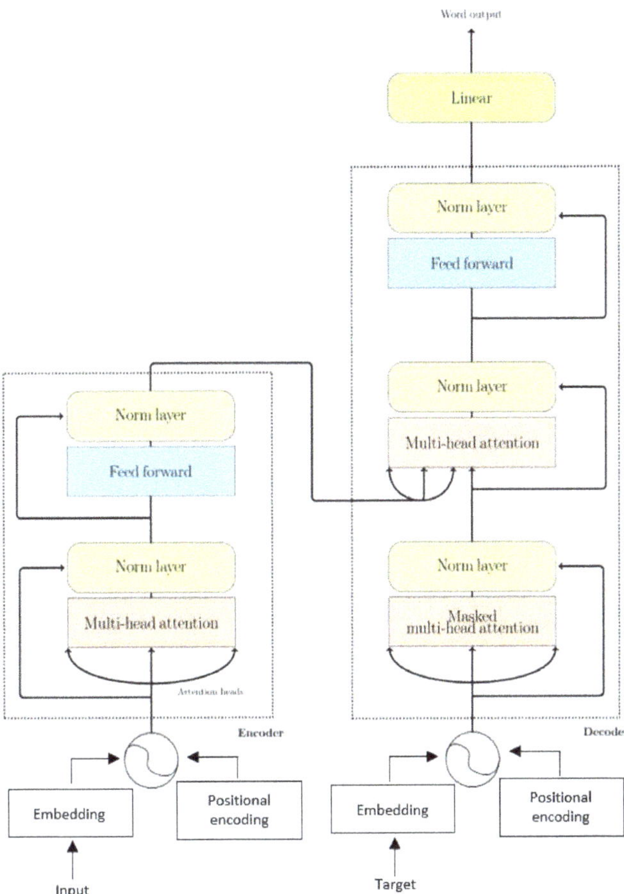

Figure 6. Structure of the Transformer model.

3.2.4. Named Entity Recognition (NER)

The NER model is used to identify the context of the recognized texts from the transcribed receipt. In this paper, a Natural Language Processing (NLP) model called spaCy [22] is applied. spaCy allows the model to recognize a wide range of words or entities. The NER model is composed of Convolutional Neural Network (CNN) and Long Short-Term Memory (LSTM) models [23], and it is processed based on a transition-based model described in the paper by Lample et al. [24].

A transition-based model directly identifies the suitable representations of a multi-token. The model is built as a stacked data structure to obtain the input's chunks in predicting the following actions. Lample et al. used a stacked LSTM model, enabling the stacked object embedding through the push and pop operations [24]. Stack LSTM is used to compute the dimensional embedding of the stack content, buffer, output, and actions taken at each time step, which represents the distribution of the possible action at each time step. Thus, the model aims to maximize the conditional probability of action sequences based on the given sentence input. The maximum probability of the action is chosen until the chunking algorithm meets the termination condition.

3.3. Implementation Details

This section describes the training procedure of the proposed pipeline. Firstly, the YOLO-v5 model is used for multi-text type classification. In this study, the YOLO-v5x model was chosen as it has shown a very promising result from the established results. An annotation software called LabelImg [25] was used to annotate the collected dataset to train the YOLOv5 model. The image is segmented into three classes (printed, handwritten, and non-text), as shown in Figure 7. The non-text category contains unreadable texts such as logos, signatures, and others. Three hundred receipt images were used, with a train and validation split of 5:1. After that, the model was trained for 100 epochs, where the best model with the highest precision score was saved. The total number of characters included in the experiment was 96, with a maximum length of 64.

(a)   (b)   (c)

Figure 7. Sample segmented images: (**a**) printed (**b**) handwritten (**c**) non-text.

After the handwritten segments were obtained from YOLOv5, we combined the ResNet-101 and Transformer models, named ResNet-101T, for HTR. ResNet-101 is used as the feature extractor, the backbone of the proposed architecture where the linear projection layer is excluded, and Transformer is used to analyze the extracted features, as inspired by [26]. In [26], CNN is used together with Transformer to achieve the object detection goal. In this paper, ResNet-101 is responsible for learning 2D representations that encompass shape/outline and positional information of the texts/words in the image. The last feature map by ResNet-101 serves as the input to Transformer, which is then flattened with 2D-positional encoding and passed to the encoder. The ResNet-101 feature map also serves as the input for the decoder, which is the target in this case. The output embeddings of the decoder are then passed to the final linear layer for predictions.

Both the ResNet-101 and Transformer (ResNet-101T) are jointly trained. Two inputs are fed into the model: handwritten segments and the label. The image is fed into ResNet-101T with the shape $30 \times 375 \times 3$. The label length is 64 characters. The input shape and the corresponding number of parameters for each layer of ResNet-101T are given in Table 2. The total number of text segments used was 15,297, and the train, test, validation split ratio was 8:1:1. The model was trained for 100 epochs, and the hyperparameters were tuned to find the optimal result in terms of character error rate (CER) and word error rate (WER). The hyperparameters that are tuned in the experiments include cell units of the ResNet-101 and the number of decoders, encoders, attention heads of the Transformer model.

Table 2. Output shape and number of parameters of each layer in the proposed ResNet-101T model.

Layer (Type)	Input Shape	Parameter
ResNet	[1, 3, 30, 375]	9408
Embedding layer	[1, 63]	25,344
Positional Encoding	[1, 63, 256]	0
Transformer Encoder	[12, 1, 256]	5,260,800
Transformer Decoder	[12, 1, 256], [63, 1, 256]	6,315,520
Linear	[1, 63, 256]	25,443

Total parameter: 12,151,651
Trainable parameter: 12,151,651
Non-trainable parameters: 0

In training the NER model, a tag editor is used to annotate the text data, where the text data is obtained from the annotated text labels. Unlike the labeled data for the HTR model recognition, the text labels are annotated for both printed and handwritten texts. The model is trained for 100 epochs, and the best model with the highest accuracy is saved. The dataset contains a total of 54,522 tokens, with 954 sentences. Tokens refer to chunks within a sentence, or the string between spaces and punctuation symbols, while sentences represent the sequence of tokens. Eleven attributes, such as the clinic's name, address, and contact details, were identified in the study.

4. Experimental Results

This section presents the experimental results of the proposed HTR pipeline. The experimental setup and performance of each model are provided in the following sections.

4.1. Experimental Setup

The proposed HTR pipeline consists of three funnels: ROI localization, handwritten word recognition, and context recognition. Each of these funnels is evaluated separately. The experiments are conducted using a The experiments are conducted using a laptop with a processor of Intel(R) Core (TM) i7-10875H CPU @2.30 GHZ and an GeForce RTX 3060 GPU manufactured by NVIDIA corporation, purchased from the supplier Illegear, Johor, Malaysia.

4.2. You Only Look Once v5 (YOLOv5)

The YOLOv5 model has demonstrated promising results in accurately identifying the ROIs of the printed text, handwritten text, and non-text. Figure 8 shows a sample output of the detected regions from a receipt, along with the confidence score. The model has a very high mean average precision (mAP) of 0.5 confidence, at 91.78%, where the mAP of confidence score of 0.5 and above show a lower expectancy, at 52.75%. Moreover, the model can achieve a precision of 91.61% and a recall of 86.24%. Throughout the training, the loss graphs of the model in detecting the bounding boxes, the error between the detected objects, and the classification loss showed a steady decreasing trend. The learning rate also decreased steadily after 20 iterations. Figure 9 illustrates the metric changes throughout the training period.

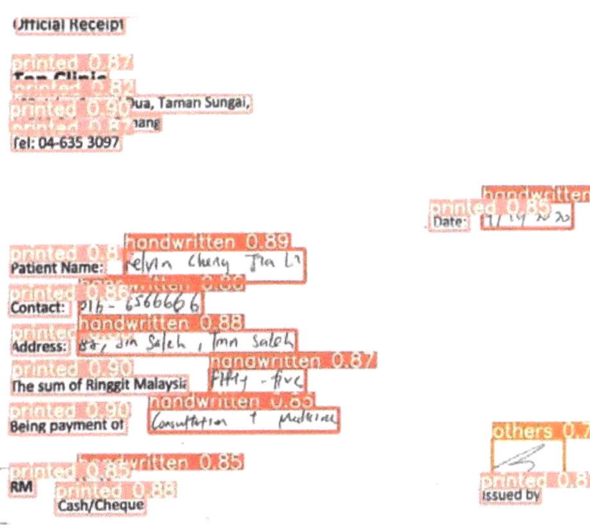

Figure 8. Sample detected output of the YOLOv5 model.

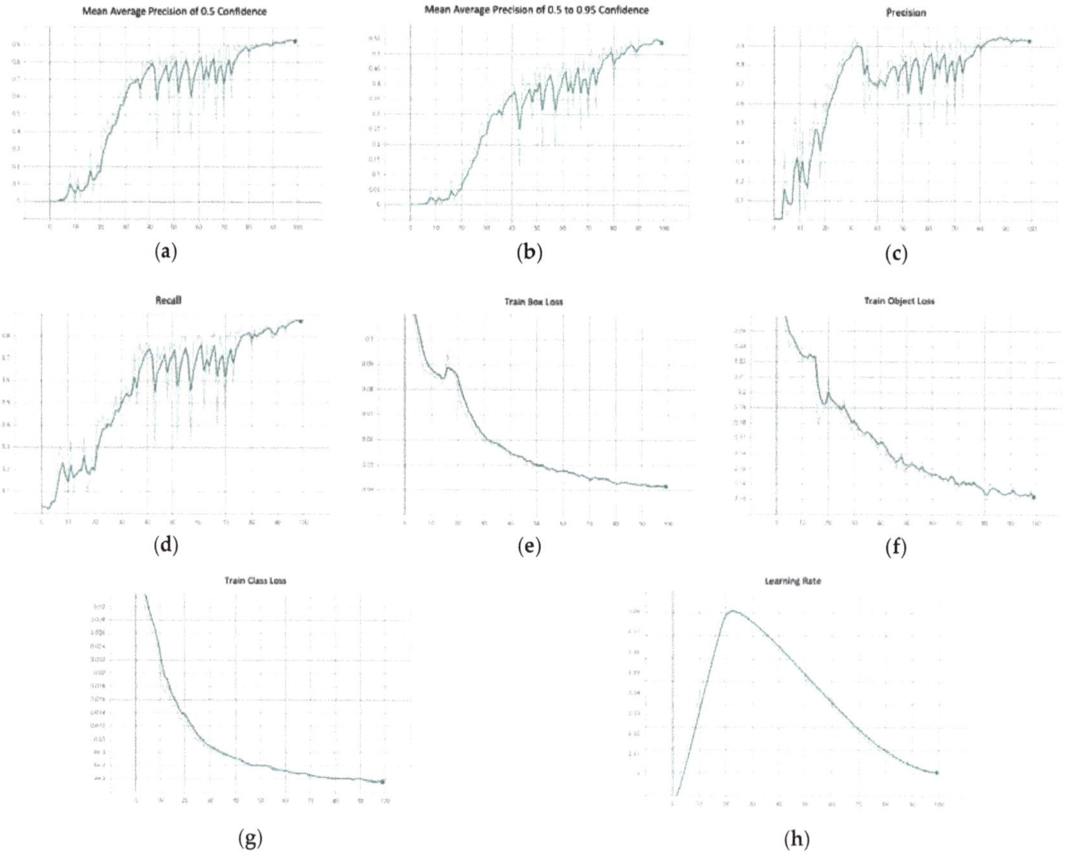

Figure 9. Experimental results of the Yolov5 model: (**a**) mean average precision of 0.5 confidence; (**b**) mean average precision of 0.5 to 0.95 confidence; (**c**) precision; (**d**) recall; (**e**) train box loss; (**f**) train object loss; (**g**) train class loss; (**h**) the learning rate.

4.3. ResNet-101 with Transformer (ResNet-101T)

The proposed ResNet-101T model is trained for 100 epochs. ResNet-101 is used for feature extraction, where the extracted features are fed as input to the Transformer to identify the underlying information based on the image pixels. The ResNet output is in a 2D format as the last two layers are dropped. Figure 10 shows some examples of the extracted feature maps from the first layer of ResNet. The topology of the words is still visible in the ResNet output, which encodes positional information, i.e., word sequence. After that, a 2D positional encoding is used to flatten the 2D representation into a 1D sequence. We believe the feature vector possesses some sequential information. This is where the Transformer model plays a role in processing the sequential data.

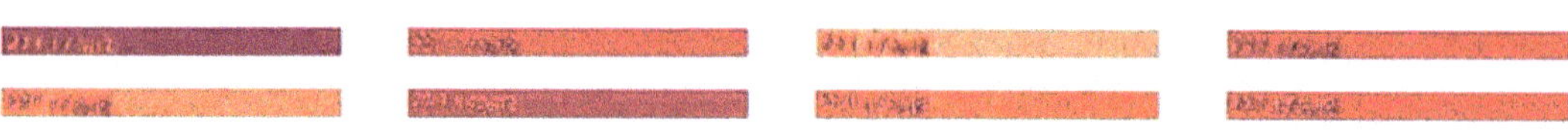

Figure 10. Example of extracted features of the first layer of ResNet-101.

The model achieved a character error rate (CER) and word error rate (WER) of 7.77% and 10.77% on the testing data. In addition, the model demonstrated a stable decrease in both the training and validation losses. To improve the performance of the proposed model, hyperparameter tuning was carried out. The hyperparameters being tuned included the number of heads, encoders, decoders, and the hidden dimension size of the ResNet-101T. The hyperparameter tuning process took a very long time, as a more complex structure is more computationally intensive. Due to training time constraints, the training epoch was fixed at 100 for each setting. Table 3 shows the experimental result of hyperparameter tuning in terms of CER, WER, and time in seconds. The initial model, with the setting of 4 encoders, decoders, and attention heads, yields the best performance among the competing models. This is possible as models with a more complex structure would require a longer convergence time. Additionally, a too-complex structure might lead to overfitting. Therefore, the two models with the settings of the number of encoders, decoders, attention heads of (6, 6, 4) and (8, 8, 8), and cell units of 1024 were terminated earlier due to the extremely long training time, and no significant loss reduction was noticed after ten epochs.

Table 3. Hyperparameter tuning results of the proposed ResNet-101T model.

Number of Encoders, Decoders, Attention Heads	Unit Dimension		
	256	512	1024
4, 4, 4			
CER	7.77	8.66	20.93
WER	10.77	11.81	29.02
Times (Second)	350,864	392,254	592,497
6, 6, 4			
CER	9.15	11.87	86.76
WER	13.08	16.93	87.39
Times (Second)	388,399	487,445	73,817 (Stopped at 11)
8, 8, 8			
CER	12.96	13.35	1.0
WER	17.15	19.25	1.0
Times (Second)	429,401	539,650	87,148 (Stopped at 11)

4.4. Named Entity Recognition (NER)

There are a total of 11 attributes contained in the NER model, including medicine, payment, clinic name, address, contact, website, receipt number, name, email, and service. The model's performance is measured in terms of accuracy, entities precision, recall, and f1-score. Table 4 summarizes the model performance of each entity. Generally, the trained NER model achieved a promising result, with a full score for accuracy, precision, recall, and f1-score for all the attributes except payment. The results demonstrate that the NER model can perform well on the context recognition task.

Table 4. NER model performance on entities recognition.

Entities	Precision	Recall	F1-Score
Medicine	1	1	1
Payment	0.9967	1	0.9983
Clinic Name	1	1	1
Address	1	1	1
Contact	1	1	1
Website	1	1	1
Receipt Number	1	1	1
Name	1	1	1
Email	1	1	1
Service	1	1	1

4.5. Comparisons with State-of-the-Art Methods

To validate the effectiveness of the proposed model, a comparison is made with LSTM [27] and Visual Transformer (ViT) [28] models. LSTM was selected for benchmark comparison as it is the state-of-the-art sequential processing method in HTR. Moreover, ViT was investigated to demonstrate the superiority of the proposed ResNet-101T method over the sole Transformer model. For a fair comparison, the same setting is imposed in all the experiments, such as the dataset split ratio. ViT was trained with a total of 1600 steps and evaluated at every 200 steps. The LSTM model achieved a CER and WER of 11.55 and 26.64, and ViT had a CER and WER of 10.60 and 18.41, where the performance of both models was inferior to ResNet-101T, as presented in Table 5. In terms of speed comparison, the LSTM model used the least computational time for training while ViT took a much longer time to train. Note that although LSTM has a faster training speed due to its simpler architecture, its accuracy is much lower than the proposed ResNet-101T model. Figure 11 shows the training and validation loss of the models throughout training.

Table 5. Result comparison of Transformer and LSTM.

Model	CER	WER	Computational Time for Model Training (s)
Proposed ResNet-101T	7.77	10.77	350,864
LSTM [28]	11.55	26.64	87,148
ViT [29]	12.47	20.18	571,428

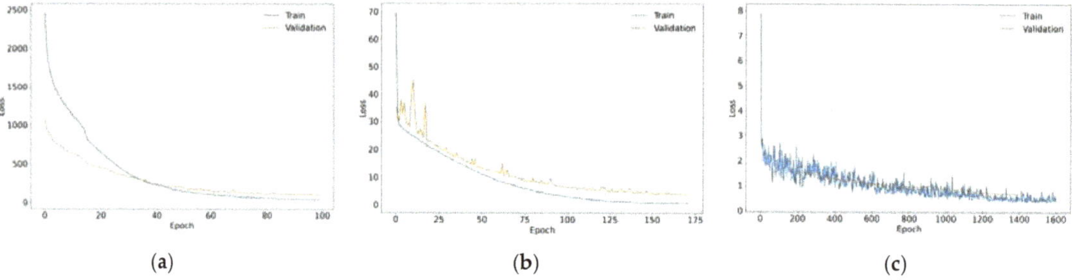

(a)   (b)   (c)

Figure 11. Experimental results of the ResNet-101T: (**a**) training and validation loss of the Transformer model; (**b**) training and validation loss of the LSTM model; (**c**) training and validation loss of ViT.

4.6. Demonstration

This section demonstrates some of the sample input and output of the proposed method. Figures 12 and 13 show the transcription results of the proposed HTR pipeline for different receipt templates. The printed text region is highlighted in blue, while the handwritten text region is bounded with a red square. A label is displayed at the top left corner of the square if the NER model identifies any underlying information from the text regions. We observe that the proposed model can correctly recognize the context of the transcribed texts. However, the application of OCR to printed texts sometimes failed to recognize the texts appropriately. This is considered a future work to enhance the model's performance.

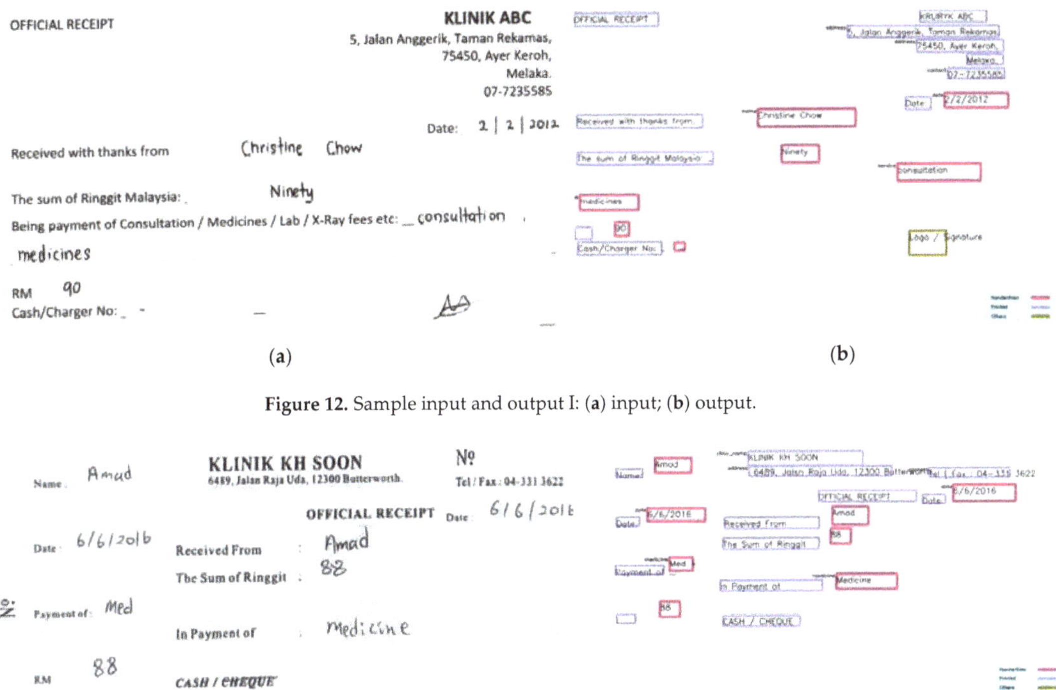

Figure 12. Sample input and output I: (**a**) input; (**b**) output.

Figure 13. Sample input and output II: (**a**) input; (**b**) output.

5. Discussion

Some interesting findings have been discovered in this study. Real-life documents contain a substantial amount of noise (the noise can occur before and after the digitization process). Thus, the document layout should be properly analyzed, and data pre-processing plays an important role in treating different types of documents. In contrast to the line segmentation technique [3] and the full-page document recognition method [4], we applied an object detection approach for implicit ROI localization. The proposed approach was able to perform text type classification at the same time. There generally exist different text types in handwritten receipts, such as printed, handwritten, and non-text. Different text types should be treated individually to ensure optimal performance, rather than considering all text types in one go, which might result in inferior performance. Thus, explicitly segmenting different ROIs according to the text types would help to increase recognition accuracy and is more suitable for real-life applications.

Along this line, we find that a single HTR model cannot cope with the different text types. For example, a model that is good at recognizing printed texts does not necessarily work well with handwritten texts. Moreover, a separate model needs to be developed to recognize non-text attributes such as signature, which is considered vital for documentation. Therefore, a pipeline approach was proposed by training different models to deal with different text types. The transcribed texts are not directly useful for practical applications. Hence, a NER module was introduced to assign the transcribed texts into their corresponding entities/groups (e.g., name, date, address, phone number). The pipeline approach ensures a fully automated workflow with better efficiency, recognition accuracy, and data management ability in a practical application.

We also wish to highlight that the ResNet-101T model is proposed due to the following reasons. First, ResNet-101 has the advantage of being able to minimize negative effects when the depth of the network is increased. Second, the Transformer model can better model the words input with the attention mechanism compared to the other RNN variations. Moreover, it also has a relatively low inductive bias compared to its RNN counterpart. In addition, the experimental result suggests the application of ResNet can effectively extract the 2D representative features, such as the shape/outline and positional information of the words for training the Transformer model. The result of feeding ResNet output to the Transformer model is better than using raw text input for the Transformer model. Empirical results show that the proposed model has clearly outperformed LSTM and ViT in terms of CER and WER. Although LSTM takes less time for training due to its relatively simpler architecture, its accuracy is far inferior to the proposed ResNet-101T method.

6. Conclusions and Future Works

A system that can recognize human handwritten text is significantly essential in automatic information storage and management. This paper presents a pipeline approach towards HTR. The proposed approach is composed of ROI localization, text type classification, text recognition, and context recognition funnels. A ResNet-101T model is introduced to recognize handwritten texts. The proposed model, trained using a self-collected clinical receipts dataset containing 15,297 text segments, achieves a CER and WER of 7.77% and 10.77%, respectively. In addition, more experimental studies can be carried out to investigate the use of ViT for HTR, such as fine-tuning and cross-validation.

For future endeavors, more training data will be collected to enhance the system's efficiency and accuracy. More receipt samples will be distributed to collect different handwritten styles, such as the clinicians and medical staff in the hospital, who might have messier handwritten styles. In this way, the proposed approach can be fine-tuned and applied in real-life scenarios. The HTR pipeline can be further improved and extended to different domains, such as clinical reports and receipts updates, insurance records, industrial documents, and others. In addition, Explainable AI techniques [29,30] are considered for future works to enhance the model's explainability and to learn meaningful representations to improve the model's performance.

Author Contributions: Conceptualization, Y.F.T. and T.C.; Methodology, Y.F.T., T.C. and M.K.O.G.; Software, M.K.O.G.; Validation, A.B.J.T.; Writing—original draft, Y.F.T. and T.C.; Writing—review and editing, A.B.J.T. All authors have read and agreed to the published version of the manuscript.

Funding: This work was supported by the National Research Foundation of Korea (NRF) grant funded by the Korea government (MSIP) (NO. NRF-2019R1A2C1003306).

Institutional Review Board Statement: Not applicable.

Informed Consent Statement: Informed consent was obtained from all subjects involved in the study.

Data Availability Statement: The data presented in this study are available at https://github.com/yeefantan/ResNet-101T-for-HCR/blob/main/README.md (accessed on 22 January 2022).

Conflicts of Interest: The authors declare no conflict of interest.

References

1. Chowdhury, A.; Vig, L. An Efficient End-to-End Neural Model for Handwritten Text Recognition. *arXiv* **2018**, arXiv:1807.07965. Available online: http://arxiv.org/abs/1807.07965 (accessed on 22 October 2021).
2. Chung, J.; Delteil, T. A Computationally Efficient Pipeline Approach to Full Page Offline Handwritten Text Recognition. *arXiv* **2020**, arXiv:1910.00663. Available online: http://arxiv.org/abs/1910.00663 (accessed on 22 October 2021).
3. Ingle, R.R.; Fujii, Y.; Deselaers, T.; Baccash, J.; Popat, A.C. A Scalable Handwritten Text Recognition System. In Proceedings of the 2019 International Conference on Document Analysis and Recognition (ICDAR), Sydney, Australia, 20–25 September 2019; IEEE: New York, NY, 2019; pp. 17–24. [CrossRef]
4. Singh, S.S.; Karayev, S. Full Page Handwriting Recognition via Image to Sequence Extraction. *arXiv* **2021**, arXiv:2103.06450. Available online: https://arxiv.org/abs/2103.06450 (accessed on 22 October 2021).
5. Zhang, X.; Yan, K. An Algorithm of Bidirectional RNN for Offline Handwritten Chinese Text Recognition. In *Intelligent Computing Methodologies*; Huang, D.-S., Huang, Z.-K., Hussain, A., Eds.; Springer International Publishing: Cham, Switzerland, 2019; pp. 423–431.

6. Hassan, S.; Irfan, A.; Mirza, A.; Siddiqi, I. Cursive Handwritten Text Recognition using Bi-Directional LSTMs: A Case Study on Urdu Handwriting. In Proceedings of the 2019 International Conference on Deep Learning and Machine Learning in Emerging Applications (Deep-ML), Istanbul, Turkey, 26–28 August 2019; IEEE: New York, NY, USA, 2019; pp. 67–72. [CrossRef]

7. Nogra, J.A.; Romana, C.L.S.; Maravillas, E. LSTM Neural Networks for Baybáyin Handwriting Recognition. In Proceedings of the 2019 IEEE 4th International Conference on Computer and Communication Systems (ICCCS), Singapore, 23–25 February 2019; IEEE: Singapore, 2019; pp. 62–66. [CrossRef]

8. Yu, Y.; Si, X.; Hu, C.; Zhang, J. A Review of Recurrent Neural Networks: LSTM Cells and Network Architectures. *Neural Comput.* **2019**, *31*, 1235–1270. [CrossRef] [PubMed]

9. Vaswani, A.; Shazeer, N.; Parmar, N.; Uszkoreit, J.; Jones, L.; Gomez, A.N.; Kaiser, L.; Polosukhin, I. Attention Is All You Need. *arXiv* **2017**, arXiv:1706.03762. Available online: http://arxiv.org/abs/1706.03762 (accessed on 22 October 2021).

10. Marti, U.-V.; Bunke, H. The IAM-database: An English sentence database for offline handwriting recognition. *Int. J. Doc. Anal. Recognit. IJDAR* **2002**, *5*, 39–46. [CrossRef]

11. Bluche, T. Joint Line Segmentation and Transcription for End-to-End Handwritten Paragraph Recognition. *arXiv* **2016**, arXiv:1604.08352.

12. Wigington, C.; Tensmeyer, C.; Davis, B.; Barrett, W.; Price, B.; Cohen, S. Start, Follow, Read: End-to-End Full-Page Handwriting Recognition. In *Computer Vision—ECCV 20180*; Ferrari, V., Hebert, M., Sminchisescu, C., Weiss, Y., Eds.; Springer International Publishing: Cham, Switzerland, 2018; pp. 372–388. [CrossRef]

13. Asha, K.; Krishnappa, H. Kannada Handwritten Document Recognition using Convolutional Neural Network. In Proceedings of the 2018 3rd International Conference on Computational Systems and Information Technology for Sustainable Solutions (CSITSS), Bengaluru, India, 20–22 December 2018; IEEE: New York, NY, USA, 2018; pp. 299–301. [CrossRef]

14. Wu, K.; Fu, H.; Li, W. Handwriting Text-line Detection and Recognition in Answer Sheet Composition with Few Labeled Data. In Proceedings of the 2020 IEEE 11th International Conference on Software Engineering and Service Science (ICSESS), Beijing, China, 16–18 October 2020; IEEE: New York, NY, USA, 2020; pp. 129–132. [CrossRef]

15. Jocher, G.; Stoken, A.; Chaurasia, A.; Borovec, J.; Xie, T.; Kwon, Y.; Michael, K.; Changyu, L.; Fang, J.; Abhiram, V.; et al. Ultralytics/Yolov5: v6.0—YOLOv5n "Nano" Models, Roboflow Integration, TensorFlow Export, OpenCV DNN Support, Zenodo. 2021. Available online: https://zenodo.org/record/5563715#.YgYOB-pByUk (accessed on 22 October 2021).

16. Thakare, S.; Kamble, A.; Thengne, V.; Kamble, U. Document Segmentation and Language Translation Using Tesseract-OCR. In Proceedings of the 2018 IEEE 13th International Conference on Industrial and Information Systems (ICIIS), Rupangar, India, 1–2 December 2018; IEEE: New York, NY, USA, 2018; pp. 148–151. [CrossRef]

17. Cantoni, V.; Mattia, E. Hough Transform. In *Encyclopedia of Systems Biology*; Dubitzky, W., Wolkenhauer, O., Cho, K.-H., Yokota, H., Eds.; Springer: New York, NY, USA, 2013; pp. 917–918. [CrossRef]

18. Galamhos, C.; Matas, J.; Kittler, J. Progressive probabilistic Hough transform for line detection. In Proceedings of the 1999 IEEE Computer Society Conference on Computer Vision and Pattern Recognition (Cat. No PR00149), Fort Collins, CO, USA, 23–25 June 1999; Volume 1, pp. 554–560. [CrossRef]

19. Bradski, G. The openCV library. *Dr. Dobb's J. Softw. Tools Prof. Program.* **2000**, *25*, 120–123.

20. Redmon, J.; Farhadi, A. YOLOv3: An Incremental Improvement. *arXiv* **2018**, arXiv:1804.02767. Available online: http://arxiv.org/abs/1804.02767 (accessed on 9 November 2021).

21. Tzutalin, LabelImg. Available online: https://github.com/tzutalin/labelImg (accessed on 22 October 2021).

22. He, K.; Zhang, X.; Ren, S.; Sun, J. Deep Residual Learning for Image Recognition. In Proceedings of the 2016 IEEE Conference on Computer Vision and Pattern Recognition (CVPR), Las Vegas, NV, USA, 27–30 June 2016; IEEE: New York, NY, USA; pp. 770–778. [CrossRef]

23. Carion, N.; Massa, F.; Synnaeve, G.; Usunier, N.; Kirillov, A.; Zagoruyko, S. End-to-End Object Detection with Transformers. *arXiv* **2020**, arXiv:2005.12872. Available online: http://arxiv.org/abs/2005.12872 (accessed on 1 December 2021)

24. Honnibal, M.; Montani, I.; Van Landeghem, S.; Boyd, A. spaCy: Industrial-Strength Natural Language Processing in Python, (2020). Available online: https://zenodo.org/record/5764736#.YgYOaupByUk (accessed on 1 December 2021).

25. Tarcar, A.K.; Tiwari, A.; Dhaimodker, V.N.; Rebelo, P.; Desai, R.; Rao, D. NER Models Using Pre-Training and Transfer Learning for Healthcare. *arXiv* **2019**, arXiv:1910.11241. Available online: http://arxiv.org/abs/1910.11241 (accessed on 22 October 2021).

26. Lample, G.; Ballesteros, M.; Subramanian, S.; Kawakami, K.; Dyer, C. Neural Architectures for Named Entity Recognition. *arXiv* **2016**, arXiv:1603.01360. Available online: http://arxiv.org/abs/1603.01360 (accessed on 23 October 2021).

27. Wigington, C.; Stewart, S.; Davis, B.; Barrett, B.; Price, B.; Cohen, S. Data Augmentation for Recognition of Handwritten Words and Lines Using a CNN-LSTM Network. In Proceedings of the 2017 14th IAPR International Conference on Document Analysis and Recognition (ICDAR), Kyoto, Japan, 9–15 November 2017; pp. 639–645. [CrossRef]

28. Li, M.; Lv, T.; Cui, L.; Lu, Y.; Florencio, D.; Zhang, C.; Li, Z.; Wei, F. TrOCR: Transformer-based Optical Character Recognition with Pre-trained Models. *arXiv* **2021**, arXiv:2109.10282. Available online: http://arxiv.org/abs/2109.10282 (accessed on 6 December 2021).

29. Zhou, X.; Jin, K.; Shang, Y.; Guo, G. Visually Interpretable Representation Learning for Depression Recognition from Facial Images. *IEEE Trans. Affect. Comput.* **2018**, *11*, 542–552. [CrossRef]

30. Zhou, B.; Khosla, A.; Lapedriza, A.; Oliva, A.; Torralba, A. Learning deep features for discriminative localization. In Proceedings of the IEEE Conference on Computer Vision and Pattern Recognition, Las Vegas, NV, USA, 27–30 June 2016; pp. 2921–2929.

applied sciences

Article

Constrained Backtracking Matching Pursuit Algorithm for Image Reconstruction in Compressed Sensing

Xue Bi [1,*], Lu Leng [2,3,*], Cheonshik Kim [4,*], Xinwen Liu [5], Yajun Du [6] and Feng Liu [5]

1   School of Electrical Engineering and Electronic Information, Xihua University, Chengdu 610039, China
2   School of Software, Nanchang Hangkong University, Nanchang 330063, China
3   School of Electrical and Electronic Engineering, College of Engineering, Yonsei University, Seoul 05006, Korea
4   Department of Computer Engineering, Sejong University, Seoul 05006, Korea
5   School of Information Technology and Electrical Engineering, The University of Queensland,
    Brisbane 4072, Australia; xinwen.liu@uq.net.au (X.L.); feng@itee.uq.edu.au (F.L.)
6   Information and Network Center, Xihua University, Chengdu 610039, China; duyajun@mail.xhu.edu.cn
*   Correspondence: bixue@mail.xhu.edu.cn (X.B.); leng@nchu.edu.cn (L.L.); mipsan@sejong.ac.kr (C.K.)

Abstract: Image reconstruction based on sparse constraints is an important research topic in compressed sensing. Sparsity adaptive matching pursuit (SAMP) is a greedy pursuit reconstruction algorithm, which reconstructs signals without prior information of the sparsity level and potentially presents better reconstruction performance than other greedy pursuit algorithms. However, SAMP still suffers from being sensitive to the step size selection at high sub-sampling ratios. To solve this problem, this paper proposes a constrained backtracking matching pursuit (CBMP) algorithm for image reconstruction. The composite strategy, including two kinds of constraints, effectively controls the increment of the estimated sparsity level at different stages and accurately estimates the true support set of images. Based on the relationship analysis between the signal and measurement, an energy criterion is also proposed as a constraint. At the same time, the four-to-one rule is improved as an extra constraint. Comprehensive experimental results demonstrate that the proposed CBMP yields better performance and further stability than other greedy pursuit algorithms for image reconstruction.

Keywords: constrained backtracking matching pursuit; sparse reconstruction; compressed sensing; greedy pursuit algorithm; image processing

Citation: Bi, X.; Leng, L.; Kim, C.; Liu, X.; Du, Y.; Liu, F. Constrained Backtracking Matching Pursuit Algorithm for Image Reconstruction in Compressed Sensing. *Appl. Sci.* **2021**, *11*, 1435. https://doi.org/10.3390/app11041435

Academic Editor: Andrés Márquez

Received: 17 January 2021
Accepted: 2 February 2021
Published: 5 February 2021

Publisher's Note: MDPI stays neutral with regard to jurisdictional claims in published maps and institutional affiliations.

1. Introduction

Image reconstruction is a significant application of multimedia signal processing. Compressed sensing (CS) is a technique that reconstructs sparse, compressible signals from under-determined random linear measurements. Over the past few decades, CS has been widely applied to image processing, including image reconstruction [1–5] and acquisition [6–8].

Various algorithms have been proposed for CS-based signal reconstruction with sparse constraints [9], which can be categorized into three classes. The first class is the non-convex optimization [10], such as re-weighted l_1 norm minimization [11] and l_q norm minimization [12]. However, non-convex optimization is a non-deterministic polynomial (NP)-hard problem, which is hard to solve. The second class focuses on convex optimization based on the minimization of the l_1 norm. The basis pursuit (BP) algorithm is typically used for convex optimization, but its l_1 norm-based cost function is sometimes not differentiable. It also involves high computational complexity, thus limiting its practical applications [13–15].

The third category includes a set of greedy pursuit algorithms, which are to easy implement and have low computational complexity [13–21]. Specifically, orthogonal matching pursuit (OMP) [15–17], stage-wise OMP (StOMP) [18], and regularized orthogonal matching pursuit (ROMP) [19,20] have been proposed. The reconstruction complexity of basic greedy pursuit algorithms is roughly about O(kMN), which is much lower than that of BP algorithm.

While the greedy pursuit algorithms show superiority in easy implementation and computational efficiency, they typically require additional measurements for reconstruction and lack stable reconstruction capability. The problem is alleviated when backtracking is introduced. For example, the subspace pursuit (SP) algorithm [21] and compressive sampling pursuit (CoSaMP) algorithm [22] have been proposed based on the backtracking scheme. The difference between SP and CoSaMP is that the latter chooses $2k$ indices to combine the estimated support set from the previous iteration. However, it is necessary to estimate the sparsity level of signal k before applying SP and CoSaMP. Indeed, it is impractical to know the accurate sparsity level k of unknown signals in advance.

Then, sparsity adaptive matching pursuit (SAMP), which can recover signals without knowing the sparsity level, was proposed by Do et al. [23]. It alternatively estimates the sparsity level when the residue's energy increases between two consecutive stages and updates the support set size of the signal using a fixed and small step size. SAMP has apparent advantages when processing one-dimensional sparse signals. However, since one is used as the initial step size, when processing high-dimensional signals, the small step size significantly affects the result and efficiency of reconstruction. To further improve the reconstruction performance, an energy-based adaptive matching pursuit (EAMP) has been proposed [24]. One limitation of EAMP is that it only focuses on the binary signal reconstruction. Rasha et al. used the structured Wilkinson matrix as the measurement matrix to improve the efficiency of SAMP [25]. More recently, the improved generalized sparsity adaptive matching pursuit (IGSAMP) algorithm has been proposed. This algorithm uses a nonlinear step size to approximate the sparsity level, and only a small initial step size can be selected. Meanwhile, it requires carefully choosing the parameters without referring to the sensitivity of a large step size [26].

To improve the reconstruction performance of the sparsity adaptive matching pursuit algorithm and make it less sensitive to the step size, we propose a compositely constrained backtracking matching pursuit (CBMP) algorithm for image reconstruction. The main contributions of this paper are summarized as follows.

(1) The restricted isometry property (RIP) is analyzed, and the relationship between observed values and signals is derived and demonstrated.

(2) The reconstruction process is divided into three stages, including the large step size stage, small step size stage, and support set update stage. Different step sizes are used in these stages.

(3) A backtracking threshold operation is proposed, which adopts a composite strategy and uses dedicated parameters to control the different step sizes in the reconstruction process.

(4) The proposed algorithm can achieve satisfactory reconstruction performance and overcome the sensitivity to step size.

2. Preliminaries

2.1. A Review of Compressed Sensing

CS compresses the signal at the time of sampling while maintaining the ability to reconstruct the original signal. For a signal $x \in \mathbb{R}^N$ that has at most k terms as nonzero components in some bases Ψ, the compressed signal y is obtained through the following linear transform:

$$y = \Phi\Psi x \tag{1}$$

where y is an $M \times 1$ vector and Φ denotes an $M \times N$ random measurement matrix with $M \ll N$.

In general, M is much larger than N, so the reconstruction x from the measurements y can be solved by forming an underdetermined set of linear equations. Thus, the CS reconstruction is generally an ill-posed problem. To guarantee an exact reconstruction of every k sparse signal, one of the most important assumptions of CS is that the measurement matrix Φ satisfies the restricted isometry property (RIP) [27,28] with parameters (k, δ) [29–31].

$$(1 - \delta_k)\|x\|_2^2 \leq \|\Phi x\|_2^2 \leq (1 + \delta_k)\|x\|_2^2 \tag{2}$$

where δ_k is the RIP constant and $0 < \delta_k < 1, k < M$.

When a matrix satisfies the RIP, the lengths of all sufficiently sparse vectors are approximately preserved under the matrix transformation [29]. In [19,21], it was demonstrated that if $\delta_{2K} < \sqrt{2} - 1,$, then the signal can be exactly reconstructed via a finite number of iterations.

The CS reconstruction aims to find the sparsest possible solution that satisfies Equation (1). Then, the CS model [1,31] is represented as:

$$\min \|\Psi x\|_0 \quad \text{subject to} \quad y = \Phi\Psi x \tag{3}$$

where $\|\Psi x\|_0$ is the l_0 norm and denotes the number of nonzero components in (Ψx).

2.2. A Review of the Greedy Pursuit Algorithms

Among the reconstruction algorithms used in CS, the greedy pursuit algorithms are the most widely used due to their easy implementation and low computational complexity.

The goal of greedy pursuit algorithms is to find the support set of the unknown signal. After finding the support set, the signal can be reconstructed by solving a least squares problem [31–33]. There exit the indices of the optimal support set $J \in \{1, 2, \ldots, n\}$, and z^* satisfies $y = z^* \varphi_J$. φ_J is the J-th column (index) of Φ. Then, the error function $e(j)$ is:

$$e(j) = \min_z \|z\varphi_j - y\|_2^2 = \min_z \left[\left(\varphi_j^T \varphi_j\right)z^2 - 2\left(\varphi_j^T y\right)z + y^T y \right]$$

$$= \min_z \left(\varphi_j^T \varphi_j\right)\left(z - \frac{\varphi_j^T y}{\varphi_j^T \varphi_j}\right)^2 + y^T y - \frac{\left(\varphi_j^T y\right)^2}{\varphi_j^T \varphi_j}$$

Letting $e(j) = 0$, the optimal solution:

$$z^* = \left\{ \begin{array}{l} \dfrac{\varphi_j^T y}{\|\varphi_j\|^2}, j = J \\ 0, \text{ otherwise.} \end{array} \right\} \tag{4}$$

The matching pursuit (MP) algorithm is one of the most classical and primitive greedy pursuit algorithms. As described in Equation (4), only the column J minimizing the error function is selected in each iteration of the MP algorithm [32]. Later on, the OMP algorithm [15] was developed based on the MP algorithm. As stated in OMP, some indices are searched, corresponding to the most significant correlations between the measurement matrix and the residual. In each iteration, only one or more coordinates are selected and added to the support set. These selected coordinates correspond to the columns (indices) of observation matrices with the largest correlation with the residuals. The optimization iterates until the termination condition is satisfied. Finally, the pseudo-inverse of the observation matrix corresponding to the obtained support set is used for signal reconstruction.

CS-based greedy pursuit algorithms adopted in CS include OMP [15–17], StOMP [18], ROMP [19,20], SP [21], CoSaMP [22], SAMP [23], EAMP [24], and IGSAMP [26]. Utilizing some criteria, they can approximate the sparse signals iteratively. Each of the algorithms iteratively computes the estimated support set of the signals. In each iteration, one or several coordinates are added to the support set. In particular, in OMP, only one column of Φ is added to the estimated support set. In StOMP, a hard threshold is used to choose several columns that are to be added to the support set. Both algorithms have to select these columns previously. Otherwise, these algorithms cannot be rectified.

These greedy pursuit algorithms required more measurements for exact reconstruction and lacked stable reconstruction capability until the backtrackingidea was introduced in SP [21] and CoSaMP [22]. Refining the last estimated support set, the backtracking scheme allows eliminating the wrong coordinates, which are selected in the previous iterations. The candidate set is introduced into the greedy pursuit algorithm, which is the key point of the backtracking. However, both SP [21] and CoSaMP [22] require prior knowledge about the sparsity level k, which is impractical to know previously. SAMP [23], on the other hand, was put forward to gradually approach the sparsity level by accumulation with a step size. The SAMP algorithm is shown in Figure 1 and Algorithm 1.

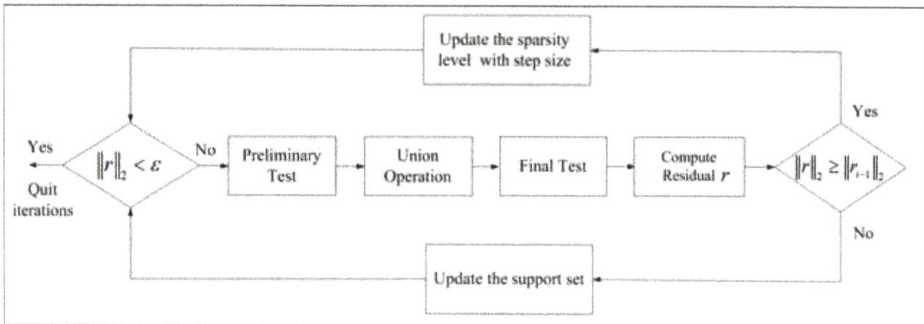

Figure 1. The pipeline of sparsity adaptive matching pursuit (SAMP) [23].

Algorithm 1 Sparsity adaptive matching pursuit algorithm

Input:

$M \times N$ measurement matrix Φ, measurement vector y, step size s

Initialization:

$\hat{x} = 0$ {Trivial Initialization}, $r^0 = y$ {Initial residue}, $U^0 = \varnothing$ {the estimated support set},

$L = s$ {size of the support set}, $j = 1$ { stage index}, $i = 1$ { iteration index}.

Repeat:

1. Preliminary test: find the matched L indices from Φ based on the correlation between Φ and r^{i-1}, that is $D^i = max(|\Phi^T r^{i-1}|, L)$.

2. Make the candidate list: $U^i = T^{i-1} \bigcup D^i, x_{U^i} = \Phi_{U^i}^\dagger y$.

3. Final test: $F = max(|x_{U^i}|, L), x_F = \Phi_F^\dagger y$.

4. Compute residual: $r = y - \Phi_F x_F$.

   if the halting condition is true, then quit the iteration;

   else if $\|r\|_2 \geq \|r_{i-1}\|_2$, then

   $j = j + 1$ {update the stage index}, $L = j \times s$ {update the size of support set};

   else $T^i = F$ {update the support set}, $r^i = r$ { update the residual}, $i = i + 1$.

   end if

   Until the halting condition is true;

Output: $\hat{x} = \Phi_T^\dagger y$ {update the stage index}, $L = j \times s$ {a sparse reconstruction computed by the least squares algorithm}

SAMP uses the "divide and conquer" principle stage-by-stage to estimate the sparsity level and the true support set of the target signals. SAMP applies two tests, namely the preliminary test and the final test, to estimate the signal's support set. The preliminary test is used to implement the selection of the L largest elements corresponding to the most considerable correlation between the residual and the measurement matrix, denoted by $D^i = \max(\Phi^T r^{i-1} |, L)$. After the preliminary test, a candidate list U is created by the union of the chosen list in the preliminary test and the support set in the previous iteration, represented by $U^i = T^{i-1} \cup D^i$. The final test firstly solves a least squares problem to obtain x_{U^i}, and then chooses a subset of the L largest elements from x_{U^i}. This subset of coordinates serves as the support set of the current iteration. The residual is finally updated by subtracting the measured vector y from its projection onto the subspace spanned by the columns in the support set. The pseudo-code of SAMP is summarized below.

$\Phi^\dagger = (\Phi^T \Phi)^{-1} \Phi^T$ represents the pseudo-inverse of Φ, in which Φ^T denotes the transposition of Φ. The main innovation of SAMP is that the increment of the residual is used as the criterion to judge the sparsity level by accumulating with the step size. As previously mentioned, SAMP uses a fixed step size that is sensitive to the reconstruction performance [23]. Specifically, when SAMP is applied to two-dimensional images, the selection of the step size seriously affects the image reconstruction performance due to the lack of flexibility and adaptation in the sparsity level update stage. As shown in Figure 2, the reconstruction performance is affected by the step size. When the step size $s = 64$, the peak signal-to-noise ratio (PSNR) is 24.04 dB, whereas when $s = 512$, the PSNR is 28.44 dB.

(**a**) $s = 64$, PSNR $= 24.04$ dB   (**b**) $s = 512$, PSNR $= 28.44$ dB

Figure 2. Performance of SAMP vs. different step sizes.

Then, a variable step size was proposed in EAMP [24], but it focuses on one-dimensional sparse binary signal reconstruction. Recently, IGSAMP [26] was proposed to improve SAMP. Furthermore, it requires one to carefully choose the parameters and control the variable nonlinear step size in the reconstruction process and does not refer to the sensitivity to the step size. In this paper, we propose an improved adaptive greedy algorithm, whose signal reconstruction performance is relatively insensitive to the step size.

3. The Constrained Backtracking Matching Pursuit Algorithm for Image Reconstruction

To overcome the sensitivity to the step size and improve the reconstruction performance of greedy pursuit algorithms, we propose the CBMP algorithm, which introduces restrictions to the backtracking stage, which provides more flexibility as the algorithm gradually approaches the true sparsity level of the unknown signal. The main steps of CBMP are described as follows:

Considering the signal to be reconstructed is a two-dimensional image, the sparsity level is relatively large; the process of sparsity level estimation is divided into large and small step size estimation stages. In the large step size stage, the increment of the step size is $s^j = 2 \times s^{j-1}$. j denotes the stage iteration index. The increment of the sparsity level in the stage of the small step size is fixed and equals the step size of the previous stage.

Due to the advantages of combing information and improving accuracy [34,35], a composite strategy is proposed to effectively control the increment of the estimated sparsity level in the two stages. It includes two constraints controlled by parameters a and b, which are required in the backtracking threshold operation of CBMP, as described in Algorithm 2. The theoretical support for the composite strategy is clarified as follows:

Theorem 1. *Let $x \in R^N$ be a sparse signal and y be a measurement vector. If the measurement matrix Φ satisfies the RIP, then $\|x\|_2^2 > \frac{1}{\sqrt{2}}\|y\|_2^2$. The proof is presented in Appendix A.*

Algorithm 2 The proposed CBMP algorithm

Input:

$M \times N$ measurement matrix Φ, measurement vector y, step size s^0

Initialization:

$x = 0\{$trivial Initialization $\}; y_r^0 = y\{$initial residue $\}; T^0 = \varnothing\{$the estimated support set $\}; L^0 = s^0\{$size of the support set (sparsity level)$\}; j = 1\{$stage index$\};$

i=1$\{iteration index\}$; $U^0 = \varnothing\{$ union set $\}$

Repeat the following steps until the stopping condition holds:

1. Preliminary test: $v = \Phi^T y_r^{i-1}$, find the matched set $D^i = \{L^{j-1}$ indices corresponding to the largest absolute values of $v\}$, that is $D^i = \max(|\Phi^T y_r^{i-1}|, L^{j-1})$.

2. Union operation: to broaden the selection space and make candidate list U^i : $U^i = T^{i-1} \cup D^i, x_{U^i} = \Phi_{U^i}^\dagger y$.

3. Final test: to obtain the vector x_F^i : find the matched indices F^i based on the largest absolute values of x_{U^i}, that is $F^i = \max(|x_{U^i}|, L^{j-1})$, $\quad x_F^i = \Phi_{F^i}^\dagger y$.

4. Compute residual: $r_F^i = y - \Phi_F x_F^i$.

5. Backtracking threshold operation:

if $\|x_F^i\|_2^2 \leq a\|y\|_2^2$ and size$(U_i) < b \times M$, then shift into the large step size estimation stage: $\quad s^j = 2 \times s^{j-1}, \quad L^j = L^{j-1} + s^j, \quad y_r^i = y_r^{i-1}, j = j+1, i = i+1$, then shift into 1.

if $\|r_F^i\|_2 > \|y_r^{i-1}\|_2$, then shift into the small step size estimation stage: $s^j = s^{j-1}, L^j = L^{j-1} + s^j, \quad y_r^i = y_r^{i-1}, j = j+1, \quad i = i+1$, then shift into 1.

Otherwise, shift into the stage that updates the support set based on the current estimated sparsity level: $y_r^i = r_F^i, L^j = L^{j-1}, \quad T^i = F^i, \quad i = i+1$, then shift into 1.

Output: $x = \Phi_T^\dagger y$ { a sparse reconstruction computed by the least squares algorithm }

According to Theorem 1, the energy of the original signal x is greater than the square root of one half of that of the measurement vector y, that is $\|x\|_2^2 > \frac{1}{\sqrt{2}}\|y\|_2^2$. Different step sizes are used in CBMP. Specifically, the estimated sparsity level is far smaller than the true one at the early stage. Based on this theorem, the energy criterion can be improved by introducing a parameter a to constrain the reconstruction stages.

Inspired by the "four-to-one" practical rule proposed in [27], the measurement number should be four times the signal sparsity level for signal reconstruction. In CBMP, U_i is the union of a new matched set and the estimated support set of the previous iteration. M is the row number of the measurement matrix. We introduce the "four-to-one" rule to CBMP and use the parameter b to constrain the estimation stage. The relationship between the parameters a and b is analyzed in Section 4.

Figure 3 shows the flowchart of the CBMP algorithm. The reconstruction process is divided into the sparsity level update stage and the support set update stage. As for the details, the sparsity level update stage includes both the large and small step size update stages. In the early stage of reconstruction, the estimated sparsity level is far less than the true one, so large step sizes are adopted to estimate the sparsity level. As the iteration goes on, after the threshold condition is satisfied, it enters a small step size stage. The reason why CBMP can achieve better reconstruction performance than SAMP is attributed to its superior capability in handling the wrong indices (atoms). When the current obtained sparsity level is far less than the true one, those false indices can be easily added into the candidate support set. However, these false indices are difficult to eliminate in the later iteration. Therefore, at the beginning of the iteration, a large step size allows those false indices to be filtered out.

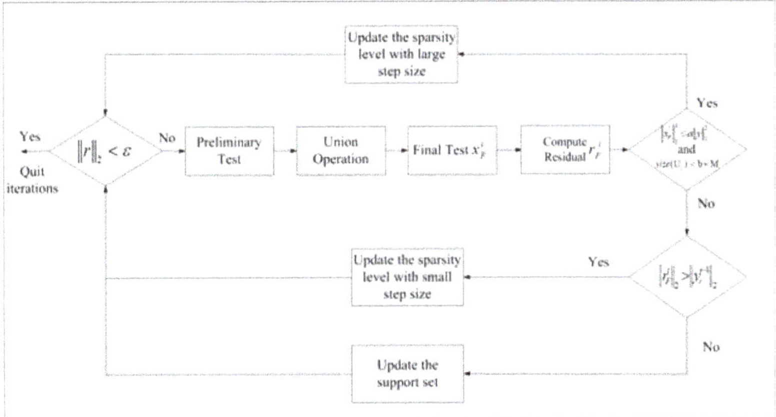

Figure 3. The flowchart of the constrained backtracking matching pursuit (CBMP) algorithm.

4. Experimental Results

Several experiments were conducted to illustrate the performance of the CBMP algorithm. The proposed CBMP was compared with SAMP [23] and IGSAMP [26]. The halting condition used by these algorithms was $\|y_r\| \leqslant 10^{-5}$. For a fair comparison, the same initial step size was used by CBMP, SAMP, and IGSAMP. It should be noted that in SAMP and IGSAMP, the reconstruction results shown in their simulation experiments [23,26] are obtained by a small step size (s = 1). In practical applications, when two-dimensional images are stacked into long one-dimensional vectors, the sparsity level in the transform domain is far greater than one. Correspondingly, the step sizes of the proposed algorithm were relatively large. The step sizes used in the experiment were 64, 128, 256, and 512, respectively. Different sampling rates were used to demonstrate the reconstruction performance of CBMP. The wavelet transform was chosen as the sparse basis to represent images. The quality of recovered images was measured by the peak signal-to-noise ratio (PSNR), which is expressed as:

$$\text{MSE} = \frac{1}{M \times N} \sum_{i=0}^{M-1} \sum_{j=0}^{N-1} |I(i,j) - \hat{I}(i,j)|^2 \tag{5}$$

$$\text{PSNR}(I, \hat{I}) = 10 \log_{10}\left(\frac{\text{MAX}}{\text{MSE}}\right) \tag{6}$$

where $M = N = 512$, $I(i,j)$ denotes the original value of the test image at the position (i,j) and $\hat{I}(i,j)$ denotes the reconstructed value at the position (i,j). The maximum pixel intensity is given as MAX. All images here are expressed using 8 bit intensity values per pixel, so the peak intensity is 255. The experiment configuration is as follows: the CPU was an Intel® Core™ i5-7200U at 2.50 GHz, and the size of the RAM was 8 GB. The programming language used to perform the experiments was MATLAB. Several experiments were conducted to validate the advantages of CBMP.

According to Theorem 1, $\|x\|_2^2 > \frac{1}{\sqrt{2}}\|y\|_2^2$. In CBMP, x_F should gradually approach the true one and x_F is much smaller than x at the beginning. Simultaneously, there are two update stages, and then, the threshold parameter a is contracted within $\frac{1}{4\sqrt{2}}\left(\frac{1}{2} \times \frac{1}{2} \times \frac{1}{\sqrt{2}}\right)$. Our experiments demonstrate that the threshold parameters a and b do not distinctively affect the reconstruction performance if the parameter satisfies $a \leq \frac{1}{4\sqrt{2}}$. In CBMP, the support set of the signal obtained by the current iteration is constrained by the parameter a in the step size update stage, while U_i is the union of the estimated support set of the previous iteration and the currently selected support set. Therefore, the relationship between these two parameters is set as $b = 2a$. These two parameters play different roles, as a is used after the final test, while b corresponds to the union operation after the preliminary test.

The relationship between the threshold parameters and reconstruction performance is shown in Figures 4 and 5. Meanwhile, SAMP and IGSAMP are both tested. Two standard images, "Lena" and "Peppers", were reconstructed to test the reconstruction performance of different parameter pairs (a,b). For a fair comparison, the sampling rate was 0.4, and the same initial step sizes were used. The initial step sizes were chosen from 64 to 512. From Figure 4, we can see that the reconstruction performance of CBMP with different threshold parameters is better than that of SAMP and IGSAMP. For example, when $a = \frac{1}{16\sqrt{2}}$, all the PSNR values of CBMP with different initial step sizes are greater than 33.5. While the initial step size is 512, SAMP achieves the maximum PSNR value, which is less than 32, and IGSAMP offers the maximum PSNR value, which is less than 32.5. Therefore, CBMP offers better reconstruction performance than SAMP and IGSAMP. From Figure 5, it is noticed that if $a \leq \frac{1}{4\sqrt{2}}$, all the PSNR values of CBMP with different step sizes are greater than 31. Therefore, the introduction of the threshold operation is necessary for the improvement of greedy pursuit algorithms. At the same time, threshold parameters do not distinctively affect the reconstruction performance if they are satisfied with the constrained condition in CBMP. Meanwhile, the reconstruction performance of CBMP with $a = \frac{1}{16\sqrt{2}}$ is better than the others; thus, this a value is regarded as the optimal value in the CBMP.

Tables 1 and 2 compare CBMP, SAMP, and IGSAMP in terms of the reconstruction performance (PSNR) on the Lena image with different sampling ratios and initial step sizes. Tables 3 and 4 compare CBMP, SAMP, and IGSAMP in terms of the reconstruction performance (PSNR) on the Peppers image with different sampling ratios and initial step sizes.

In Table 1, when the sampling ratio is 0.3, each PSNR value of the CBMP algorithm is greater than that of SAMP and IGSAMP. For example, with the initial step size of 64, the PSNR value of SAMP and IGSAMP is 25.45 dB and 26.23 dB, respectively, but the PSNR value of CBMP is 32.13 dB. Table 2 shows the PSNR values of SAMP, IGSAMP, and CBMP with the same sampling ratio of 0.4. Different step sizes are used. The PSNR values of SAMP with different initial step sizes range from 26.99 dB to 31.60 dB. The PSNR values of IGSAMP are increased from 27.32 dB to 32.46 dB. However, the PSNR values of CBMP are greater than those of SAMP and IGSAMP, achieving 33.9675 dB as the average value.

Similarly, Table 3 shows the PSNR values of the Peppers image by SAMP, IGSAMP, and CBMP when the sampling ratio is 0.3. Each PSNR value of the CBMP algorithm is greater than that of SAMP and IGSAMP. Table 4 shows the PSNR values of the Peppers image by SAMP, IGSAMP, and CBMP when the sampling ratio is 0.4. Different step sizes are used. For example, as the initial step size is 512, the PSNR value of SAMP and IGSAMP is 30.67 dB and 32.75 dB, respectively, while the PSNR value of CBMP is 32.84 dB. Therefore, CBMP can achieve better reconstruction performance with different sampling ratios and initial step sizes.

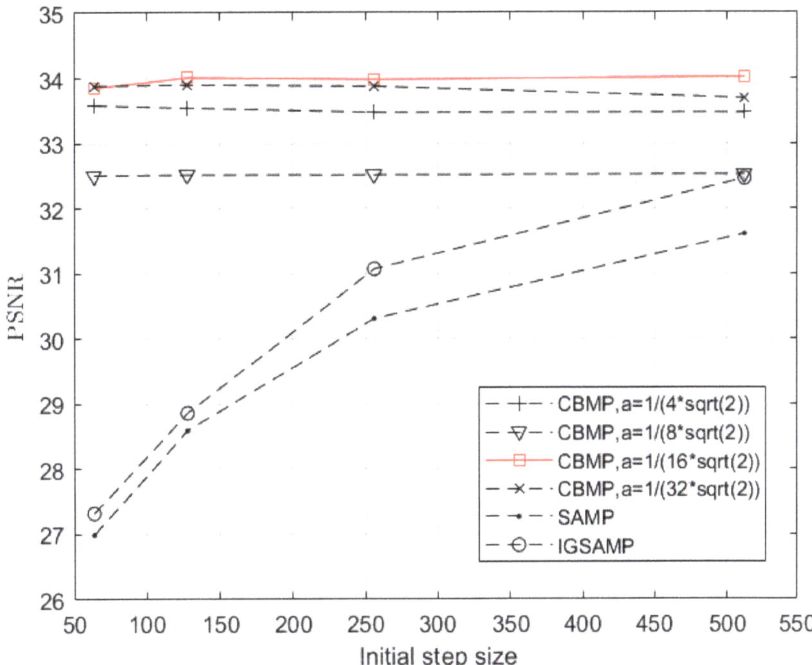

Figure 4. PSNR (dB) under different initial step sizes of the Lena image.

Table 1. PSNR (dB) comparison of the Lena image when the sampling ratio is 0.3.

Initial Step Size	PSNR of SAMP	PSNR of IGSAMP	PSNR of CBMP
64	25.45	26.23	32.13
128	26.41	28.45	31.91
256	27.62	29.89	31.87
512	28.98	31.76	31.83

Table 2. PSNR (dB) comparison of the Lena image when the sampling ratio is 0.4.

Initial Step Size	PSNR of SAMP	PSNR of IGSAMP	PSNR of CBMP
64	26.99	27.32	33.85
128	28.59	28.86	34.01
256	30.31	31.07	33.99
512	31.60	32.46	34.02

Table 3. PSNR (dB) comparison of the Peppers image when the sampling ratio is 0.3.

Initial Step Size	PSNR of SAMP	PSNR of IGSAMP	PSNR of CBMP
64	24.04	25.33	31.46
128	25.19	27.85	31.42
256	26.87	30.54	31.40
512	28.44	31.10	31.38

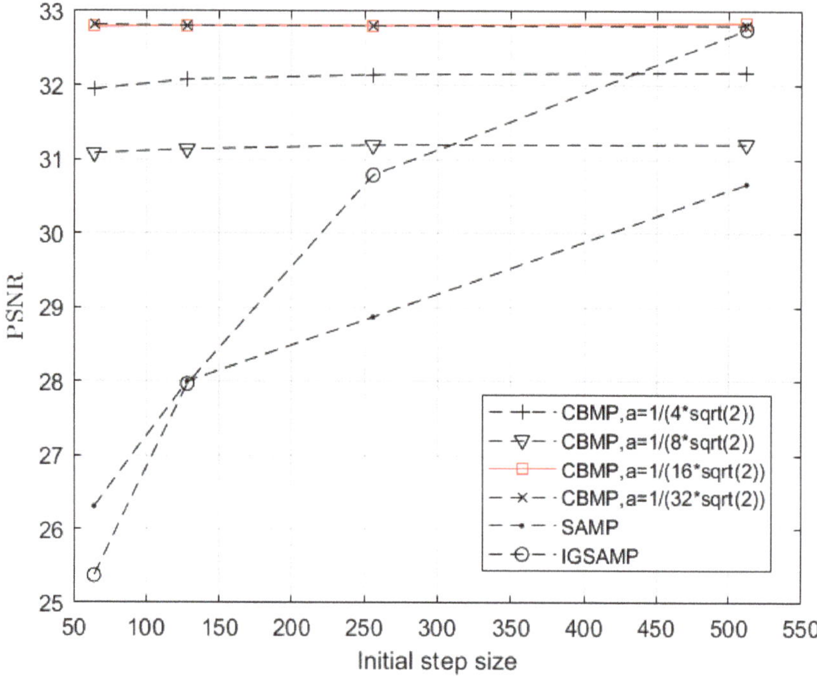

Figure 5. PSNR (dB) under different initial step sizes of the Peppers image.

Table 4. PSNR (dB) comparison of the Peppers image when the sampling ratio is 0.4.

Initial Step Size	PSNR of SAMP	PSNR of IGSAMP	PSNR of CBMP
64	26.30	25.37	32.80
128	28.00	27.96	32.81
256	28.87	30.79	32.81
512	30.67	32.75	32.84

Finally, the reconstructed results of the Lena image using SAMP, IGSAMP, and CBMP are shown in Figures 6 and 7. The sampling rate is 0.3; different step sizes are used. The reconstructed results of the Peppers image using SAMP, IGSAMP, and CBMP are shown in Figures 8 and 9.

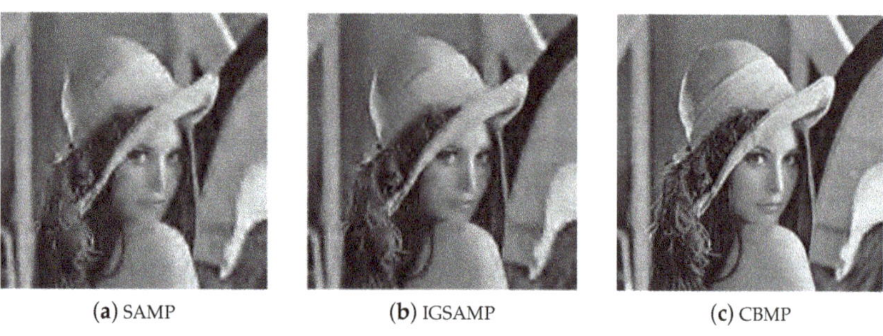

(**a**) SAMP   (**b**) IGSAMP   (**c**) CBMP

Figure 6. Reconstructed results of the Lena image by SAMP, IGSAMP, and CBMP with the initial step size of 64.

(**a**) SAMP   (**b**) IGSAMP   (**c**) CBMP

Figure 7. Reconstructed results of the Lena image by SAMP, IGSAMP, and CBMP with the initial step size of 512.

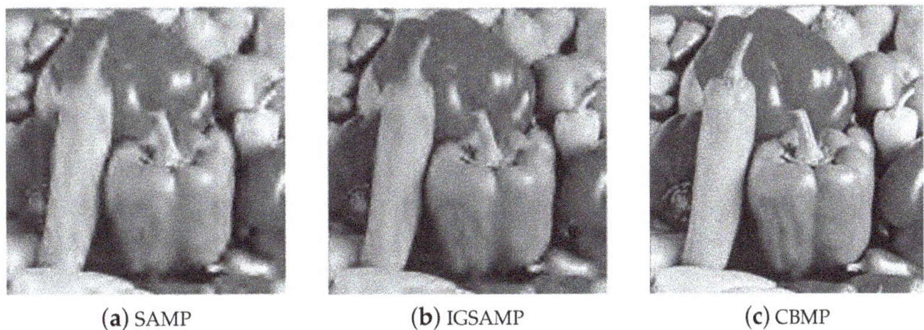

(**a**) SAMP   (**b**) IGSAMP   (**c**) CBMP

Figure 8. Reconstructed results of the Peppers image by SAMP, IGSAMP, and CBMP with the initial step size of 64.

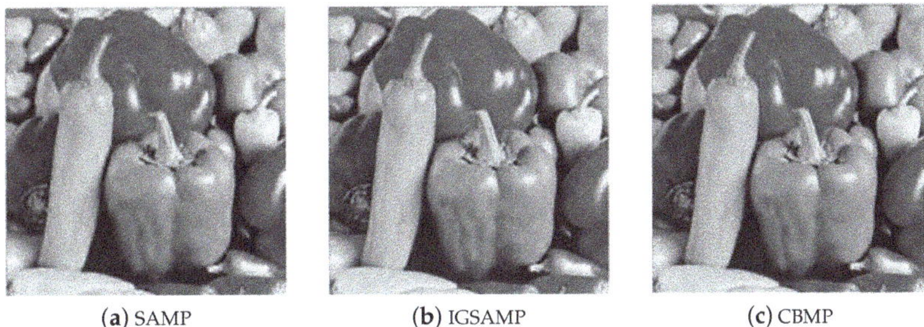

(**a**) SAMP   (**b**) IGSAMP   (**c**) CBMP

Figure 9. Reconstructed results of the Peppers image by SAMP, IGSAMP, and CBMP with the initial step size of 512.

In our test, CBMP outperforms SAMP and IGSAMP in terms of visual effect and PSNR, which is irrelevant to the setup of the initial step size value. At the same time, with different step sizes, the reconstruction performance of CBMP is stable. For example, Figures 8a and 9a show different visual reconstruction effects when the initial step size is 64 and 512, individually, and the same conclusion can be made from Figures 8b and 9b. It is noted that the visualization effect is not obvious in Figures 8c and 9c when the initial step size is 64 and 512, respectively. Therefore, it can be concluded that the CBMP algorithm is relatively insensitive to the step size.

5. Conclusions

In this paper, a constrained backtracking matching pursuit algorithm is proposed for image reconstruction using compressed sensing. A composite strategy, including two constraints, is adopted to effectively control the estimated sparsity level's increment at different stages and accurately estimate the true support set of the image to be reconstructed. On the one hand, the energy criterion between the estimated signal and the measurement is used as a constraint. On the other hand, the four-to-one practical rule is considered and improved as another control. Due to the introduction of these composite mechanisms, the reconstruction performance of the proposed algorithm outperforms the greedy pursuit algorithms, including SAMP and IGSAMP. In particular, CBMP offers a stable reconstruction performance, which is insensitive to the initial step size. In our future works, the CBMP algorithm will be applied to neural network framework-based signal reconstruction, including medical image reconstruction.

Author Contributions: Conceptualization, X.B. and L.L.; Formal analysis, L.L.; Funding acquisition, L.L., X.B. and Y.D.; Investigation, X.B. and L.L.; Methodology, X.B.; Supervision, C.K., Y.D. and F.L.; Writing original draft, X.B.; Writing review—editing, X.B., L.L., C.K., X.L. and F.L. All authors have read and agreed to the published version of the manuscript.

Funding: The work is supported by the National Natural Science Foundation of China (Nos. 61866028, 61872298), the Chunhui Project of the Ministry of Education Project Foundation of China (No. Z2017075), the Sichuan Provincial Department of Education Foundation (No. 17ZB0416) and the Key Project Foundation of Xihua University (No. Z1520908).

Data Availability Statement: Not applicable.

Conflicts of Interest: The authors declare no conflict of interest.

Appendix A

Let $x \in R^N$ be a sparse signal and y be a measurement vector. If the measurement matrix Φ satisfies the RIP, then $\|x\|_2^2 > \frac{1}{\sqrt{2}}\|y\|_2^2$.

Proof. From the right-hand side of the RIP, one has:

$$\|\Phi x\|_2^2 \leq (1 + \delta_k)\|x\|_2^2 \tag{A1}$$

Furthermore, $\|y\|_2^2 \leq (1 + \delta_k)\|x\|_2^2$. and:

$$\frac{\|y\|_2^2}{(1 + \delta_k)} \leq \|x\|_2^2 \tag{A2}$$

According to the monotonicity of δ_k [21], for two integers $k < k'$:

$$\delta_k < \delta_{k'}$$

Furthermore, $\delta_k < \delta_{2k}$:

$$1 + \delta_k < 1 + \delta_{2k}$$

and:

$$\frac{1}{1 + \delta_{2k}} < \frac{1}{1 + \delta_k}$$

Then:

$$\frac{\|y\|_2^2}{1 + \delta_{2k}} < \frac{\|y\|_2^2}{1 + \delta_k} \tag{A3}$$

Combining (A2) with (A3):

$$\frac{\|y\|_2^2}{1 + \delta_{2k}} < \frac{\|y\|_2^2}{1 + \delta_k} \leq \|x\|_2^2 \tag{A4}$$

Based on the demonstration in SP [21] and RIP [30], $0 < \delta_{2k} < \sqrt{2} - 1$ is the sufficient condition for signal reconstruction in CS.

Then:

$$1 < 1 + \delta_{2k} < \sqrt{2}$$
$$\frac{\|y\|_2^2}{\sqrt{2}} < \frac{\|y\|_2^2}{1+\delta_{2k}} < \|x\|_2^2$$

Therefore:

$$\|x\|_2^2 > \frac{1}{\sqrt{2}}\|y\|_2^2.$$

$\square$

References

1. Candès, E.J.; Romberg, J.; Tao, T. Robust uncertainty principles: Exact signal reconstruction from highly incomplete frequency information. *IEEE Trans. Inf. Theory* **2006**, *5*, 489–509. [CrossRef]
2. Wei, Z.R.; Zhang, J.L.; Xu, Z.Y.; Liu, Y. Optimization methods of compressively sensed image reconstruction based on single-pixel imaging. *Appl. Sci.* **2020**, *10*, 3288. [CrossRef]
3. Hashimoto, F.; Ote, K.; Oida, T.; Teramoto, A.; Ouchi, Y. Compressed sensing magnetic resonance image reconstruction using an iterative convolutional neural network approach. *Appl. Sci.* **2020**, *10*, 1902. [CrossRef]
4. Jiang, M.F.; Lu, L.; Shen, Y.; Wu, L.; Gong, Y.L.; Xia, L.; Liu, F. Directional tensor product complex tight framelets for compressed sensing MRI reconstruction. *IET Image Process.* **2019**, *13*, 2183–2189. [CrossRef]
5. Zhang, Z.M.; Liu, X.W.; Wei, S.S.; Gan, H.P.; Liu, F.F.; Li, Y.W.; Liu, C.Y.; Liu, F. Electrocardiogram reconstruction based on compressed sensing. *IEEE Access* **2019**, *7*, 37228–37237. [CrossRef]
6. Bi, X.; Chen, X.D.; Zhang, Y. Image compressed sensing based on wavelet transform in contourlet domain. *Signal Process.* **2011**, *91*, 1085–1092. [CrossRef]
7. Ye, J.C. Compressed sensing MRI: A review from signal processing perspective. *BMC Biomed. Eng.* **2019**, *1*, 8. [CrossRef]
8. Sandino, C.M.; Cheng, J.Y.; Chen, F.; Mardani, M.; Pauly, J.M.; Vasanawala, S.S. Compressed sensing: From research to clinical practice with deep neural networks: Shortening scan times for magnetic resonance imaging. *IEEE Signal Process. Mag.* **2020**, *37*, 117–127. [CrossRef]
9. Iwen, M.A.; Spencer, C.V. A note on compressed sensing and the complexity of matrix multiplication. *Inf. Process. Lett.* **2012**, *109*, 468–471. [CrossRef]
10. Chartrand, R.; Staneva, V. Restricted isometry properties and nonconvex compressive sensing. *Inverse Probl.* **2008**, *24*, 1–14. [CrossRef]
11. Candès, E.J.; Wakin, M.B.; Boyd, S.P. Enhancing sparsity by reweighted l1 minimization. *J. Fourier Anal. Appl.* **2008**, *14*, 877–905. [CrossRef]
12. Foucart, S.; Lai, M.J. Sparsest solutions of underdetermined linear systems via lq minimization for $0 < q <= 1$. *Appl. Comput. Harmonic Anal.* **2009**, *26*, 395–407.
13. Nam, N.; Needell, D.; Woolf, T. Linear Convergence of Stochastic Iterative Greedy Algorithms with Sparse Constraints. *IEEE Trans. Inf. Theory* **2017**, *63*, 6869–6895.
14. Tkacenko, A.; Vaidyanathan, P.P. Iterative greedy algorithm for solving the FIR paraunitary approximation problem. *IEEE Trans. Signal Process.* **2006**, *54*, 146–160. [CrossRef]
15. Tropp, J.A.; Gilbert, A.C. Signal recovery from random measurements via orthogonal matching pursuit. *IEEE Trans. Inf. Theory* **2007**, *5*, 4655–4666. [CrossRef]
16. Li, H.; Ma, Y.; Fu, Y. An improved RIP-based performance guarantee for sparse signal recovery via simultaneous orthogonal matching pursuit. *Signal Process.* **2017**, *144*, 29–35. [CrossRef]
17. Needell, D.; Vershynin, R. Greedy signal recovery and uncertainty principles. *Proc. SPIE* **2008**, *6814*, 68140J.
18. Donoho, D.L.; Tsaig, Y.; Starck, J.L. Sparse solution of under-determined linear equations by stage-wise orthogonal matching pursuit. *IEEE Trans. Inf. Theory* **2006**, *58*, 1094–1121. [CrossRef]
19. Needell, D.; Vershynin, R. Signal recovery from incomplete and inaccurate measurements via regularized orthogonal matching pursuit. *IEEE J. Sel. Top. Signal Process.* **2010**, *4*, 310–316. [CrossRef]
20. Zhang, H.F.; Xiao, S.G.; Zhou, P. A matching pursuit algorithm for backtracking regularization based on energy sorting. *Symmetry* **2020**, *12*, 231. [CrossRef]
21. Dai, W.; Milenkovic, O. Subspace pursuit for compressive sensing signal reconstruction. *IEEE Trans. Inf. Theory* **2009**, *55*, 2230–2249. [CrossRef]
22. Needell, D.; Tropp, J.A. CoSaMP: Iterative Signal Recovery from Incomplete and Inaccurate Samples. *Appl. Comput. Harmon. Anal.* **2009**, *26*, 301–321. [CrossRef]
23. Do, T.T.; Gan, L.; Nguyen, N.; Tran, T.D. Sparsity adaptive matching pursuit algorithm for practical compressed sensing. In Proceedings of the 42nd Asilomar Conference on Signals, Systems, and Computers, Pacific Grove, CA, USA, 26–29 October 2008; pp. 581–587.
24. Bi, X.; Chen, X.D.; Leng, L. Energy-based adaptive matching pursuit algorithm for binary sparse signal reconstruction in compressed sensing. *Signal Image Video Pocess.* **2014**, *8*, 1039–1048. [CrossRef]

25. Shoitan, R.; Nossair, Z.; Ibrahim, I.I.; Tobal, A. Improving the reconstruction efficiency of sparsity adaptive matching pursuit based on the Wilkinson matrix. *Front. Inf. Technol. Electron. Eng.* **2018**, *19*, 503–512. [CrossRef]
26. Zhao, L.Q.; Ma, K.; Jia, Y.F. Improved generalized sparsity adaptive matching pursuit algorithm based on compressive sensing. *J. Electr. Comput. Eng.* **2020**, *4*, 1–11.
27. Candès, E.J.; Wakin, M.B. An introduction to compressive sampling. *IEEE Signal Process. Mag.* **2008**, *25*, 21–30. [CrossRef]
28. Kutyniok, G.; Eldar, Y.C. *Compressed Sensing: Theory and Applications*; Cambridge University Press: Cambridge, UK, 2012.
29. Baraniuk, R.; Davenport, M.; DeVore, R.; Wakin, M. A simple proof of the restricted isometry property for random matrices. *Constr. Approx.* **2008**, *28*, 253–263. [CrossRef]
30. Candès, E.J. The restricted isometry property and its implications for compressed sensing. *Comptes Rendus Math.* **2008**, *346*, 589–592. [CrossRef]
31. Needell, D. Topics in Compressed Sensing. Ph.D. Dissertation, University of California, Berkeley, CA, USA, 2009.
32. Mallat, S.G.; Zhang, Z. Matching pursuits with time-frequency dictionaries. *IEEE Trans. Signal Process.* **1993** , *41*, 3397–3415. [CrossRef]
33. Elad, M. *Sparse and Redundant Representations—From Theory to Applications in Signal and Image Processing*; Springer: New York, NY, USA, 2010.
34. Leng, L.; Zhang, J.S.; Khan, M.K.; Chen, X.; Alghathbar, K. Dynamic weighted discrimination power analysis: A novel approach for face and palmprint recognition in DCT domain. *Int. J. Phys. Sci.* **2010**, *5*, 2543–2554.
35. Leng, L.; Li, M.; Kim, C.; Bi, X. Dual-source discrimination power analysis for multi-instance contactless palmprint recognition. *Multimed. Tools Appl.* **2017**, *76*, 333–354. [CrossRef]

Article

Self-Embedding Fragile Watermarking Scheme to Detect Image Tampering Using AMBTC and OPAP Approaches

Cheonshik Kim [1,*,†] and Ching-Nung Yang [2,*,†]

1   Department of Computer Engineering, Sejong University, Seoul 05006, Korea
2   Department of Computer Science and Information Engineering, National Dong Hwa University, Hualien 97401, Taiwan
*   Correspondence: mipsan@sejong.ac.kr (C.K.); cnyang@gms.ndhu.edu.tw (C.-N.Y.)
†   These authors contributed equally to this work.

Abstract: Research on self-embedding watermarks is being actively conducted to solve personal privacy and copyright problems by image attack. In this paper, we propose a self-embedded watermarking technique based on Absolute Moment Block Truncation Coding (AMBTC) for reconstructing tampered images by cropping attacks and forgery. AMBTC is suitable as a recovery bit (watermark) for the tampered image. This is because AMBTC has excellent compression performance and image quality. Moreover, to improve the quality of the marked image, the Optimal Pixel Adjustment Process (OPAP) method is used in the process of hiding AMBTC in the cover image. To find a damaged block in a marked image, the authentication data along with the watermark must be hidden in the block. We employ a checksum for authentication. The watermark is embedded in the pixels of the cover image using 3LSB and 2LSB, and the checksum is hidden in the LSB. Through the recovering procedure, it is possible to recover the original marked image from the tampered marked image. In addition, when the tampering ratio was 45%, the image (Lena) could be recovered at 36 dB. The proposed self-embedding method was verified through an experiment, and the result was the recovered image showed superior perceptual quality compared to the previous methods.

Keywords: watermarking; self-embedding; digital signature; AMBTC; fragile watermarking

Citation: Kim, C.; Yang, C.-N. Self-Embedding Fragile Watermarking Scheme against Tampering Image by Using AMBTC and OPAP Approaches. *Appl. Sci.* **2021**, *11*, 1146. https://doi.org/10.3390/app11031146

Academic Editor: Francesco Bianconi
Received: 6 January 2021
Accepted: 22 January 2021
Published: 27 January 2021

Publisher's Note: MDPI stays neutral with regard to jurisdictional clai-ms in published maps and institutio-nal affiliations.

1. Introduction

With the advanced high-speed communication technology, recently, many SNS subscribers freely share the digital contents they have created, and with useful image processing software, digital contents are easily manipulated to create interesting images. In addition, images are deliberately or unintentionally manipulated during transmission, causing many social problems. For this reason, the problem of verifying the integrity of an image is becoming an important area of image security.

To solve such a problem, in the past, several signature-based image authentication schemes [1,2] were proposed for integrity verification. Digital signatures are always stored by third parties in a digital signature-based method. In this approach, the digital signature extracted from the image is compared to a digital signature stored by a third party. Comparing the two signatures may detect if the image has been tampered with [3–5]. This method makes it easy to determine whether an image is authentic or not, but they cannot find the tampered area. Besides, adding signatures requires additional bandwidth and storage space. Digital signatures have these obvious limitations. First of all, to recover a marked image with high quality, a method is required that can accurately detect the tampered area of the marked image. As an alternative to digital signatures, the watermarking technology not only detects tampered areas using watermarks, but it also suggests an alternative to recover marked images and is currently being actively studied.

Watermarking methods are classified as strong watermarking [6–9], semi-fragile watermarking [10–12] and fragile watermarking [13–16]. The strong watermarking method allows you to extract hidden watermarks from watermarked images, even after image processing (e.g., image compression and filtering). Thus, it can be exploited to verify copyright and intellectual property rights. The fragile watermarking technique can be easily destroyed by simple image processing; thus, it can accurately detect the tampered area. There are currently two types of fragile watermarking techniques. The first type detects only the tampered area from the cover image. The second can detect and find the tampered area as well as recover the area on the image.

Self-embedding is a way of recovering the tampered area with the recovered bits, which are embedded in the pixels of the cover image, where the recovering bits are composed of the feature of the original image. The performance of the self-embedding method based on watermarking technology is generally evaluated by the quality of the recovered image. In most self-embedding methods, the recovery bits of a specific block are always hidden in the other block of the image. A method like this can fail if the block containing the recovery bit has been tampered with. This is called the tampering coincidence problem [17].

The most important factor for image recovery depends on the ability to detect forged areas. Walton [5] proposed the first fragile watermarking method for detection of tampered areas based on inserting checksums in gray levels. Fridrich et al. [18,19] introduced a self-embedding method based on DCT. Here, the DCT is converted to a bitstream, and then it is embedded to pixels of the distant block. The reconstruction quality using Algorithm 1 in this method is 50% quality, which is significantly worse than that for a JPEG compressed image. He et al. [19] proposed an adjacent block-based statistical detection method to accurately identify the tampered block, and they provided an analysis of the tampered detection performance. It has been shown that the statistical detection method can identify the tampered block of the host image. However, there may also be a recovering problem due to statistical error.

Lin et al. [20] introduced a hierarchical-based watermarking method to detect and recover cover image damage. It is effective because the detection is based on a hierarchical structure so that the accuracy of tamper localization can be ensured. The drawback is that it is not possible to check whether the location of the error is an error of a lower block of the current block or an error of a lower block within the same block. Therefore, the scheme [21] proposed a new mechanism to facilitate recovery with a higher probability by inserting a double copy of the watermark into two different blocks.

Zhang and Wang (2008) [22] proposed a new vector coding-based fragile watermarking method that can recover the tampered area without error, as long as it is not too serious. For restoration, recovery and authentication, bits are compressed losslessly and then are embedded in the cover image. The mean PSNR is about 28.70 dB. Moreover, if the tampering rate is less than 3.2%, the tampered area may be totally recovered. To improve the tampering rate, Zhang et al. (2009) [23] proposed another fragile watermarking scheme that restores the content of the original image. While the reference bits are embedded into the entire cover image, the hash bits are embedded into the local blocks.

Qian et al. [24] proposed a fragile watermarking method for high-quality restoration. They first categorized the image into one of six types depending on the degree of smoothness. Complex blocks were compressed into more bits for recovery and smooth blocks were needed fewer bits for recovery. Finally, the recovery bit and authentication bit were embedded into the three number of LSBs of every pixel for the image.

Luo et al. [25] proposed a self-contained watermarking scheme for digital images. The host image was converted into a halftone image using a digital halftone technique, and the converted pixels were used as recovery bits. Halftone can preserve the characteristics of the host image in the most compressed type. The halftoning watermark was used for tamper detection, and the tampered area can be approximately recovered using the extracted watermark. They adopted a simple low-pass filtering approach for inverse halftoning. The reconstructed image based on halftone is not satisfactory from the perspective of the image quality. Hsu and Tu's study [26] used the degree of smoothness to distinguish the types of image blocks, and they employed different watermark embedding, tamper detection and recovery strategies for different block types to enhance hiding efficiency, authentication and recovery effects.

Yang & Shen [27] proposed a method to detect and recover images tampered with by integrating Wong's watermarking method [28] and vector quantization (VQ). This integration also required a little extra cost, i.e., an increase in codebook size. However, with a codebook, the image is recovered by using VQ if the mapping information for recovery is lost. Due to the limitation of the quality of VQ, the restored image is not of high quality.

In [29–31], they proposed a fragile watermarking technique based on Block Truncation Coding (BTC). In this method, the bitstream compressed with BTC [32] or AMBTC [33] of the original image was hidden in the LSB and 2LSB of the cover image to store the features of the original image. When a part of the image has been tampered, the location is detected, and the information on the tampered area is recovered with AMBTC. Kim et al. [29] adopted a method of improving the image quality through Gaussian filtering after using the reconstructed bits for image restoration. Hemida et al. [30] used a quantum chaos map to escape the tampering attack of the mark. The error rate of tamper detection using the XOR operation between the bitmap of each block and the binary random number may higher than that of using the decimal number. Chang et al. [31] proposed a method to improve the quality of marked images by enhancing the compression performance of AMBTC encoding bits for the original image.

The quality of the reconstructed image was not good due to the loss of recover bits according to compress the bitmaps. They employed an inpainting technique to improve the quality of the recovered image. For the tampered block, the most important thing to recover depends on how to exactly find the tampered location. In this paper, we propose a fragile watermark technique based on self-embedding using AMBTC to restore the tampered cover image. To improve the quality of the cover image, it uses Optimal Pixel Adjustment Process (OPAP) [34] to encode self-embedded data (Watermark). OPAP is introduced to optimize the error in the DH process using LSB replacement, and it is a coding method with excellent performance. In this scheme, we used checksum for authentication of every block and embedded authentication bits in every block to detect forgery. Although the checksum is a simple method, it guarantees relatively accurate performance in detecting whether or not it is a tampered block through threshold comparison. This improves the accuracy of tamper detection and localization.

This scheme has several advantages: (1) high accuracy of tampered detection; (2) the quality of the recovered marked image is guaranteed by the use of high-quality compression bits generated by AMBTC; (3) the quality of marked images is guaranteed because the original image encoded with AMBTC is hidden in the cover-image using OPAP; (4) recovering bits for a block are concealed at a far distance from the current block to prepare for cropping attacks. Experimental results show that the proposed scheme allows high-quality recovery up to a modulation rate of 45%.

The rest of this paper is organized as follows. Section 2 briefly introduces watermark technology. Section 3 presents a review of current and related work. Section 4 introduces the proposed self-embedding watermarking scheme. Section 5 explains the experimental results, and Section 6 provides the conclusions and future work.

2. Conspectus of Watermarking Technology

In this section, an overview of watermarking technology and its main terms is introduced.

2.1. Watermark Requirements

The critical requirements that a watermarking method must have are as follows [35]. First, embedding capacity: it must have capable of storing data of sufficient capacity to protect the copyright of digital contents. Second, robustness: the embedded watermark in a cover image needs to be able to resist various attacks such as compression of images and image processing. Third, security: attackers must not be able to easily access the embedded watermark. Fourth, unrecognizable: It should be possible to hide the presence of watermarks by preventing distortion of marked images in case of embedding a watermark in the cover image. Fifth, blind: it should be possible to recover the watermark without reference to the original image.

2.2. Watermarking Techniques Classification

Watermarking technology is the most basic application used for copyright protection of digital content such as (color) images [35], video [36] and 3D mesh [37]. That is, the ownership information of the watermark is exploited for identifying copyright ownership and preventing fraud and theft of digital content. In fact, marked digital images can prove ownership when someone claims it by a legitimate owner. In addition, authentication is another watermarking application that aims to verify the integrity of the watermarked digital images and detect attempts to alter the original images. These watermarks are designed to be subject to signal manipulation and are used to indicate the authenticity of digital content.

Watermarking technology can be divided according to several perspectives. First, according to human perception, it is divided into two types: visible watermarking technology and invisible technology. The former means that the watermark is visible from digital images, and the latter means that the watermark cannot be recognized by the human eye. Second, watermarking technology is divided into non-blind and blind, depending on whether the original image is needed for watermark recovery. In the non-blind technique, both the original image and watermark are required during the authentication of the watermark. The blind technique does not require a watermark or original image. Third, it is divided into a spatial domain and a frequency domain based on the work area. The former is done by directly manipulating the pixel values in the original image. The merit of the work based on the spatial domain is its simple implementation and low computational complexity, while its demerit is its weak robustness to compression. In the latter case, you need to convert the host image to an appropriate frequency working domain. Then, the coefficient is adjusted according to the values of a watermark. In general, domain transformation techniques are Discrete Cosine Transform (DCT), Singular Value Decomposition (SVD), and Discrete Wavelet Transform (DWT) [18,38,39]. The frequency domain-based approach is more resilient to compression attacks and image conversion attacks [36].

Fourth, depending on whether the watermark can withstand various attacks, it is classified into a strong watermark, a fragile watermark or a semi-fragile watermark. The strong watermark-based method provides the performance to withstand compression and various image manipulations. To do this, it is characterized by converting the image into the frequency domain. The fragile watermark-based method is vulnerable to image compression and image processing, so the use of this method may be different. This is because the hidden information cannot be restored even with trivial image processing. The semi-fragile watermark-based method provides selective robustness for specific manipulations.

There are two ways to watermark a color image in the conversion domain. The first uses gray level techniques to process each channel individually, and the second treats each pixel in the color image into a quaternion vector to which the transformation is applied.

3. Preliminaries

3.1. AMBTC

The Block Truncation Coding (BTC) [32] is a simple lossy compression method based on moment preserving quantization for blocks of pixels in a grayscale image. Since BTC produces a set of bitmap, mean and standard deviation to represent a block, it gives a CR (size of the original image/size of the compressed image) of 4; hence, the bit rate is 2 bits per pixel for a 4×4 block. Though the BTC method provides good compression without much degradation on the reconstructed images, it shows some artifacts like the staircase effect. Absolute Moment Block Truncation Coding (AMBTC) [33] preserves the higher mean and lower mean of each of the blocks and improves the staircase effect of the conventional BTC method. Besides, AMBTC is simpler than BTC, thus the computation speed is very fast. The AMBTC algorithm involves the following steps:

Step 1: The original image of size $N \times N$ is divided into non-overlapping blocks (C) of the size $m \times m$ (let $m = 4$), and each block is processed separately. Let $m^2 = k$.

Step 2: For each block, the average pixel value is calculated by Equation (1).

$$\bar{x} = \frac{1}{k} \sum_{i=1}^{k} x_i \tag{1}$$

where x_i represents the ith pixel value of this block with the size of k. All pixels in the block are quantized into a bitmap b_i (0 or 1) using Equation (2). That is, if the corresponding pixel x_i is greater than or equal to the average ($\bar{x}$), it is assigned with '1', otherwise it is '0'. Pixels in each block are divided into two groups of '1' or '0'.

$$b_i = \begin{cases} 1, & \text{if } x_i \geq \bar{x}, \\ 0, & \text{if } x_i < \bar{x}. \end{cases} \tag{2}$$

Step 3: The block $\mathcal{M}$ is partitioned into two sets of pixels $\mathcal{M}_0$ and $\mathcal{M}_1$ such that $\mathcal{M} = \mathcal{M}_0 \cup \mathcal{M}_1$ and $\mathcal{M}_0 \cap \mathcal{M}_1 = \phi$ where $\mathcal{M}_0 = \{0_0, 0_1, \ldots, 0_t\}$ and $\mathcal{M}_1 = \{1_1, 1_2, \ldots, 1_{k-t}\}$,, and t and $k - t$ refer to the numbers of pixels in the '0' and '1' groups, respectively. The means Q_1 and Q_2 of the two groups indicate the quantization levels of the groups '0' and '1'. The two quantization levels are calculated by Equations (3) and (4).

$$Q_1 = \left\lfloor \frac{1}{t} \sum_{x_i < \bar{x}} x_i \right\rfloor \tag{3}$$

$$Q_2 = \left\lfloor \frac{1}{k-t} \sum_{x_i \geq \bar{x}} x_i \right\rfloor \tag{4}$$

Step 4: To reconstruct the pixel marked by '0' it will be given the value Q_1, and that marked by '1' will be given the value Q_2. The values Q_1 and Q_2 satisfy the following relation. The compressed block is simply uncompressed by using Equation (5).

$$g_i = \begin{cases} Q_1, & \text{if } b_i = 0, \\ Q_2, & \text{if } b_i = 1. \end{cases} \tag{5}$$

The image block is compressed into two quantization levels Q_1 and Q_2, and a bitmap $\mathcal{M}$, and it can be represented as a $trio(Q_1, Q_2, \mathcal{M})$. A bitmap $\mathcal{M}$ contains the bit-planes that represent the pixels, and the values Q_1 and Q_2 are used to decode the AMBTC-compressed image by using Equation (5). If the block size is 4×4 then it will give the 32-bit compressed data (i.e., the size of the block bitmap is 16 bits; converting Q_1 and Q_1 to binary results in 16 bits), and hence the bit rate is 2 bpp. For $m = 4$, 16 pixels are represented by a $trio(Q_1, Q_2, \mathcal{M})$ of $8 + 8 + 16 = 32$ bits, so the compression ratio (CR) is $(16 \times 8)/32 = 4$. For 512×512 pixel images, the file size of 2M-bits can be reduced to 0.5 M-bits.

3.2. LSB Substitution and OPAP

LSB (Least-Significant-Bit) alternative technology is a method of directly concealing the watermark in the LSB of the pixels constituting the cover image. Wang et al. [34] introduced an optimal LSB substitution and genetic algorithms, and it was found that the Worst-case Mean Squared Error (WMSE) (which is a measurement obtained by comparing the original and marked image) is $1/2$ of that obtained with simple LSB substitution techniques. Let us look at the DH procedure for the original 8-bit grayscale represented by $x_i \in \{0, 1, \ldots, 255\}$. S denotes n-bit hidden values represented as $S = \{s_k | 0 \leq k < n, s_k \in \{0, 1\}\}$. The mapping between the n-bit secret bits $S = \{s_k\}$ and the embedded bits $S' = s'_k$ can be defined as follows: $s'_k = \sum_{j=0}^{\delta-1} s_{k\times\delta+j} \times 2^{\delta-1-j}$. The pixel value x_i for embedding the δ-bit s'_k is changed to form the stego-pixel x'_i like $x'_i = x_i - (x_i \bmod 2^\delta) + s'_k$. The δ LSBs of the pixels are extracted by $s_k = x'_i \bmod 2^\delta$.

It has been mathematically proven that OPAP can improve the quality of marked images by reducing WMSE by using LSB replacement based on the minimization rule. Let x_i be the pixel of the cover image, x'_i be the obtained pixel from pixel x_i using the LSB replacement, and x''_i is the optimized pixel derived from x'_i by the OPAP method. The value of Δ_i may be segmented into three intervals. The OPAP modifies x' to form the stego-pixel x'' as the following rules:

1. Rule 1 $(2^{\delta-1} < \Delta_i < 2^\delta)$: if $x'_i \geq 2^\delta$, then $x''_i = x'_i - 2^\delta$; otherwise $x''_i = x'_i$;
2. Rule 2 $(-2^{\delta-1} \leq \Delta_i \leq 2^{\delta-1})$: $x''_i = x'_i$;
3. Rule 3 $(-2^\delta < \Delta_i < -2^{\delta-1})$: if $x'_i < 256 - 2^\delta$, then $x''_i = x'_i + 2^\delta$; otherwise $x''_i = x'_i$;

3.3. Luo et al.'s Method

Luo et al. [25] proposed a self-embedding watermarking scheme by using the digital halftoning technique. Here, the tampered image is restored by converting the original image's features into the halftone image and secretly embedding them in the pixels of cover image. If the halftone image composed of 1's and 0's is used as the restoration bits, the size of the watermark is small, but the quality of the restored image is low because it cannot retain sufficient features for the original image.

Suppose I and W denote the host image and the watermark image, respectively, and both are of size $N \times N$. W obtains the enhanced edge by using the error diffusion halftoning algorithm. The watermark W permutes the locations of all pixels constituting the watermark using the key K. The permuted watermark W_p is embedded into the pixels' LSBs in the cover image I (Figure 1).

For reconstruction, recover W' from the marked image I', then divide I' and W' into non-overlapping $m \times m$ block BI_l and $BW_l(l = 1, 2, \ldots, N \times N/m)$ respectively. Compute the difference D between pixel values of each block BI_l and the corresponding block BW_l. If the difference is smaller than threshold T, it is the authenticated block; otherwise, it is not the authenticated block. For the tampered block, it is replaced with the same block of W'.

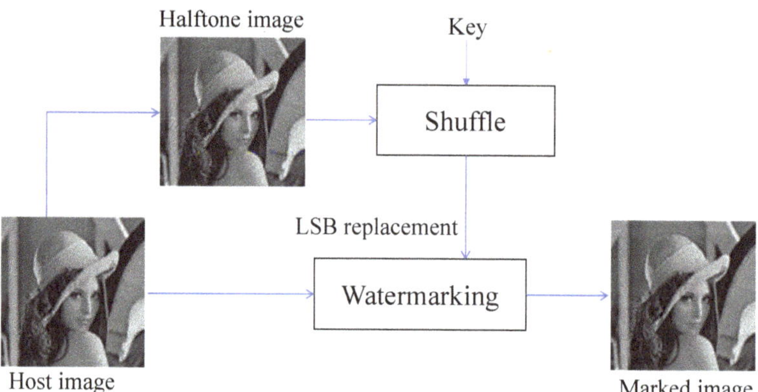

Figure 1. Block diagram of the proposed scheme—watermark generation and embedding.

3.4. Fridrich and Goljan's Method

Fridrich and Goljan [18] described a self-embedding technique. First, the cover image is divided into blocks of 8×8 pixels. Set the LSB of each pixel in a given block to 0, and convert the block into DCT. The quantized matrix is encoded with 64 bits, and the bits are embedded into the LSBs of a distant block. After embedding the watermark, on average it modifies 5% of pixels in a block, and the quality of the reconstructed image is somewhat worse than 50% of JPEG quality. The following three steps are carried out for each block B:

Step 1: The original image is divided into blocks of 8×8 pixels. All blocks are transformed into the interval $[-127, 128]$, and the LSBs of all pixels are set to zero.

Step 2: Each 8×8 block is transformed into the frequency domain using DCT. The first 11 coefficients (in zig-zag order) are quantized with the following quantization table Q (Fridrich and Goljan [18]) that corresponds to 50% JPEG quality: The quantized values are further binary encoded. The bit lengths of their codes (including the signs) are shown in matrix L (Fridrich and Goljan [18]). Coding based on L will guarantee that the first 11 coefficients from each block will be coded using exactly 64 bits.

Step 3: The binary sequence obtained in Step 2 (e.g., the 64-bit string) is encrypted and inserted into the LSB of the block $B + \vec{p}$, where $\vec{p}$ is a vector of length approximately $3/10$ of the image size with a randomly chosen direction.

4. Proposed Scheme for Self-Embedding

This section introduces an efficient self-embedding method based on AMBTC. First, the proposed method obtains the feature information of the original image by converting the original image into AMBTC. A basic configuration of the AMBTC image is a trio composed of a bitmap and two quantization levels. Next, the encrypted trios are embedded in their own pixels in the cover image, and the quality of the cover image is somewhat reduced by this procedure.

Figure 2 schematically shows the procedure of obtaining two bitmaps from the original image and the procedure of embedding the checksum and two bitmaps into the cover image after completing the mapping process. The compression method in Section 3.1 is the procedure to obtain the bitmap $\mathbb{M}_1$ and two quantization levels Q_1 and Q_2 per block (Figure 2).

Both quantization levels are converted to binary bitmap $\mathbb{M}_2$ for DH. In preparation for the cropping attack, every block of two $\mathbb{M}_1$ and $\mathbb{M}_2$ moves as far away as possible from the original location of the block using a scramble (mapping) algorithm. Afterward, the watermarks are embedded in the 2LSB and 3LSB of the pixels in the cover image. After creating a checksum for block authentication, it secretly inserts it into the pixels of the cover block. It is used for authentication of the block during the restoration process.

Figure 3 shows a simple schematic explaining the recovery procedure of a tampered watermarked image. To recover the tampered image, the concealed watermarks $\mathbb{M}_1$ and $\mathbb{M}_2$ must first be extracted from 3LSB and 2LSB (OPAP method). The original bitmap is reconstructed by applying Equation (8) to each block of $\mathbb{M}_1$ and $\mathbb{M}_2$. After converting $\mathbb{M}_2$ to a quantization level per block, a grayscale image is restored by performing a decoding process. To check the forged block, we should restore the checksum V'_{sum} hidden in the LSB and then generate the checksum V_{sum} for the block by using the key.

The two generated checksums are compared, and the authentication process is executed in block units. In other words, if the two checksums match, it means that there is no forgery attack on the block. If not, forgery has occurred. Therefore, the block is replaced with the corresponding block of the AMBTC image created as a watermark.

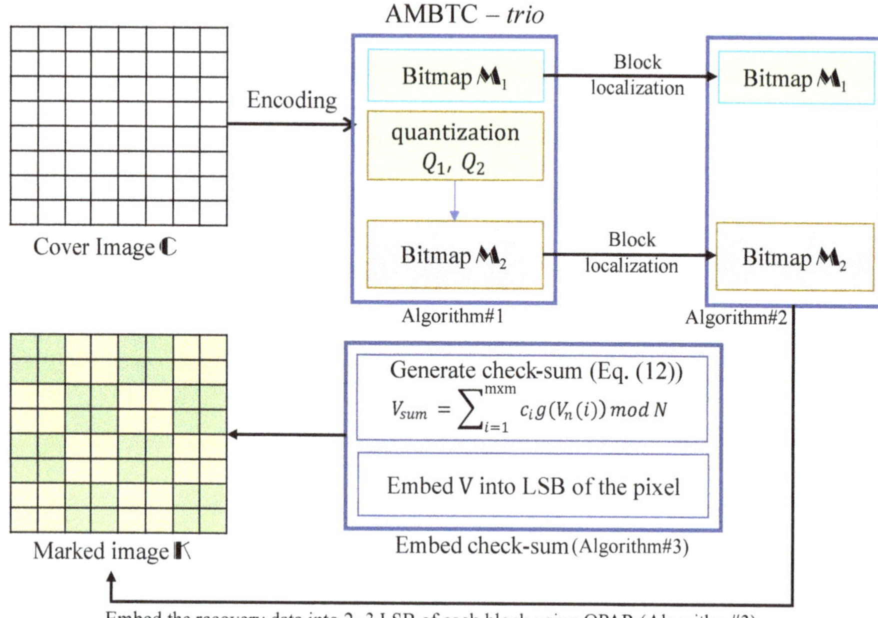

$$V_{sum} = \sum_{i=1}^{mxm} c_i\, g(V_n(i))\, mod\ N$$

Figure 2. Block diagram for generating a watermarked image using AMBTC and the proposed embedding scheme (See Algorithms 1–3).

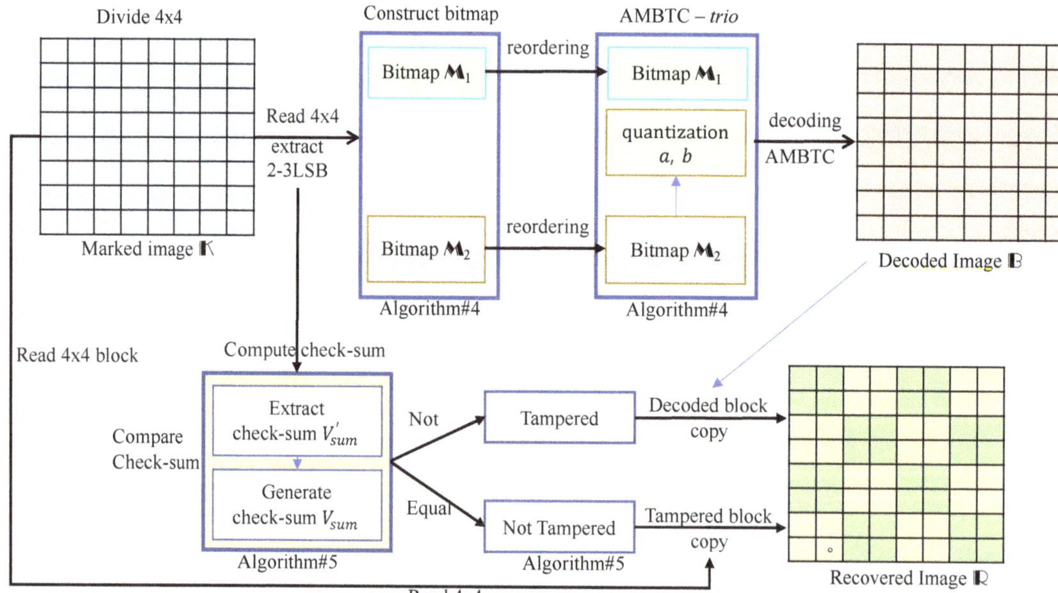

Figure 3. Block diagram for watermark extraction, tampering detection, localization and image recovery (See Algorithms 4 and 5).

4.1. Watermark Generation, Localization and Authentication

Let us suppose that $\mathbb{C}$ and $\mathbb{M}$ denote the cover image and watermark (two maps of AMBTC), respectively, and both have a size of $N \times N$. The two watermark $\mathbb{M}_{1,2}$ is taken by the previously mentioned AMBTC algorithm. The watermark $\mathbb{M}$ is randomly permuted by a key and then is subjected to the embedding procedure.

The step-by-step embedding procedure is as follows:

Algorithm 1 Watermark Generation

Input: A original image $\mathbb{O}$ with a size of $N \times N$

Output: Two bitmap $\mathbb{M}_1$ and $\mathbb{M}_2$ sized $N \times N$.

Step 1: A compressed set $trio(Q_1, Q_2, \mathcal{M})$ is obtained from the original image $\mathbb{O}$ by using the AMBTC algorithm (Section 3.1). The sized $N \times N$ bitmap $\mathbb{M}_1$ and $\mathbb{M}_2$ are initialized with zeros.

Step 2: For given $trio(Q_1, Q_2, \mathcal{M})$, the bitmap $\mathbb{M}_1$ is constructed by adding the block bitmap $\mathcal{M}_n$ to the $\mathbb{M}_1$ using Equation (6), where $n \in \{1 \leq n \leq (N \times N)/(m \times m)\}$ and $m = 4$.

$$\mathbb{M}_1^n = \sum_{n=1} \mathcal{M}_n, \text{ where } \mathcal{M} \in trio(Q_1, Q_2, \mathcal{M})_n, \tag{6}$$

Step 3: After converting the two quantization levels Q_1^n and Q_2^n into 8 bits using Equation (7) respectively, the $b_{i,t}$ are assigned to the block $\mathcal{M}_n$ sequentially. Then, $\mathcal{M}_n$ is added to the bitmap $\mathbb{M}_2$ like Equation (6), i.e., $\mathbb{M}_2^n = \sum_{n=1} \mathcal{M}_n$, where $\mathcal{M} \in trio(Q_1, Q_2, \mathcal{M})_n$.

$$b_{i,t} = \left\lfloor \frac{Q_{1,2}}{2^t} \right\rfloor \bmod 2, t = 0, 1, \ldots, 7. \tag{7}$$

If the watermark is concealed in the same order as the original image, the tampered area cannot be restored when the marked image is damaged. Therefore, the recovery bits, watermarks, are not embedded into the block itself. These watermarks are embedded in LSBs of the mapped block $\mathcal{M}_j$. Here, blocks $\mathcal{M}_i$ and $\mathcal{M}_j$ are chosen such that $\{i \neq j | (i,j) \in [1, 2, \ldots, R]\}$, where R is the total number of blocks in the cover image.

The procedure of watermark localization is as follows:

Algorithm 2 Block Mapping

Input: Two watermark bitmaps, $\mathbb{M}_1$ and $\mathbb{M}_2$ in Algorithm 1, Key ξ

Output: Mapped (scrambled) bitmaps $\mathbb{M}_1$, and $\mathbb{M}_2$

Step 1: Divides the bitmap $\mathbb{M}_1$ into non-overlapping blocks of $m \times m$ pixels. Read a block $\mathcal{M}_n$ from $\mathbb{M}_1$, where $n \in \{1 \leq n \leq R\}$ and $R = (N \times N)/(m \times m)$ and $m = 64$. The optimal position j for the block $\mathcal{M}_n$ is obtained by using Equation (8), where ξ is prime number. Swap the block $\mathcal{M}_n$ with the block $\mathcal{M}_j$ in the map $\mathbb{M}_1$.

$$j = \begin{cases} f(i) = (\xi \times n) \bmod R \\ \quad j = f(n) + 1; \end{cases} \tag{8}$$

When dividing the image into 16 areas, the block location n and j must not be in the same, and the key ξ must search for its location, which can be the most distant from each other.

Step 2: Read a block $\mathcal{M}_n$ from the bitmap $\mathbb{M}_2$. The optimal position j for the block $\mathcal{M}_n$ is obtained by using Equation (8) and swap the block $\mathcal{M}_n$ with the block $\mathcal{M}_j$ in the map $\mathbb{M}_2$.

Step 3: Repeat Step-1 and Step-2 for all blocks.

4.2. Watermark Embedding Procedure

In Section 4.1, we introduced obtaining bitmaps $\mathbb{M}_1$ and $\mathbb{M}_2$ for the restoration of marked images using Algorithm 1. However, if one block of the marked image has tampered with, $\mathbb{M}_1$ and $\mathbb{M}_2$ of the block have tampered with, so localization is required. Algorithm 2 introduced obtaining localized maps $\mathbb{M}_1$ and $\mathbb{M}_2$. Section 3.2 introduces the procedure of hiding the two maps and the authentication bits in the cover image. Here, the three LSB layers of the cover image are replaced with watermark bits and authentication bits.

The watermark embedding procedure is as follows:

Algorithm 3 Watermark Embedding

Input: cover image $\mathbb{C}$

Output: marked image $\mathbb{K}$

Step 1: Divide the cover image $\mathbb{C}$ and two maps ($\mathbb{M}_1$ and $\mathbb{M}_2$) in Algorithm 2 into non-overlapping blocks sized $m \times m$, where $m = 4$.

Step 2: Read three blocks from the cover image $\mathbb{C}$, two maps $\mathbb{M}_1$ and $\mathbb{M}_2$, respectively, then these blocks are assigned to $\mathcal{P}_n$, $\mathcal{M}_1$, and $\mathcal{M}_2$, where $n \in \{1 \le n \le (N \times N)/(m \times m)\}$ and n is a block index number.

Step 3: Obtain 1LSB (b_i^1), 2LSB (b_i^2), and 3LSB (b_i^3) using Equation (9), where n is the index, and i is the pixel index. After that, OPAP is applied according to the rule of Equation (10), and $\mathcal{M}_1^n$ is hidden in the 3LSB of $\tilde{\mathcal{P}}$. That is, $\tilde{\mathcal{P}}_n' = \sum_{i=1}^{m \times m} f(\tilde{\mathcal{P}}_{n,i}, \mathcal{M}_1^{n,i})$, here, f is a function representing the logic of Equation (10).

$$
\begin{cases}
\tilde{\mathcal{P}}_{n,i} \xleftarrow[\text{LSB}]{\text{removed}} \lfloor \mathcal{P}_{n,i}/2 \rfloor \\
b_i^1 \xleftarrow{\text{1LSB}} \mathcal{P}_{n,i} \bmod 2 \\
b_i^2 \xleftarrow{\text{2LSB}} \tilde{\mathcal{P}}_{n,i} \bmod 2 \\
b_i^3 \xleftarrow{\text{3LSB}} \lfloor \tilde{\mathcal{P}}_{n,i}/2 \rfloor \bmod 2
\end{cases}
\tag{9}
$$

$$
\tilde{\mathcal{P}}_n' =
\begin{cases}
\tilde{\mathcal{P}}_{n,i} - 1, & \text{if } (b_i^3 = 0 \text{ and } b_i^2 = 0) \text{ and } \mathcal{M}_1^{n,i} = 1, \\
\tilde{\mathcal{P}}_{n,i} + 1, & \text{if } (b_i^3 = 0 \text{ and } b_i^2 = 1) \text{ and } \mathcal{M}_1^{n,i} = 1, \\
\tilde{\mathcal{P}}_{n,i} - 1, & \text{if } (b_i^3 = 1 \text{ and } b_i^2 = 0) \text{ and } \mathcal{M}_1^{n,i} = 0, \\
\tilde{\mathcal{P}}_{n,i} + 1, & \text{if } (b_i^3 = 1 \text{ and } b_i^2 = 1) \text{ and } \mathcal{M}_1^{n,i} = 0, \\
\text{no change}, & \text{otherwise}
\end{cases}
\tag{10}
$$

Step 4: Embed $\mathcal{M}_2^{n,i}$ into $(b_i^3 \oplus b_i^2)$ of $\tilde{\mathcal{P}}_n'$ using OPAP rule (Equation (11)), where f is the function represented the logic of Eqaution (11). Before applying OPAP, b_i^3 and b_i^2 need to be re-calculated using Equation (9). That is why the LSBs are updated values by using Equation (10).

$$
\tilde{\mathcal{P}}_n'' =
\begin{cases}
\tilde{\mathcal{P}}_n'(i) - 1, & \text{if } (b_i^3 = 1 \text{ and } b_i^2 = 1) \text{ and } \mathcal{M}_2^n(i) = 1, \\
\tilde{\mathcal{P}}_n'(i) + 1, & \text{if } (b_i^3 = 1 \text{ and } b_i^2 = 0) \text{ and } \mathcal{M}_2^n(i) = 0, \\
\tilde{\mathcal{P}}_n'(i) - 1, & \text{if } (b_i^3 = 0 \text{ and } b_i^2 = 1) \text{ and } \mathcal{M}_2^n(i) = 0, \\
\tilde{\mathcal{P}}_n'(i) + 1, & \text{if } (b_i^3 = 0 \text{ and } b_i^2 = 0) \text{ and } \mathcal{M}_2^n(i) = 1, \\
\text{no change}, & \text{otherwise}
\end{cases}
\tag{11}
$$

In order to reflect the changed pixel block $\tilde{\mathcal{P}}_n''$ to $\mathcal{P}_n$, the following calculation must be applied. That is, $\mathcal{P}_n(i) = f(\tilde{\mathcal{P}}_n''(i) \times 2) + b_i^1$.

Algorithm 3 *Cont.*

Step 5: For image authentication, we compute a checksum for each block and hide it in a block. First, we choose a large number $\mathcal{G}$ that will be used for calculating the checksums (Equation (12)). For each block, every pixel $(\mathcal{V}_{n,i})$ is generated by a key (ξ) with a pseudo-random number. Here, $g(\mathcal{V}_{n,i})$ is the gray level of the pixel $\mathcal{V}_{n,i}$. We also generate $m \times m$ integers $c_1, c_2, \ldots, c_{m \times m}$ comparable in size to $\mathcal{G}$. The checksum K_{sum} is calculated as

$$\begin{cases} \mathcal{V}_{sum} = \sum_{i=1}^{m \times m} c_i g(\mathcal{V}_n(i)) \bmod \mathcal{G} \\ k_{i,t} = \left\lfloor \dfrac{\mathcal{V}_{sum}}{2^t} \right\rfloor \bmod 2, t = 0, 1, \ldots, m \times m. \end{cases} \tag{12}$$

Finally, the transformed bits k_i from checksum K_{sum} are acquired.

Step 6: Embed k_i into the LSBs of $\mathcal{P}_{n,i}$ using the logic of Equation (13). That is, $\mathcal{P}'_n = \sum_{i=1}^{m \times m} f(\mathcal{P}_{n,i}, k_{i,t})$, where f is the function representing the rule.

$$\mathcal{P}_n(i)' = \begin{cases} \text{no operation,} & \text{if } k_i = b_i^1 \\ \mathcal{P}_n(i) + 1, & \text{if } (k_i \neq b_i^1) \text{ and } b_i^1 = 0, \\ \mathcal{P}_n(i) - 1, & \text{if } (k_i \neq b_i^1) \text{ and } b_i^1 = 1, \end{cases} \tag{13}$$

4.3. Watermark Extraction and Reconstructing AMBTC

In the method we proposed, the information (checksum) for its authentication is secretly concealed in the LSB, so it is possible to check whether the marked image is tampered with or forged even if there is no original image. The receiver side can find the tampered and forged block by using the validity of the checksum while moving each block. If a modulated block is found, the damaged block can be recovered according to the recovery procedure. Figure 3 shows a block diagram for the content recovery procedure.

Algorithm 4 Watermark Extracting

It extracts two maps, which are watermarks hidden in the marked image $\mathbb{K}$. The restoration process is performed using two maps and the checksum.

Input: A marked image $\mathbb{K}$ (output of Algorithm 3) with a size of $N \times N$, Key ξ.

Output: A reconstructed image $\mathbb{B}$ with a size of $N \times N$.

Step 1: Divide the marked image $\mathbb{K}$ into non-overlapping blocks sized $m \times m$, where $m = 4$. The sized $N \times N$ bitmap $\mathbb{M}_1$ and $\mathbb{M}_2$ are initialized with zeros.

Step 2: Read a block from the marked image $\mathbb{K}$, then this block is assigned to $\mathcal{P}_n$, where n is a block number. After that, the embedded hidden bits in b_i^1(LSB1), b_i^2(LSB2) and b_i^3(LSB3) are obtained from the block $\mathcal{P}_n$ using Equation (9). Then, a block $(\mathcal{M}_{n,i})$ of bitmap $\mathbb{M}_1$ is restored using Equation (14).

$$\mathcal{M}_{n,i} = \sum_{i=1}^{m \times m} \lfloor \tilde{\mathcal{P}}_{n,i}/2 \rfloor \bmod 2 \tag{14}$$

The restored bitmap block $\mathcal{M}_{n,i}$ is assigned to $\mathbb{M}_1$, i.e., $\mathbb{M}_1^n = \mathcal{M}_n$, where n is a block index.

Step 3: After applying Equation (15) to block $\mathcal{P}_n$, the obtained recovery block $\mathcal{M}_{n,i}$ is assigned to $\mathbb{M}_2$. That is, $\mathbb{M}_2^n = \mathcal{M}_n$.

$$\mathcal{M}_{n,i} = \sum_{i=1}^{m \times m} (\lfloor \tilde{\mathcal{P}}_{n,i}/2 \rfloor \bmod 2) \oplus (\tilde{\mathcal{P}}_{n,i} \bmod 2) \tag{15}$$

Algorithm 4 *Cont.*

Step 4: If the procedure of Steps 2 and 3 is repeated, the number of blocks $((N \times N)/(m \times m))$, the two maps $\mathbb{M}_1$ and $\mathbb{M}_2$ are reconstructed.

Step 5: The mapped blocks $\mathbb{M}_1$ and $\mathbb{M}_2$ constructed by Equation (8) are reconstructed to have their original location. For this, first, divide the bitmap $\mathbb{M}_1$ and $\mathbb{M}_2$ into non-overlapping blocks of $m \times m$ pixels, where $m = 64$. Repeat Steps 5-1 and 5-2 until $\mathbb{M}_1$ and $\mathbb{M}_2$ are reconstructed.

Step 5-1: Read a block $\mathcal{M}_n$ from $\mathbb{M}_1$, where $n \in \{1 \leq n \leq (N \times N)/(m \times m)\}$. Obtain an index of two blocks needed to be exchanged applying $\mathbb{M}_1$ to Equation (8). That is, $j = \sum_{n=1} f(\mathcal{M}_n, \xi, n)$ where f is the function of the rule of Equation (8) and ξ is the key. Here, the indexes n and j are the indexes of the blocks to be exchanged. Swap the values of the block $\mathcal{M}_n$ and the block $\mathcal{M}_j$ in the map $\mathbb{M}_1$. When the blocks corresponding to the two positions are exchanged, the original positions are returned.

Step 5-2: Read a block $\mathcal{M}_n$ from $\mathbb{M}_2$, where $n \in \{1 \leq n \leq (N \times N)/(m \times m)\}$. Obtain an index to exchange two blocks applying Equation (8) to $\mathbb{M}_1$. That is, $j = \sum_{n=1} f(\mathcal{M}_n, \xi, n)$. Swap the values of the block $\mathcal{M}_n$ and the block $\mathcal{M}_j$ in the map $\mathbb{M}_2$.

Step 6: Divide the bitmap $\mathbb{M}_1$ and $\mathbb{M}_2$ into non-overlapping blocks of $m \times m$ pixels, where $m = 4$. The sized $N \times N$ AMBTC grayscale image $\mathbb{B}$ are initialized with zeros.

Step 6-1: Read a block $\mathcal{M}_n$ from $\mathbb{M}_2$, where $n \in \{1 \leq n \leq (N \times N)/(m \times m)\}$. The moment values ($Q_1$ and Q_2) are reconstructed from $\mathcal{M}_n$ using Equation (16), where $base = [2^7, 2^6, 2^5, 2^4, 2^3, 2^2, 2^1, 2^0]^T$.

$$\begin{cases} Q_1^n = \sum_{i=1}^{m \times m/2}(base \cdot \mathcal{M}_{n,i}) \\ Q_2^n = \sum_{i=9}^{m \times m/2}(base \cdot \mathcal{M}_{n,i}) \end{cases} \tag{16}$$

Step 6-2: Read a block $\mathcal{M}_n$ from $\mathbb{M}_1$. Equation (5) is applied to replace a bitmap block $\mathcal{M}_n$ with a grayscale block G_n. That is, $G_{n,i} = \sum_{i=1}^{m \times m} f(\mathcal{M}_n, Q_1, Q_2)$, where f is a function logic of Equation (5) and n is block number. A grayscale block G coding obtained by the decoding is assigned to $\mathbb{B}$, i.e., $\mathbb{B}(n) = G_n$.

Step 7: The image derived from AMBTC is reconstructed as repeating the procedure of Step 6 as much as the number of blocks.

Until now, we explained the restoration of a grayscale image based on AMBTC through extracting watermarks(maps) from the marked image. Next, we will explain how to restore the tampered block after finding the tampered block from the marked image.

Algorithm 5 Watermark Authentication, Tamper Detection and Reconstruct Cover Image

---

This describes the extracting checksum from the marked image and the restoration procedure of the tampered block using the checksum and the recovered trio.

Input: A marked image $\mathbb{K}$ and a grayscale image $\mathbb{B}$ (output of Algorithm 4) based on AMBTC with a size of $N \times N$, Key ξ.

Output: A reconstructed cover image $\mathbb{R}$ with a size of $N \times N$.

Step 1: Divide the images $\mathbb{K}$ and $\mathbb{B}$ into non-overlapping blocks sized $m \times m$, where $m = 4$. The sized $N \times N$ image $\mathbb{R}$ are initialized with zeros.

Step 2: Read a block of the images $\mathbb{K}$ and $\mathbb{B}$, then this block is assigned to $\mathcal{P}_n$ and $\mathcal{B}_n$, where n is a block number. $\mathcal{P}_{sum}$ embedded in the LSB of the block $\mathcal{P}_n$ is recovered by using Equation (17).

$$\mathcal{P}_{sum} = \sum_{i=1}^{m \times m} (\mathcal{P}_{n,i} \bmod 2) \times 2^i \tag{17}$$

$\mathcal{P}_{sum}$ is an embedded checksum in the block $\mathcal{P}_n$.

Algorithm 5 *Cont.*

Step 3: Generate checksum V_{sum} using Equation (12) and then discriminate whether the block has been tampered with or not using Equation (18). That is, if $V_{sum} = P_{sum}$, this block is a safe block; otherwise, it is a tampered block. If the block is safe, P_n is assigned to $\mathbb{R}_n$. Meanwhile, if it is tampered, the recovered block $\mathcal{B}_n$ is assigned to $\mathbb{R}_n$.

$$\mathbb{R}_n = \begin{cases} \mathcal{P}_n, & if \ (V_{sum} = P_{sum}), \\ \mathcal{B}_n, & otherwise, \end{cases} \tag{18}$$

Step 4: The recovered image $\mathbb{R}$ is made by repeating the procedure of Steps 2 and Step 3 as much as the number of blocks.

---

5. Experimental Results

In this section, the experiments and analysis are described to prove the performance of the proposed method. The computing platform used in the experiment has a Core i5-8250U processor, 1.60 GHz speed and 8 GB of RAM, and the software for the simulation is MATLAB R2019b. The standard USC-SIPI image database was used in the experiment for image restoration. Of these, some of the original 512×512 grayscale images were selected and used for the experiment. Figure 4 shows a set of test images (e.g., Lena, Pepper, Airplane, Boat, Goldhill, Couple, Baboon, and Zelda) used in the experiment.

For evaluation, Structural Similarity Index Metric (SSIM) and peak signal-to-noise ratio (PSNR) were used to compare the performance of the existing and proposed methods. The quality of the image was measured by the PSNR defined as

$$\text{PSNR} = 10 log_{10} \frac{255^2}{\text{MSE}} \tag{19}$$

PSNR is calculated as $10log$ (signal power/noise power), and signal power and noise power are calculated using peak power. The MSE used for PSNR calculation is the difference in average intensity between the marked image and the reference image, and a low MSE value can be evaluated as good image quality. In other words, the MSE is the mean of the squares of the errors $(p_i - p_i')^2$, where p and p' are reference and distorted images, respectively. The MSE is calculated as follows:

$$\text{MSE}(p, p') = \frac{1}{N} \sum_{i=1}^{N} (p_i - p_i')^2. \tag{20}$$

Here, the allowable pixel intensity is 255^2.

SSIM is a formula that measures the similarity between the original image and the displayed image. It consists of luminance, contrast and structure, and it measures the quality of an image. The range of the SSIM value is limited between 0 and 1, and if the value is close to 1, the image is similar to the cover image. The computation of SSIM is as follows:

$$SSIM(x, y) = \frac{(2\mu_x \mu_y + c_1)(2\sigma_{xy} + c_2)}{(\mu_x^2 \mu_y^2 + c_1)(\sigma_x^2 + \sigma_y^2 + c_2)}, \tag{21}$$

where μ_x and μ_y denote values of cover image x and the marked image y, σ_x and σ_y are standard deviation values of the cover image and the marked image, while $\sigma_{x,y}$ denotes the covariance of both two images. c_1 and c_2 are constants to stabilize the the division.

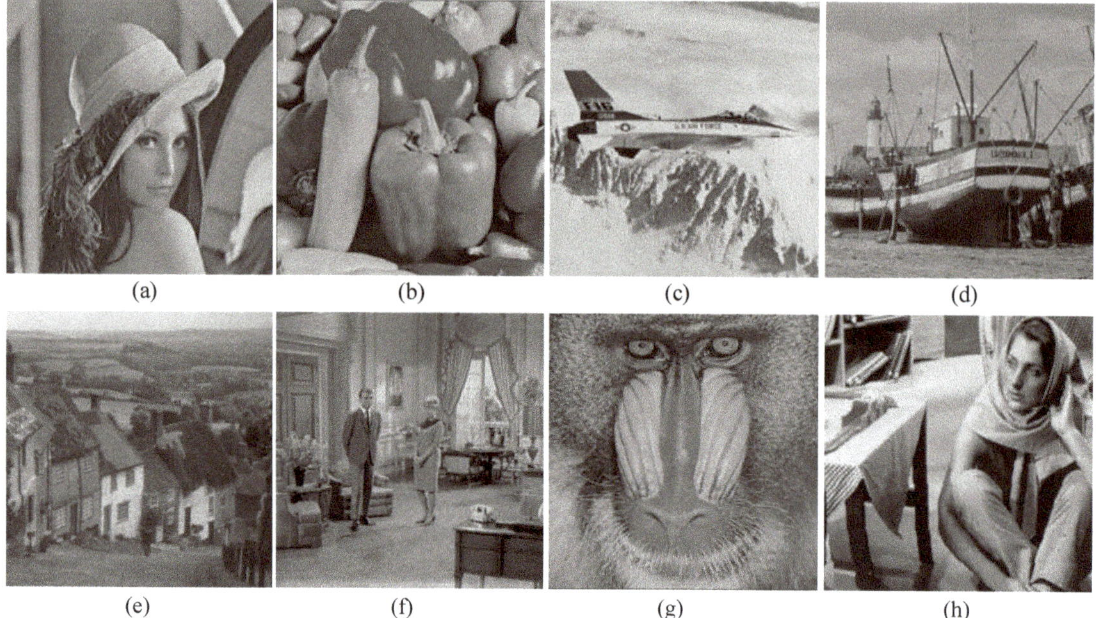

Figure 4. Test images used in our experiments: (**a**) Lena, (**b**) Pepper, (**c**) Airplane, (**d**) Boat, (**e**) Goldhill, (**f**) Couple, (**g**) Baboon and (**h**) Zelda.

Figure 5 shows a cropping attack on the marked image with limited ratios and the results of recovering the tampered images using the proposed method. The ratio was limited to 10% to 45%. For the cropping attack, the visual difference between the original image and the restored images as applying the method proposed in Section 4 (see Figure 5b) was very similar. Objective evaluations such as PSNR and SSIM of Figure 5 can be found in Tables 1 and 2.

The merit of our proposed method is that it manages PSNR (Table 1) and SSIM (Table 2) about marked images and recovered images reasonably. In the case of Lena image, when the ratio of cropping attacks is 5% and 45%, the difference between the two PSNRs is only 3.4573 dB. While, in the case of the Barbara image, the difference was highest among the comparison PSNRs in Table 1, i.e., it was 6.3909 dB.

In Table 2, the reason for introducing SSIM to measure image quality is that SSIM was high in the case of Baboon images with low PSNR (in Table 1), and subjective evaluation of image quality like the human visual system is possible, unlike PSNR measurement results. Overall, it means that SSIM was more accurate than PSNR in subjective terms. Therefore, two measurements are required for complementarity. As a result, it showed very good performance compared to other existing methods.

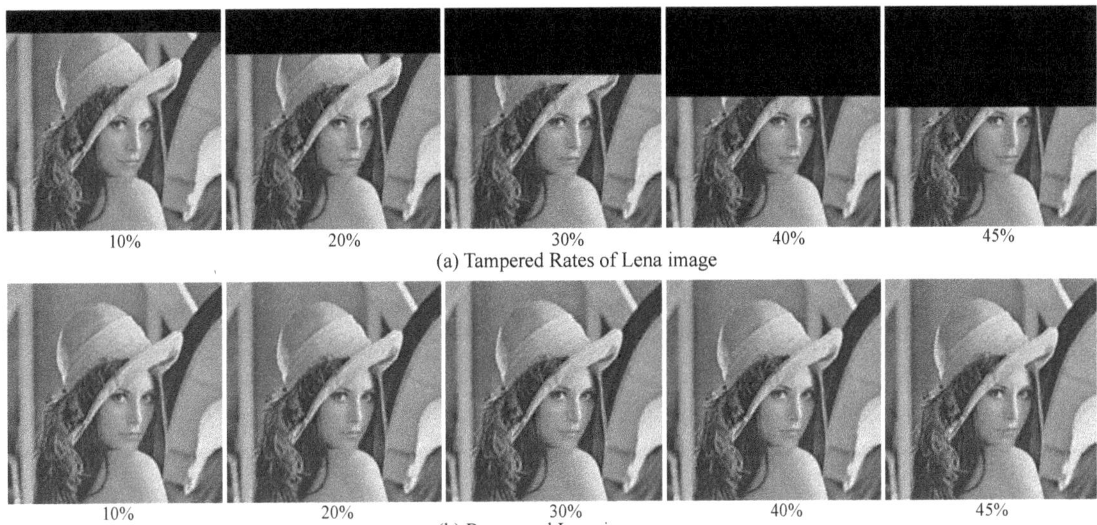

Figure 5. Cropping attack according to various ratios on Lena images and reconstructed Lena images applying the method we proposed.

Table 1. PSNR comparisons between original images and recovered images according to tampered rates.

Cover Image	Tampering Rate								
	5%	10%	15%	20%	25%	30%	35%	40%	45%
Lena	40.0894	40.0741	39.8066	39.2411	38.7654	38.2036	37.6783	36.754	36.6321
Pepper	39.1559	38.8223	38.258	37.8034	37.4989	37.0547	36.7648	36.1718	36.0495
Airplane	39.9916	40.0421	39.8804	39.0067	39.087	38.9265	39.1673	35.7464	35.7072
Boat	40.1693	39.9564	39.6507	38.9742	38.4153	37.7102	37.5841	35.8619	35.7968
Goldhill	39.7814	39.6703	39.5971	38.0752	37.8333	37.5566	37.4104	36.2497	35.8103
Couple	38.6335	37.9082	36.0183	34.9058	33.5838	33.1666	32.4731	32.1509	31.7149
Baboon	34.9313	33.6695	32.1257	31.5393	31.0723	30.4577	29.6961	29.551	28.6731
Zelda	40.0645	39.7914	39.0381	38.847	38.4935	37.7322	38.0708	36.3148	37.4341
Barbara	37.8367	37.0637	35.9489	35.8704	35.4713	34.1328	33.2672	31.9851	31.4458
Average	38.96151	38.55533	37.81376	37.14034	36.6912	36.10454	35.79023	34.53173	34.36264

Table 2. SSIM comparisons between original images and recovered images according to tampered rates.

Cover Image	Tampering Rate								
	5%	10%	15%	20%	25%	30%	35%	40%	45%
Lena	0.9508	0.9517	0.9534	0.9506	0.951	0.9516	0.9521	0.9495	0.9507
Pepper	0.9504	0.9505	0.9504	0.9495	0.9497	0.9484	0.948	0.9458	0.9463
Airplane	0.9477	0.9492	0.9509	0.953	0.955	0.9573	0.9599	0.9546	0.9571
Boat	0.9582	0.9601	0.9625	0.9638	0.9644	0.9647	0.9668	0.9608	0.9632
Goldhill	0.9665	0.9668	0.9674	0.9604	0.9593	0.958	0.9572	0.9502	0.9502
Couple	0.9664	0.9658	0.9604	0.9609	0.9552	0.9513	0.9465	0.9441	0.94
Baboon	0.9776	0.9732	0.9655	0.9615	0.9572	0.9516	0.9454	0.9421	0.9359
Zelda	0.9494	0.9486	0.947	0.9462	0.9446	0.943	0.9443	0.9398	0.9421
Barbara	0.969	0.9686	0.967	0.9689	0.9682	0.9644	0.9618	0.9585	0.9565
Average	0.959556	0.959389	0.958278	0.9572	0.956067	0.954478	0.953556	0.949489	0.949111

Figure 6 shows the quality of the restored image as PSNR when the cropping attack ratio was applied from 5% to 45% for the marked image. As the attack rate increased, PSNR decreased, showing a downward trend. However, the line was relatively smooth. In the case of the Lena image, the maximum performance was 40 dB or more at 5% and 36 dB or more at 45%.

In the case of the Pepper image, it showed about 39 dB at 5% and more than 36 dB at 45%, showing the lowest performance. The texture of the Pepper image has a smoother characteristic than that of the Lena image, and the surface of the cover image is very bright. Such features seem to degrade the quality of the recovered image during the reconstruction process using AMBTC. That can be a minor weakness of the method we have proposed.

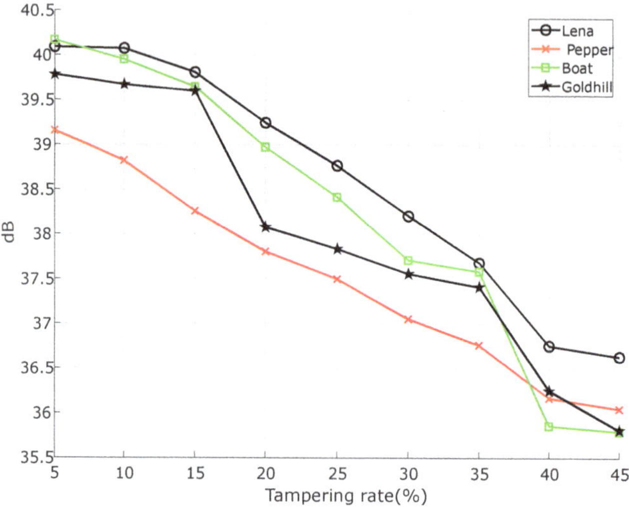

Figure 6. Tampered images at various tampering rates (%).

Since the recovered image using the proposed method used AMBTC derived from the BTC, this staircase effect could be reduced to some extent. The staircase effect may tend to appear larger in the brighter parts of the image. As a result, as the tampering rate increased, the staircase effect on the nose of Barbara's image was slightly revealed. Overall, the quality of the image restored with the method we proposed was excellent.

Figure 7 compares the performance of the existing self-embedding methods (i.e., Zhang et al. [17], Luo et al. [25], Yang & Shen [27], Hemida & He [30]) with our proposed method. The Tampering Rate (TR) for the marked cover-image (Lena) ranged from 5% to 45%. Both methods proposed by Hemida & He [30] and Luo et al. [25] measured about 35dB when the TR was 5%, and there was a slight difference in the performance of the two methods until the TR reached about 20%. However, the PSNR of the two showed similar reduction, and when TR was 20%, it was about 30 dB.

Then, the PSNR of Luo et al. [25] decreased to a smooth descending curve and was about 24 dB when TR = 45%, while the method proposed by Hemida & He [30] drastically decreased to about 11 dB when TR = 45%. One of the reasons Hemida& He's method [30] did not perform well is because it uses simple binary operations to detect the tampering area. This seems to be due to the threshold error according to the use of binary operations. Luo et al. [25] used the 7MSB binary value of each block for image authentication. Although Luo et al.'s authentication method performed better than Hemida & He [30], authentication failures may occur due to errors caused using binary numbers and may affect performance.

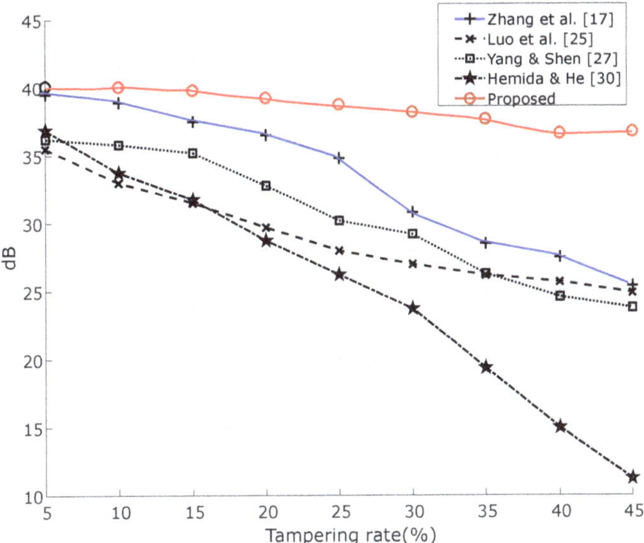

Figure 7. Comparison of the performance between previous methods and our proposed method.

Using halftones for image restoration is important to investigate. The image quality obtained by restoration using halftones was around 30 dB, which is a bit insufficient to ensure high image quality. The reason is that in the process of converting halftone to a grayscale image, it is visually observed that the quality of the image is clearly different from the texture of the original image. This means that the halftone could not be enough for the recovery bits. The PSNR of Yang & Shen [27] was limited from about 35 dB (highest) to about 24 dB (lowest), but it showed a relatively stable PSNR performance. For tampering authentication, they used Wong's watermarking technique [28], and its plan had a good impact on performance.

When the TR was 5%, the PSNR of Zhang et al.'s method [17] and the proposed method were shown at about 39 dB and about 40 dB, respectively. The proposed method decreased slowly until TR = 45%, and when TR = 45%, PSNR was 36 dB. On the other hand, the PSNR of Zhang et al. [17] was about 25 dB (TR = 45%), which appeared as a slightly steeper curve than ours. However, the advantage of this method is that it is designed to prevent tampering coincidence problems, which has a positive effect on performance and shows superior performance among existing methods. Nevertheless, the proposed method shows good performance among self-embedding methods.

The factors for improving the performance of the proposed method are as follows: first, the recovery bits for a specific block were stored in a long block located as far away as possible. Second, there was correct detection of tampered blocks. Since our proposed method was faithful in this perspective, we find that it had a positive effect on performance.

Table 3 shows a comparison of the PSNR of the marked image and the restored image obtained after applying various self-embedding watermarking methods to the Lena image. Looking at the PSNR of the reconstructed image in Table 3, we show that the proposed method was superior to the previous methods. In the case of He et al.'s method [19], some blocks at the boundary of the tampered region were erroneously identified. Zhang et al.'s method [23] showed that the quality of the recovered image was high when the tampered domain was less than 35% of the entire image. Qian et al.'s method did the authentication bit and reference bit in the three LSB layers of the image. Therefore, the restored image was excellent under limited conditions. Yang & Shen [27] produced an index table of the original image through vector quantization (VQ), and the obtained data were hidden in the cover image for image restoration. If the VQ is lost with a tamper attack, the restoration of the lost VQ area is impossible. In Kim et al.'s method [29], when the image damaged area was less than 50%, the quality of the reconstructed image was high (33.6). This was improved through image filtering after image restoration. In conclusion, our proposed method had the best restored image quality, but the quality of marked images was not the best. This is because up to 3LSB was used for restoration performance.

Table 3. Comparisons of PSNR of marked image and recovered image among different schemes.

Methods	Marked Images (PSNR)	Recovered Images (PSNR)	Criteria of Restoration
He et al. [19]	51.1	32.2	Tampered areas must be reserved
Zhang et al. [23]	37.9	29.9	<59%
Qian et al. [24]	37.9	35.0	<35%
Yang and Shen [27]	40.7	32.0	<50%
Kim et al. [29]	43.7	33.6	<50%
The proposed method	40.0	36.6	<45%

Table 4 shows the PSNR and NCC (Normalized Cross-Correlation) of the watermarked images, and it shows their embedding times (seconds). The PSNRs of the marked images were at levels difficult to discriminate with the human visual system. Therefore, the marked image made by the proposed method was very good in the aspect of the images' quality. In addition, the time performance measured with MATLAB was not bad, but the reason why the performance was not high is due to the performance of MATLAB, and it seems that there will be no problem in terms of time when developing in C language.

Table 4. Measuring PSNR, SSIM, NCC and time (second) of marked images based on the proposed method.

Cover Image	PSNR (dB)	SSIM	NCC	Embedding Time (s)
Lena	40.0076	0.9488	0.9996	1.0396
Pepper	40.0121	0.9516	0.9997	0.9516
Airplane	40.0306	0.9469	0.9996	0.8567
Boat	40.0093	0.9549	0.9996	0.9976
Goldhill	39.9983	0.9662	0.9995	1.2315
Couple	40.0196	0.9691	0.9996	1.2363
Baboon	40.0095	0.9841	0.9996	1.2099
Zelda	39.9805	0.9480	0.9994	0.8555

6. Conclusions

In this paper, we present a productive, fragile, self-embedding watermarking method based on AMBTC. Here, we concealed the recovery bits in LSB2 and LSB3 and the checksum bits in LSB in block units using the OPAP method. In addition, a checksum was introduced for accurate block authentication. In the existing method, binary bits were used for authentication, but there was a lack of precision, so checksum was used. Since the proposed method is a fragile watermarking method, the watermark information may be destroyed by image processing such as compression and filtering. However, in case of a partial cropping attack, it is possible to restore the tampered area to the level of the marked area before the forgery by using the hidden restoration information. The limitation of our proposed method is that if the damaged area of the image is large, the restoration information is also removed, so that an area that cannot be restored may occur. In the future we would like to find a way to solve the problem of corruption of recovery information that occurs when the size of the damaged image area is more than 50%.

Author Contributions: Conceptualization, C.-N.Y.; writing—original draft preparation, C.K.; Writing—review & editing, C.K.; Validation, C.-N.Y., and C.K.; formal analysis, C.-N.Y.; methodology, C.K.; data curation, C.K.; funding acquisition, C.-N.Y., C.K. All authors have read and agreed to the published version of the manuscript.

Funding: This work was partially supported by the Ministry of Science and Technology (MOST), under Grant 108-2221-E-259-009-MY2 and 109-2221-E-259-010, and by the Basic Science Research Program through the National Research Foundation of Korea (NRF) funded by (2015R1D1A1A01059253), and it was supported under the framework of international cooperation program managed by NRF (2016K2A9A2A05005255). This work was supported by the faculty research fund of Sejong University in 2020.

Institutional Review Board Statement: Not applicable.

Informed Consent Statement: Not applicable.

Data Availability Statement: Not applicable.

Acknowledgments: Thank you to the reviewers who reviewed this paper and the MDPI editor who edited it professionally.

Conflicts of Interest: The authors declare no conflicts of interest.

Abbreviations

The following abbreviations are used in this manuscript:

$\mathbb{M}_1$	first bitmap image
$\mathbb{M}_2$	second bitmap image
$\mathcal{V}_{sum}$	generated checksum
$\mathbb{C}$	cover image
$\mathbb{M}$	watermark (two maps of AMBTC)
$\mathbb{O}$	original image
$\mathcal{P}_n$	storing $m \times m$ pixel in a block
$\tilde{\mathcal{P}}_{n,i}$	a block which is removed LSB from $\mathcal{P}_n$
$trio(Q_1, Q_2, \mathcal{M})$	Q is a quantization level and $\mathcal{M}$ is a bitmap block
N	the size of original image
m	the size of a block
ξ	Key for block mapping in Algorithm 2
G_n	a grascale block
BTC	Block Truncation Coding
AMBTC	Absolute Moment Block Truncation Coding
OPAP	Optimal Pixel Adjustment Process
DH	Data Hiding
XOR	Exclusive-OR
CR	Compression Ratio
PSNR	Peak Signal-to-Noise Ratio
MSE	Mean Squared Error
WMSE	Worst-case Mean Squared Error
LSB	Least-Significant-Bit
SSIM	Structural Similarity Index Metric

References

1. Lou, D.-C.; Liu, J.-L. Fault resilient and compression tolerant digital signature for image authentication. *IEEE Trans. Consum. Electron.* **2000**, *46*, 31–39.
2. Umamageswari, A.; Suresh, G.R. Secure medical image communication using ROI based lossless watermarking and novel digital signature. *J. Eng. Res.* **2014**, *2*, 87–108. [CrossRef]
3. Tsai, P.; Hu, Y.; Chang, C. Novel image authentication scheme based on quadtree segmentation. *Imaging Sci. J.* **2005**, *53*, 14–162. [CrossRef]
4. Ababneh, S.; Ansari, R.; Khokhar, A. Iterative compensation schemes for multimedia content authentication. *J. Vis. Commun. Image Represent.* **2009**, *20*, 303–311. [CrossRef]
5. Walton, S. Image authentication for a slippery new age. *Dr. Dobb's J.* **1995**, *20*, 18–26.
6. Mishra, A.; Agarwal, C.; Sharma, A.; Bedi, P. Optimized gray-scale image watermarking using DWT-SVD and firefly algorithm. *Expert Syst. Appl.* **2014**, *41*, 7858–7867. [CrossRef]

7.   Parah, S.A.; Sheikh, J.A.; Loan, N.A.; Bhat, G.M. Robust and blind watermarking technique in DCT domain using inter-block coefficient differencing. *Digit. Signal Process.* **2016**, *53*, 11–24. [CrossRef]

8.   Di, Y.-F.; Lee, C.-F.; Wang, Z.-H.; Chang, C.-C.; Li, J. A robust and removable watermarking scheme using singular value decomposition. *KSII Trans. Internet Inf. Syst.* **2016**, *12*, 5268–5285.

9.   Zear, A.; Singh, A.K.; Kumar, P. A proposed secure multiple watermarking technique based on dwt, DCT and SVD for application in medicine. *Multimed. Tools Appl.* **2018**, *77*, 4863–4882. [CrossRef]

10.   Preda, R.O. Semi-fragile watermarking for image authentication with sensitive tamper localization in the wavelet domain. *Measurement* **2013**, *46*, 367–373. [CrossRef]

11.   Al-Otum, H.M. Semi-fragile watermarking for grayscale image authentication and tamper detection based on an adjusted expanded-bit multiscale quantization-based technique. *J. Vis. Commun. Image Represent.* **2014**, *25*, 1064–1081. [CrossRef]

12.   Qi, X.J.; Xin, X. A singular-value-based semi-fragile watermarking scheme for image content authentication with tamper localization. *J. Vis. Commun. Image Represent.* **2015**, *30*, 312–327. [CrossRef]

13.   Tong, X.J.; Liu, Y.; Zhang, M.; Chen, Y. A novel chaos-based fragile watermarking for image tampering detection and self-recovery. *Signal Process. Image Commun.* **2013**, *28*, 301–308. [CrossRef]

14.   Chen, F.; He, H.J.; Tai, H.M.; Wang, H.X. Chaos-based self-embedding fragile watermarking with flexible watermark payload. *Multimed. Tools Appl.* **2014**, *72*, 41–56. [CrossRef]

15.   Ansari, I.A.; Pant, M.; Ahn, C.W. SVD based fragile watermarking scheme for tamper localization and self-recovery. *Int. J. Mach. Learn. Cybern.* **2016**, *7*, 1225–1239. [CrossRef]

16.   Singh, D.; Singh, S.K. Effective self-embedding watermarking scheme for image tampered detection and localization with recovery capability. *J. Vis. Commun. Image Represent.* **2016**, *38*, 775–789. [CrossRef]

17.   Zhang, X.; Wang, S.; Qian, Z.; Feng, G. Reference Sharing Mechanism for Watermark Self-Embedding. *IEEE Trans. Image Process.* **2011**, *20*, 485–495. [CrossRef]

18.   Fridrich, J.; Goljan, M. Images with self-correcting capabilities. In *Proceedings of International Conference on Image Processing (ICIP)*; IEEE: Korbe, Japan, 1999; pp. 792–796.

19.   He, H.; Zhang, J.; Chen, F. Adjacent-block based statistical detection method for self-embedding watermarking techniques. *Signal Process.* **2009**, *89*, 1557–1566. [CrossRef]

20.   Lin, P.L.; Hsieh, C.K.; Huang, P.W. A hierarchical digital watermarking method for image tamper detection and recovery. *Pattern Recogn.* **2005**, *38*, 2519–2529. [CrossRef]

21.   Lee, T.-Y.; Lin, S.D. Dual watermark for image tamper detection and recovery. *Pattern Recogn.* **2008**, *41*, 3497–3506. [CrossRef]

22.   Zhang, X.; Wang, S. Fragile watermarking with error-free restoration capability. *IEEE Trans. Multimed.* **2008**, *10*, 1490–1499 [CrossRef]

23.   Zhang, X.; Wang, S.; Feng, G. Fragile Watermarking Scheme with Extensive Content Restoration Capability. In Proceedings of the IWDW 2009, Guildford, UK, 24–26 August 2009; Volume 5703.

24.   Qian, Z.; Feng, G.; Zhang, X.; Wang, S. Image self-embedding with high-quality restoration capability. *Digit. Signal Process.* **2011**, *21*, 278–286. [CrossRef]

25.   Luo, H.; Chu, S.-C.; Lu, Z.-M. Self Embedding Watermarking Using Halftoning Technique. *Circuits Syst. Signal Process.* **2008**, *27*, 155–170. [CrossRef]

26.   Hsu, C.-S.; Tu, S.-F. Image tamper detection and recovery using adaptive embedding rules. *Measurement* **2016**, *88*, 287–296. [CrossRef]

27.   Yang, C.-W.; Shen, J.-J. Recover the tampered image based on VQ indexing. *Signal. Process.* **2010**, *90*, 331–343. [CrossRef]

28.   Wong, P.W.; Memon, N. Secret and public key image watermarking schemes for image authentication and ownership verification. *IEEE Trans. Image Process.* **2001**, *10*, 1593–1601. [CrossRef]

29.   Kim, C.; Shin, D.; Yang, C.-N. Self-embedding fragile watermarking scheme to restoration of a tampered image using AMBTC. *Pers. Ubiquit. Comput.* **2018**, *22*, 11–22. [CrossRef]

30.   Hemida, O.; He, H. A self-recovery watermarking scheme based on block truncation coding and quantum chaos map. *Multimed. Tools Appl.* **2020**, *79*, 18695–18725. [CrossRef]

31.   Chang, C.; Lin, C.; Su, G. An effective image self-recovery based fragile watermarking using self-adaptive weight-based compressed AMBTC. *Multimed. Tools Appl.* **2020**, *79*, 24795–24824. [CrossRef]

32.   Delp, E.; Mitchell, O. Image compression using block truncation coding. *IEEE Trans. Commun.* **1979**, *27*, 1335–1342. [CrossRef]

33.   Lema, M.D.; Mitchell, O.R. Absolute moment block truncation coding and its application to color images. *IEEE Trans. Commun.* **1984**, *COM-32*, 1148–1157. [CrossRef]

34.   Wang, R.Z.; Lin, C.F.; Lin, J.C. Hiding data in images by optimal moderately significant-bit replacement. *IEE Electron. Lett.* **2000**, *36*, 2069–2070. [CrossRef]

35.   Hsu, L.Y.; Hu, H.T. Blind watermarking for color images using EMMQ based on QDFT. *Expert Syst. Appl.* **2020**, *149*, 1–16. [CrossRef]

36.   Hammami, A.; Hamida, A.B.; Amar, C.B. Blind semi-fragile watermarking scheme for video authentication in video surveillance context. *Multimed. Tools Appl.* **2020**. [CrossRef]

37.   Hamidi, M.; Chetouani, A.; Haziti, M.E.; Hassouni, M.E.; Cherifi, H. Blind robust 3D mesh watermarking based on mesh saliency and wavelet transform for copyright protection. *Information* **2019**, *10*, 67. [CrossRef]

38. Sadek, R.A. SVD based image processing applications: State of the art, contributions and research challenges. *Int. J. Adv. Comput. Sci.* **2012**, *3*, 26–34
39. Joshi, A.M.; Gupta, S.; Girdhar, M.; Agarwal, P.; Sarker, R. Combined DWT-DCT-based video watermarking algorithm using arnold transform technique. In *Proceedings of the International Conference on Data Engineering and Communication Technology*; Springer: Singapore, 2017; pp. 455–463.

Article

Hybrid Data Hiding Based on AMBTC Using Enhanced Hamming Code

Cheonshik Kim [1,*,†], Dongkyoo Shin [1], Ching-Nung Yang [2,†] and Lu Leng [3,4,*]

[1]   Department of Computer Engineering, Sejong University, Seoul 05006, Korea; shindk@sejong.ac.kr
[2]   Department of Computer Science and Information Engineering, National Dong Hwa University,
      Hualien 97401, Taiwan; cnyang@gms.ndhu.edu.tw
[3]   Key Laboratory of Jiangxi Province for Image Processing and Pattern Recognition,
      Nanchang Hangkong University, Nanchang 330063, China
[4]   School of Electrical and Electronic Engineering, College of Engineering, Yonsei University, Seoul 120749, Korea
*   Correspondence: mipsan@sejong.ac.kr (C.K.); leng@nchu.edu.cn (L.L.)
†   These authors contributed equally to this work.

Received: 3 July 2020; Accepted: 30 July 2020; Published: 2 August 2020

Abstract: The image-based data hiding method is a technology used to transmit confidential information secretly. Since images (e.g., grayscale images) usually have sufficient redundancy information, they are a very suitable medium for hiding data. Absolute Moment Block Truncation Coding (AMBTC) is one of several compression methods and is appropriate for embedding data due to its very low complexity and acceptable distortion. However, since there is not enough redundant data compared to grayscale images, the research to embed data in the compressed image is a very challenging topic. That is the motivation and challenge of this research. Meanwhile, the Hamming codes are used to embed secret bits, as well as a block code that can detect up to two simultaneous bit errors and correct single bit errors. In this paper, we propose an effective data hiding method for two quantization levels of each block of AMBTC using Hamming codes. Bai and Chang introduced a method of applying Hamming (7,4) to two quantization levels; however, the scheme is ineffective, and the image distortion error is relatively large. To solve the problem with the image distortion errors, this paper introduces a way of optimizing codewords and reducing pixel distortion by utilizing Hamming (7,4) and lookup tables. In the experiments, when concealing 150,000 bits in the Lena image, the averages of the Normalized Cross-Correlation (NCC) and Mean-Squared Error (MSE) of our proposed method were 0.9952 and 37.9460, respectively, which were the highest. The sufficient experiments confirmed that the performance of the proposed method is satisfactory in terms of image embedding capacity and quality.

Keywords: data hiding; AMBTC; BTC; Hamming code; LSB

---

1. Introduction

Recently, the Internet space has become like a single trading world where almost all digital content is distributed because every trading system is connected by high speed Internet, such as 5G. Many people distribute digital content in this space and are constantly consuming digital content. The problem with this digital space is that a copyright protection problem occurs because digital content is easily redistributed, copied, and modified by illegal users. There are various solutions to this problem, but the commonly used method is digital watermarking [1–3], which is used to protect the integrity and reliability of digital media.

Besides watermarking technology, Data Hiding (DH) technology is the most commonly used method of concealing information in digital media. The DH [4–6] technique can be used in various fields, such as digital signatures, fingerprint recognition, authentication, and secret communication.

It has been proven various times that DH could be used for secret communication, as well as the protection of the copyright of digital content. The people who use Internet communication know that the Internet is not a fully protected communication channel due to the many attackers. However, secret communication using DH can safely protect secret messages in digital cover media from the incomplete Internet channel.

DH may achieve the role of a secret communication strategy only when it satisfies two important criteria. First, the quality of the cover image (including data) should not be significantly different from the quality of the original image, since the cover image must not be detected by attackers while it is transmitted. Second, it must have the ability to transmit many secret data to the receiver securely.

The DH method is mainly conducted in two domains, namely the spatial domain and the frequency domain. In the spatial domain, a secret bit is concealed in the pixels of a host image directly. In the case of DH based on the spatial domain, it is applied to a grayscale image; even though the four Least Significant Bits (LSBs) [7–10] of each pixel are used for information hiding, they may not be detected by the Human Visual System (HVS).

Reversible DH [11–20] is a special case of DH in the academic community. In RDH, after the embedded bits are extracted, the stego image can be recovered back to the original image without distortion. The representative RDH methods are Difference Expansion (DE) [11,12], image compression [13], Histogram Shifting (HS) [14,15], Prediction-Error expansion (PE) [16,17], and encrypted images [18,19] for privacy preserving.

In the frequency domain, a cover image is converted into a frequency form, and then, the data are concealed in the coefficients of the frequency. The two most common methods based on the frequency domain are Discrete Cosine Transform (DCT) [21,22] and Discrete Wavelet Transform (DWT) [23]. Since changing the coefficient adversely affects the image quality, it is necessary to find and change the positions of the coefficient that have a relatively small influence on the image quality during data insertion. Spatial domain methods have the merit of the ability to conceal many secret data compared to the frequency domain methods, and the quality of the image is better, while they have the demerits of compression, noise, and filtering attacks compared to the frequency domain methods. Meanwhile, excellent compression images like JPEG are preferred as digital media, because the file size is small compared to the raw images and is well transmitted. For this reason, many researchers closely studied the watermarking and DH methods based on JPEG compression a long time ago.

Block Truncation Coding (BTC) [24] is one of the compression methods, and the configuration of the BTC is very simple compared to conventional JPEG. Thus, the computation time of BTC is much shorter than that of JPEG, and the quality of an image based on BTC is not significantly deteriorated compared to that of the original image. For this reason, it seems many researchers are interested in DH based on Absolute Moment Block Truncation Coding (AMBTC) [25–27], originated from BTC recently. Chuang and Chang [28] proposed a DH method based on AMBTC replacing the bitmaps of smooth blocks with the secret bits after dividing the blocks of an image into smooth blocks and complex blocks directly. It is called the Direct Bitmap Substitution (DBS) method. The merit of this method is that it may control the quality of the stego image by adjusting the threshold $T(= b - a)$ because the number of blocks using DH is decided according to the threshold value T. Here, a and b are quantization levels for each block in AMBTC. With the increase of the threshold T, the embedding capacity will be increased, while the image quality will be worse. In the case of decreasing the threshold T, the quality of the image will be improved, but the embedding capacity may be reduced.

Ou and Sun [29] introduced a way to embed data in the bitmaps of smooth blocks and proposed a method to reduce the distortions of the image by adjusting two quantization levels through re-computation, but the original image is required for re-calculation. Bai and Chang [30] proposed a way to embed secret data by applying a Hamming Code, i.e., HC(7,4) [7], to two quantization levels and bitmaps of AMBTC, respectively. When HC (7,4) is used for a complex block of AMBTC, it may be undesirable for high image quality. Kumar et al. [31] used two threshold values to increase the capacity of DH without significantly improving the image quality. Chen et al. [32] proposed a lossless

DH method using the order of two quantization levels in *trio*. This method is named the Order of Two Quantization Level (OTQL) method, which can conceal one bit per a block. For example, to store the bit "1", the order of two quantization levels, a and b, is reversed as $trio(b, a, BM)$. This method does not change the coefficients of both quantization levels, so it does not affect the quality of the image.

Hong [33] proposed a DH using Pixel Pair Matching (PPM) [34], where PPM is applied to the quantization levels; while the existing OTQL and DBS are used together for complex and smooth blocks, respectively. In 2017, Huang et al. [35] proposed a scheme for hiding data using pixel differences (hidden bits = $log_2 T$: derived from the difference expansion method) at two quantization levels and introduced a method to adjust the differences in the quantization levels to maintain image quality. This method is also a hybrid method by using OTQL and DBS as well. Chen and Chi [36] sub-divided less complex blocks and highly complex blocks. In 2016, Malik et al. [37] introduced a DH based on AMBTC using a two bit plane and four quantization levels. The merit of this method is the high payload, and the demerit is the decrease in the compression rates.

The motivations to propose a DH method using the Hamming code based on the image compressed with AMBTC are as follows. First, AMBTC is suitable for DH because it has reasonable compression performance, very low computational complexity, and (although not many) redundant bits. In addition, DH is relatively less studied for grayscale images. Second, the Hamming code is very efficient for redundant bits, such as for grayscale images. This has been demonstrated in previous studies [7,10]. However, since the image compressed with AMBTC has fewer redundant bits than the grayscale image, the embedding of enough secret bits at two quantization levels results in a negative effect on the image in the decoding of the bitmap. Third, Bai and Chang [30] attempted to conceal data at two quantization levels, but this did not achieve optimized performance. Therefore, it is essential to develop an optimized method in the DH process.

The main contributions of this paper are summarized as follows:

(i)   We introduce a general framework for DH based on AMBTC with the minimal squared error by the optimal Hamming code using a Lookup Table (LUT).

(ii)   Our method calculates the codeword corresponding to the minimum distance from the standard array of the (7,4) Hamming code table and then extracts the corresponding code. The method has little effect on program performance and can be easily conducted.

(iii)   We provide a comparative analysis and evaluate the efficiency based on the specified criteria.

(iv)   Sufficient experimental results are used to show the effectiveness and advantages of the proposed method.

The rest of this paper is organized as follows. Section 2 gives the introduction of the background research. The proposed method is described in detail in Section 3. The experimental results are analyzed in Section 4. Section 5 draws the conclusions.

2. Preliminaries

2.1. AMBTC

Absolute Moment Block Truncation Coding (AMBTC) [25] efficiently improves the computation time of Block Truncation Coding (BTC) and improves the image quality over BTC. The basic configuration of one block in AMBTC is two quantization levels and one bitmap, while one block is compressed by preserving the moment. Here, the two quantization values are obtained by calculating the higher mean and the lower mean of each block. For AMBTC compression, the grayscale image is first divided into $(k \times k)$ blocks without overlapping, where k can determine the compression level by (4×4), (6×6), (8×8), etc. AMBTC adopts block-by-block operations. For each block, the average pixel value is calculated by:

$$\bar{x} = \frac{1}{k \times k} \sum_{i=1}^{k^2} x_i \tag{1}$$

where x_i represents the ith pixel value of this block with a size of $k \times k$. All pixels in this block are quantized into a bitmap b_i (zero or one); that is, if the corresponding pixel x_i is greater than or equal to the average ($\bar{x}$), it is replaced with "1", otherwise it is replaced with "0". Pixels in each block are divided into two groups, "1" and "0". The symbols t and $k^2 - t$ refer to the numbers of pixels in the "0" and "1" groups, respectively. The means a and b of the two groups indicate the quantization levels of the groups "0" and "1". The two quantization levels are calculated by Equations (2) and (3).

$$a = \left\lfloor \frac{1}{t} \sum_{x_i < \bar{x}} x_i \right\rfloor \tag{2}$$

$$b = \left\lfloor \frac{1}{k^2 - t} \sum_{x_i \geq \bar{x}} x_i \right\rfloor \tag{3}$$

where a and b are also used to reconstruct AMBTC.

$$b_i = \begin{cases} 1, & \text{if } x_i \geq \bar{x}, \\ 0, & \text{if } x_i < \bar{x}. \end{cases} \tag{4}$$

$$g_i = \begin{cases} a, & \text{if } b_i = 0, \\ b, & \text{if } b_i = 1. \end{cases} \tag{5}$$

The bitmap is obtained from Equation (4), and the compressed block is simply uncompressed by using Equation (5); that is, the compressed code unit, $trio(a, b, BM)$, may be obtained by using Equations (2)–(5). The image block is compressed into two quantization levels a, b, and a Bitmap (BM) and can be represented as a $trio(a, b, BM)$. A BM contains the bit-planes that represent the pixels, and the values a and b are used to decode the AMBTC compressed image by using Equation (5).

Example 1. *Here, we describe the encoding and decoding procedure of one block of a grayscale image using an example. Figure 1a is a grayscale block, and the mean value of the pixels is 106. By applying Equations (2)–(4) on (a), we can obtain the bitmap as shown in (b) and two quantization levels ($a = 102; b = 107$). The basic unit of each block is trio(a, b, BM) = (102, 107, 0101111111011001). Using the information of the trio and Equation (5), the decoded grayscale block in (c) is reconstructed.*

Compressed unit: $trio\,(a, b, BM)\ =\ (102, 107, 0101111111011001)$

102	107	104	110
109	106	107	106
110	112	104	106
106	101	103	107

(a) a natural block

0	1	0	1
1	1	1	1
1	1	0	1
1	0	0	1

(b) bitmap

102	107	102	107
107	107	107	107
107	107	102	107
107	102	102	107

(c) A reconstructed block

Figure 1. An example of AMBTC: (**a**) a natural block; (**b**) a bitmap block; (**c**) a reconstructed block. BM, Bitmap.

2.2. Hamming Code

The Hamming Code (HC) [38] is a single error-correcting linear block code with a minimum distance of three for all the codewords. In HC(n, k), n is the length of the codeword, k is the number of information bits, and $(n - k)$ is the number of parity bits.

Let x be a k bit information word. The n bit codeword y is created by using $y = xG$, where G is the $k \times n$ generator matrix. Let $e = y - \tilde{y}$ be the error vector that determines whether an error occurred while sending y. If $e = 0$, no error occurs, and $\tilde{y} = y$.

Otherwise, the weight of e represents the number of errors. Let H be a $(n - k) \times n$ parity matrix with the relation of $G \cdot H^T = [0]_{k \times (n-k)}$. Let us assume that the codeword $\tilde{y}$ has an error like $e = (y - \tilde{y})$. In this case, we could correct one error $(e = y \oplus \tilde{y})$ from the codeword $\tilde{y}$ by using the syndrome $S = \tilde{y} \cdot H^T$, where the syndrome denotes the position of the error in the codeword. As show in Equation (6), the error e can be obtained.

$$
\begin{cases}
\tilde{y} \cdot H^T = (e \oplus y) \cdot H^T = e \cdot H^T + y \cdot H^T \\
(y \cdot H^T) = (x \cdot G) \cdot H^T = x \cdot (G \cdot H^T) = 0) \\
\qquad = e \cdot H^T + 0 = e \cdot H^T
\end{cases}
\tag{6}
$$

Consider HC(7,4) with the following parity matrix.

$$
H = \begin{bmatrix}
0 & 0 & 0 & 1 & 1 & 1 & 1 \\
0 & 1 & 1 & 0 & 0 & 1 & 1 \\
1 & 0 & 1 & 0 & 1 & 0 & 1
\end{bmatrix}
\tag{7}
$$

For example, assuming that one error bit occurred in y (e.g., the second bit from the left in $e = (e_1, e_2, \ldots, e_7) = (0\underline{1}00000)$, we may obtain the error position and recover the one bit error from the codeword y by calculating the syndrome $S(= y \cdot H^T = (010))$.

2.3. Bai and Chang's Method

For DH, the AMBTC algorithm is applied to the original cover image to obtain a low mean, a high mean, and a bitmap for every block. Then, the secret message is concealed in the AMBTC compressed $trio(a, b, BM)$. The merit of AMBTC is that it achieves a higher payload compared to other DH schemes performed in the compression domain. Here, it performs AMBTC DH in two phases. The method proposed by Bai and Chang is composed of two stages. One of them is to embed three bits in two quantization levels in $trio(a, b, BM)$ by using HC(7, 4). The detailed process of this method is as follows.

Step 1: For each *trio*, obtain seven bits from the two pixels at two quantization levels, and rearrange the seven bits to form a seven bit unit. Let $a = (a_8 \ a_7 \ a_6 \ a_5 \ a_4 \ a_3 \ a_2 \ a_1)$ and $b = (b_8 \ b_7 \ b_6 \ b_5 \ b_4 \ b_3 \ b_2 \ b_1)$ be the two original pixels. The rearranged seven bit unit is obtained by $y = (a_4, a_3, a_2, a_1 || b_3, b_2, b_1)$, where the symbol $||$ denotes that the four bits from a are concatenated with the three bits from b. Three secret message bits $(m = (m_1, m_2, m_3))$ are read from the secret bit set M.

Step 2: Compute the syndrome $S(= Hy \oplus m)$ of the codeword y, and then, the value is changed into a decimal value and is assigned to the variable i. To obtain the stego codeword $\hat{y} = (y_1, y_2, y_3, y_4, y_5, y_6, y_7)$, flip the ith bit of the codeword y.

Step 3: To reconstruct two quantization levels with the codeword y, (y_7, y_6, y_5, y_4) is replaced with four LSBs of the low-mean value a, and (y_3, y_2, y_1) is replaced with three LSBs of the high-mean value b.

Step 4: It is possible to hide an additional bit by using the order of two quantization levels and the difference between them. In this case, it may be acceptable to embed an additional bit when the criterion $(b - a \geq 8)$ is satisfied. Otherwise, it is not accepted to embed an additional bit. If the bit to be embedded is "1", swap the order of the two quantization levels as $(trio(b, a, BM))$, otherwise no change is conducted.

In Step 4, it is possible to embed an additional bit only under the given condition $(b - a \geq 8)$. The reason for the condition is necessary; if the difference between the values of a and b is small, the order of the two values may be reversed as a result of the computation of the Hamming code. An ambiguous result in the decoding procedure may occur.

3. The Proposed Scheme

In this section, we introduce a DH to embed secret data in bitmaps and the quantization levels of AMBTC using optimized the Hamming code and DBS method. First, compressed blocks, *trios*, are classified into smooth blocks and complex blocks. Then, DBS is applied to the bitmaps of the smooth blocks, while the Hamming code may be applied to the quantization levels regardless of the block characteristics. The method proposed by Bai and Chang results in the large distortion of the cover image. In Section 3.1, we introduce a way to solve this problem.

3.1. Embedding Procedure

We introduce a way of DH using the Hamming code, DBS, and OTQL based on AMBTC and explain the details of the procedure step-by-step as follows. Additionally, the flowchart of the embedding process is described in Figure 2.

Input: Original grayscale image with a size of $N \times N$, threshold T, and secret data $M = (m_1, m_2, \ldots, m_n)$.

Output: Stego AMBTC *trios*.

Step 1: The original image G is divided into 4×4 non-overlapping blocks.

Step 2: The $trio(a, b, BM)$ of the AMBTC, i.e., the compressed codes, is obtained according to Equations (1)–(4), where a and b are the low mean and the high mean quantization levels, respectively, and BM is the bitmap.

Step 3: The quantization levels are $a = (a_8 a_7 \ldots a_1)$ and $b = (b_8 b_7 \ldots b_1)$, where a_8 is the Most Significant Bit (MSB) of a and a_1 is the LSB of a. Similarly, b_8 is the MSB of b, and b_1 is the LSB of b. The rearranged seven bit codeword is obtained by:

$$y = (a_4 a_3 a_2 a_1 || b_3 b_2 b_1) \tag{8}$$

where the symbol $||$ denotes concatenation. Note that a_4 and b_1 are the MSB and LSB of the rearranged pixel y, respectively.

Step 4: In Figure 3b, the location of the coset leader that matches the decimal number d for m_i^{i+2} bits is retrieved from the Lookup Table (LUT) using the procedure in Figure 4. Assuming that $x_i = (x_7\, x_6\, x_5\, x_4\, x_3\, x_2\, x_1)$, the codewords corresponding to the retrieved coset reader are converted to $(\alpha'\, \beta')$. That is, $\alpha' = bin2dec(x_7\, x_6\, x_5\, x_4)$ and $\beta' = bin2dec(x_3\, x_2\, x_1)$. Meanwhile, for codeword y generated in Step 3, $\alpha = bin2dec(y_7\, y_6\, y_5\, y_4)$ and $\beta = bin2dec(y_3\, y_2\, y_1)$ are converted; that is, $y' = (\alpha\, \beta)$. The distances for x and y' are calculated using Equation (9). After calculating $min((\alpha - \alpha')^2 + (\beta - \beta')^2)$ for all codewords, the value with the minimum distance among them is obtained. The obtained minimum distance codeword is $h = (\alpha'\, \beta')$.

$$\epsilon = min((\alpha - \alpha')^2 + (\beta - \beta')^2) \tag{9}$$

For the codeword h, two quantization levels, a and b, are constructed as follows:

$$\begin{cases} a = (a_8 a_7 a_6 a_5 || h_7 h_6 h_5 h_4) \\ b - (b_8 b_7 b_6 b_5 b_4 || h_3 h_2 h_1) \end{cases} \tag{10}$$

Before next step, three is added to the index variable i.

Step 5: If $|a - b| \leq T$, it may be a smooth block. For the smooth block, we use DBS. The m_i^{i+15} bits replace the pixels of the BM. Fifteen is added to the index variable i before the next step. If $|a - b| > T$ and $|a - b| \geq 8$, OTQL is launched. If $m_i^i = 1$, transpose the order of two quantization levels, a and b, of the *trio*, otherwise put the *trio* as the original state.

Step 6: Repeat Steps 2~5 until all image blocks are processed. Then, the stego AMBTC compressed codes' *trio* is constructed.

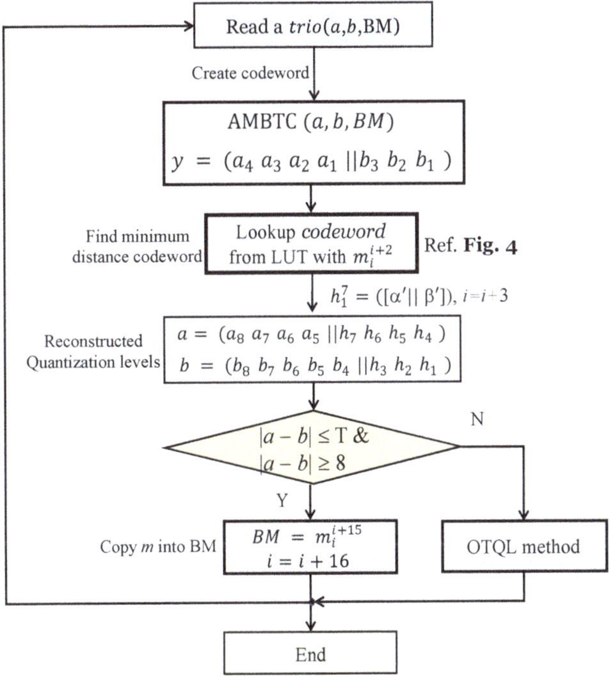

Figure 2. The flowchart of the data embedding process. OTQL, Order of Two Quantization Level.

Standard array of a (7,4) Hamming code for DH (binary)

Coset leader								
0000000	(0001101)	(0011010)	(0010111)	(0110100)	(0111001)	(0101110)	(1000011)	(1101000)
	(1100101)	(1110010)	(1111111)	(1011100)	(1010001)	(1000110)	(1001011)	
0000001	(0001100)	(0011011)	(0010110)	(0110101)	(0111000)	(0101111)	(0100010)	(1101001)
	(1100100)	(1110011)	(1111110)	(1011101)	(1010000)	(1000111)	(1001010)	
0000010	(0001111)	(0011000)	(0010101)	(0110110)	(0111011)	(0101100)	(0100001)	(1101010)
	(1100111)	(1110000)	(1111101)	(1011110)	(1010011)	(1000100)	(1001001)	
0000011	(0001110)	(0011001)	(0010100)	(0110111)	(0111010)	(0101101)	(0100000),	(1101011)
	(1100110)	(1110001)	(1111100)	(1011111)	(1010010)	(1000101)	(1001000)	
0000100	(0001001)	(0011110)	(0010011)	(0110000)	(0111101)	(0101010)	(0100111)	(1101100)
	(1100001)	(1110110)	(1111011)	(1011000)	(1010101)	(1000010)	(1001111)	
0000101	(0001000)	(0011111)	(0010010)	(0110001)	(0111100)	(0101011)	(0100110)	(1101101)
	(1100000)	(1110111)	(1111010)	(1011001)	(1010100)	(1000011)	(1001110)	
0000110	(0001011)	(0011100)	(0010001)	(0110010)	(0111111)	(0101000)	(0100101)	(1101110)
	(1100011)	(1110100)	(1111001)	(1011010)	(1010111)	(1000000)	(1001101)	
0000111	(0001010)	(0011101)	(0010000)	(0110011)	(0111110)	(0101001)	(0100100)	(1101111)
	(1100010)	(1110101)	(1111000)	(1011011)	(1010110)	(1000001)	(1001100)	

Standard array of a (7,4) Hamming code for DH (decimal)

MSB 4-BIT to DECIMAL NUMBER, LSB 3-BIT to DECIMAL NUMBER
Ex) ($a_1 = 1$, $b_1 = 5$) $\in$ {(1 5) in coset leader 0}

Coset leader								
0	(1 5)	(3 7)	(2 7)	(6 4)	(7 1)	(5 6)	(8 3)	(13 0)
	(12 5)	(14 2)	(15 7)	(11 4)	(10 1)	(8 6)	(9 3)	
1	(1 4)	(3 3)	(2 6)	(6 5)	(7 0)	(5 7)	(4 2)	(13 1)
	(12 4)	(14 3)	(15 6)	(11 5)	(10 0)	(8 7)	(9 2)	
2	(1 7)	(3 0)	(2 5)	(6 6)	(7 3)	(10 4)	(4 1)	(13 2)
	(12 7)	(14 0)	(15 5)	(11 6)	(10 3)	(8 4)	(9 1)	
3	(1 6)	(3 1)	(2 4)	(6 7)	(7 2)	(10 5)	(4 0),	(13 3)
	(12 6)	(14 1)	(15 4)	(11 7)	(10 2)	(8 5)	(9 0)	
4	(1 1)	(3 6)	(2 3)	(6 0)	(7 5)	(5 2)	(4 7)	(13 4)
	(12 1)	(14 6)	(15 3)	(11 0)	(10 5)	(8 2)	(9 7)	
5	(9 0)	(3 7)	(2 2)	(6 1)	(7 4)	(5 3)	(4 6)	(13 5)
	(12 0)	(14 7)	(15 2)	(13 1)	(10 4)	(8 3)	(9 6)	
6	(1 3)	(3 4)	(2 1)	(6 2)	(7 7)	(5 0)	(4 5)	(13 6)
	(12 3)	(14 4)	(15 1)	(11 2)	(10 7)	(8 0)	(9 5)	
7	(1 2)	(3 5)	(2 0)	(6 3)	(7 6)	(5 1)	(4 4)	(13 7)
	(12 2)	(14 5)	(15 0)	(11 3)	(10 6)	(8 1)	(8 4)	

(a)　　　　(b)

Figure 3. Standard array of HC(7,4) for Data Hiding (DH): (**a**) binary presentation and (**b**) decimal presentation.

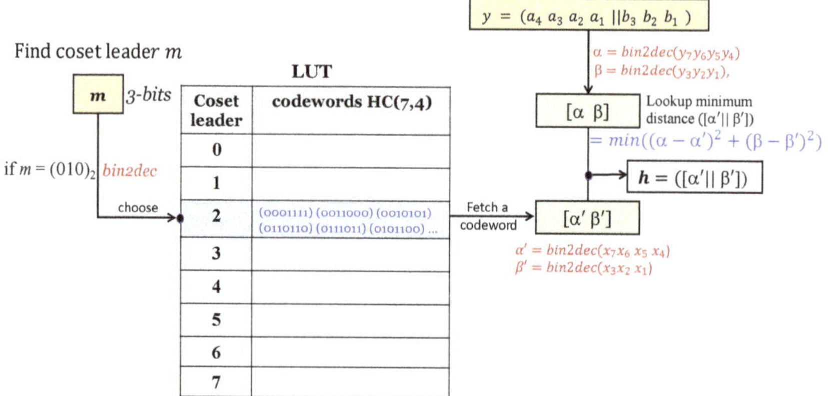

Figure 4. The flowchart of the lookup codeword with m. HC, Hamming Code.

3.2. Extraction Procedure

The procedure for extracting the hidden secret bits is shown in Figure 5. The process is explained in detail according to the following procedure.

Input: Stego AMBTC compressed codes *trios*, matrix $H = \begin{bmatrix} 0 & 0 & 0 & 1 & 1 & 1 & 1 \\ 0 & 1 & 1 & 0 & 0 & 1 & 1 \\ 1 & 0 & 1 & 0 & 1 & 0 & 1 \end{bmatrix}$, and threshold T.

Output: Secret data $M = (m_1, m_2, \ldots, m_n)$.

Step 1:   Read one block of $trio(a, b, BM)$ from a set of *trios* as a defined order, where the *trio* consists of two quantization levels and one bitmap.

Step 2:   The quantization levels are $a = (a_8 a_7 \ldots a_1)$ and $b = (b_8 b_7 \ldots b_1)$, where a_8 is the MSB of a and a_1 is the LSB of a. Similarly, b_8 is the MSB of b and b_1 is the LSB of b. The rearranged seven bit codeword is $y = (a_4 a_3 a_1 a_1 || b_3 b_2 b_1)$ by Equation (8).

Step 3:   Obtain the syndrome $S = y \cdot H^T$. Then, assign S to m_i^{i+2}, and add three to i.

Step 4:   If $|a - b| \leq T$, it is a smooth block *trio*. In this case, this means that the hidden bits were embedded in the BM in the form of pixels. Therefore, by assigning the pixels of the BM to m in order, all the values concealed in the BM can be obtained. That is, $m_i^{i+15} = BM_1^{16}$ and $i = i + 15$. If $|a - b| > T$ and $|a - b| \geq 8$, one bit is hidden in the *trio* by using the order of two quantization levels. If the order of two quantization levels is $trio(b, a, BM)$, this means that $m_i^i = 1$, otherwise $m_i^i = 0$.

Step 5:   Repeat Steps 1 $\sim$ 4 until all the *trios* are completely processed, and the extracted bit sequence constitutes the secret data m.

3.3. Examples

Here, we will show how to minimize the errors in the encoding process through an optimized method rather than the existing method. The detailed procedure of our proposed DH is explained by the process shown in Figure 6 using $trio(103, 109, 0000010001110111)$ and secret bits $m = (1011010111100001100)$. Since $b - a = 109 - 103 = 6 \leq T(7)$, the *trio* is classified as a smooth block. Therefore, in Figure 2, the data concealment process proceeds according to the processing corresponding to the smooth block of the *trio*. From now on, the process shown in Figure 6a will be explained step-by-step.

(1)   The two quantization levels of a given *trio* are assigned to variables a and b and then converted to binary, i.e., $a = 103 = 01100111_2$ and $b = 109 = 01101101_2$.

(2)   For the two converted binary numbers, the four LSB ($a = (0110\underline{0111}_2)$) of a and the three LSB ($b = (01101\underline{101}_2)$) of b are extracted.

(3)   To form a codeword, the extracted binary numbers are combined; that is, $y = (0111||101)_2$.

(4)   Calculate $y' = (bin2dec(0111)\ bin2dec(101)) = (7\ 5)$.

(5)   After converting the bit $m(= 101)$ to decimal, the value $d = 5$ is retrieved from the coset leaders of the standard array of HC (7,4).

(6)   Using Equation (9), the codeword having the minimum distance from the given codeword is retrieved from the table. Here, $(a - \underline{7})^2 + (b - \underline{4})^2 = 1$ corresponds to the minimum distance.

(7)   The new codeword is $h = (7||4) = (0111||100)_2 = (0111100)_2$.

(8)   Two quantization levels embedding three secret bits are recovered by using the codeword h. A new quantization level is obtained by replacing the upper four bits and the lower three bits of h obtained in the process of (7), respectively, with four LSB and three LSB of two quantization levels. That is, the recovered codewords are $a = 103$ and $b = 108$.

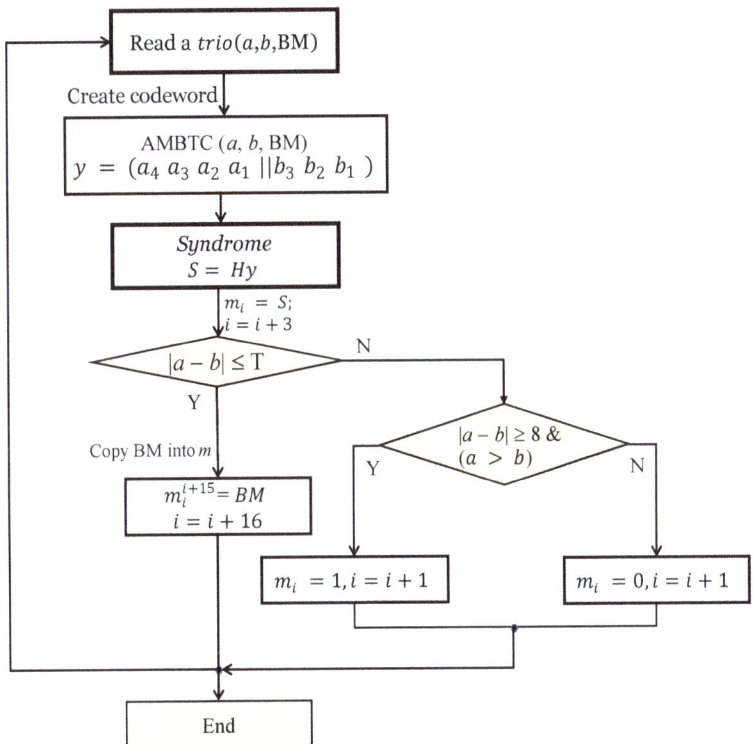

Figure 5. The flowchart of the extracting procedure.

In Figure 6b, we explain a way of embedding secret bits into the bitmap.

(1)   First, it is necessary to check whether a given block belongs to a smooth block. That is, if the difference between the absolute values of two given quantization levels is less than the threshold T, it is a smooth block, otherwise it belongs to a complex block. In Figure 6b, the difference between two given quantization levels is less than the defined threshold T, so it belongs to a smooth block.

(2)   Since the block in the given example is a smooth block, sixteen bits are concealed in the bitmap by replacing the 16 bit secret bits (m = (1010 1111 0000 1100)) directly with the bitmap.

To extract secret bits from two quantization levels, we need to construct a codeword using the quantization levels. To construct the codeword, the procedure of Figure 6a is followed. That is, the codeword ($y = 0111100$) is obtained by extracting four LSB (0111) and three LSB (100) from two quantization levels $a(= 103)$ and $b(= 108)$ and combining them. Here, we obtain the hidden secret bits, $\overline{m} = (101)$, by using the equation, $S = y \cdot H^T$, to the codeword. The decoding of the secret bits in the BM extracts the hidden bits by moving all pixels in the BM into a variable $\overline{m}$ array directly.

(a) Hamming code for DH

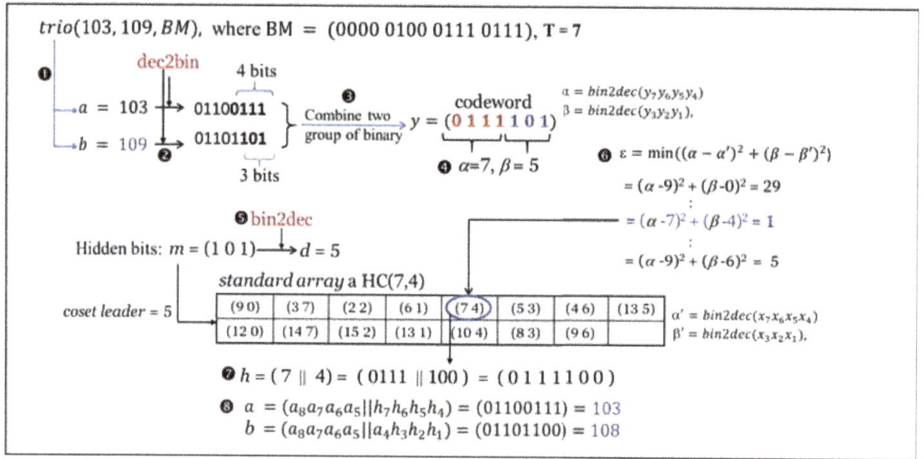

(b) Direct Bitmap Substitution (DBS) for DH

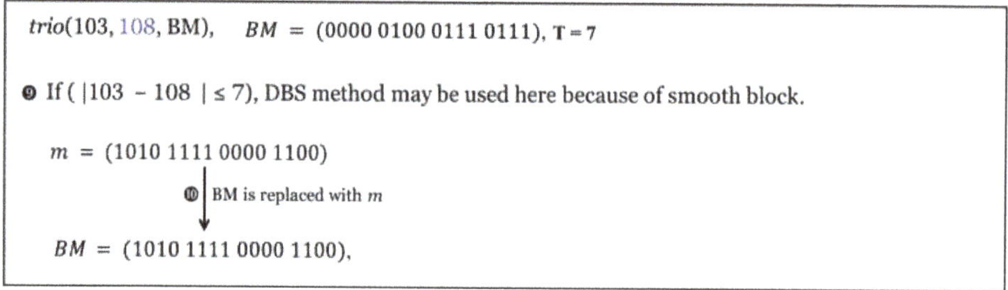

Figure 6. Illustration of data embedding.

4. Experimental Results

In this section, we prove the performance of our proposed scheme by comparing with the existing methods, such as Bai and Chang [30], W Hong [33], and Chuang et al. [28]. As shown in Figure 7, six grayscale images sized 512×512 are used for our experiments. In addition, the block size of AMBTC is set to 4×4, and the secret bits are generated by a pseudo-random number generator. Embedding Capacity (EC) and the Peak Signal-to-Noise Ratio (PSNR) are widely used as objective image evaluation indices. Here, EC is used as an indicator for the number of secret bits that can be embedded in a cover pixel. The relatively high PSNR value means that the quality of the stego image is good. The DH capacity is the size of the secret bit that is embedded in the cover image. The quality of the image is measured by the PSNR defined as:

$$PSNR = 10log_{10}\frac{255^2}{MSE} \tag{11}$$

The Mean-Squared Error (MSE) used in the PSNR denotes the average intensity difference between the stego and reference images.

Figure 7. Test images: (a~f) 512×512.

The lower the MSE value of a stego image, the better the quality of the image. The MSE is calculated using the reference image p and the distorted image $p^{'}$ as follows.

$$MSE = \frac{1}{N \times N} \sum_{i=1}^{N} \sum_{j=1}^{N} (p_{ij} - p^{'}_{ij})^2 \tag{12}$$

The error value $\epsilon = p_{ij} - p^{'}_{ij}$ indicates the difference between the original and the distorted pixels. The 255^2 means the allowable pixel intensity in Equation (11). A typical value for the PSNR in a lossy image is from 30 dB to 50 dB for an eight bit depth; the higher the better. Structural SIMilarity (SSIM) [39] estimates whether changes such as image brightness, photo contrast, and other residual errors are identified as structural changes. The SSIM values is limited to a range between zero and one. If the SSIM value is close to one, it means that the stego image is similar to the cover image and of high quality. The equation of SSIM is as follows:

$$SSIM(x,y) = \frac{(2\mu_x\mu_y + c_1)(2\sigma_{xy} + c_2)}{(\mu_x^2\mu_y^2 + c_1)(\mu_x^2\mu_y^2 + c_2)} \tag{13}$$

where μ_x, μ_y are the mean values of the cover image (x) and stego image (y), $\sigma_x, \sigma_y, \sigma_x^2, \sigma_y^2$, and σ_{xy} are the standard deviation, variances, and covariance of the cover image and stego image, and c_1, c_2, c_3 are constant values to avoid the division by zero problem.

The Normalized Cross-Correlation (NCC) has been commonly used as a metric to evaluate the degree of similarity (or dissimilarity) between two compared images. The main advantage of the NCC is that it is less sensitive to linear changes in the amplitude of illumination in the two compared images. Furthermore, the NCC is confined to the range between -1 and one. NCC is calculated by the formula given in Equation (14).

$$NCC = \frac{\sum_{x=1}^{M} \sum_{y=1}^{N} (S(x,y) \times C(x,y))}{\sum_{x=1}^{M} \sum_{y=1}^{N} (S(x,y))^2} \tag{14}$$

Table 1 represents the comparison of EC and PSNR between the proposed scheme and existing methods, i.e., Ou and Sun [29], Bai and Chang [30], and W Hong et al. [33]. Specifically, we compare the performance between our scheme and the existing methods using six images when the threshold value $T(= b - a)$ is 5, 10, and 20. The evaluation of EC and the PSNR based on threshold values is necessary for objectivity and fairness for comparative evaluation of the performance; that is, the data measured under the same threshold value may be evaluated as a more meaningful comparison. One important point for EC and PSNR is that there is a trade-off between the two assessments. That is, if EC is higher, the PSNR is reduced, and vice versa. However, in the case that the proposed method has very good performance, the deviation from the trade-off may not be large. The EC of our proposed method is efficient with respect to the EC as 151,173 bits when $T = 5$.

Table 1. PSNR and Embedding Capacity (EC) according to different thresholds T.

Images	T	Ou and Sun [29]		Bai and Chang's [30]		W Hong [33]		The Proposed	
		EC (bits)	PSNR (dB)	EC (bits)	PSNR (dB)	EC (bits)	PSNR (dB)	EC (bits)	PSNR (dB)
Boats		129,249	31.3506	64,011	31.2928	149,368	31.3203	166,176	31.2846
Goldhill		53,873	32.7028	21,291	32.7076	73,408	32.6373	100,853	31.4917
Airplane		154,545	31.7405	78,477	31.6604	175,203	31.7181	187,268	31.2018
Lena	5	135,089	33.1929	67,400	33.1760	155,889	33.1454	168,498	33.1059
Peppers		100,977	33.6253	48,164	33.6888	121,072	33.5682	138,714	33.4999
Zelda		109,585	35.7438	53,096	35.8680	128,346	35.6618	145,526	35.5624
Average		113,886	33.0593	55,407	33.0656	133,881	33.0085	151,173	32.6911
Images	T	Ou and Sun [29]		Bai and Chang's [30]		W Hong [33]		The Proposed	
		EC (bits)	PSNR (dB)	EC (bits)	PSNR (dB)	EC (bits)	PSNR (dB)	EC (bits)	PSNR (dB)
Boats		160,913	31.0204	82,644	31.1508	186,330	30.9774	201,272	30.9316
Goldhill		127,409	31.6842	64,721	32.2382	150,349	31.6372	169,667	30.7147
Airplane		194,897	31.3173	102,018	31.4875	221,824	31.2796	232,682	30.8072
Lena	10	193,249	32.3724	101,530	32.7961	220,205	32.3277	231,077	32.2792
Peppers		200,369	32.2246	106,357	32.9962	227,287	32.1842	236,657	32.1617
Zelda		212,753	33.6013	113,380	34.7771	240,483	33.5452	247,727	33.5333
Average		164,799	31.8075	85,108	32.5743	190,170	31.7623	219,847	31.7379
Images	T	Ou and Sun [29]		Bai and Chang's [30]		W Hong [33]		The Proposed	
		EC (bits)	PSNR (dB)	EC (bits)	PSNR (dB)	EC (bits)	PSNR (dB)	EC (bits)	PSNR (dB)
Boats		205,809	29.5664	110,433	30.7138	233,709	29.5557	243,122	29.5228
Goldhill		212,193	29.1224	117,286	31.2597	240,231	29.1121	249,392	28.5724
Airplane		226,977	30.1906	122,037	31.1269	256,110	30.1792	262,802	29.8300
Lena	20	233,697	30.7508	126,667	32.2383	264,366	30.7356	269,432	30.7074
Peppers		240,977	30.691	131,514	32.4373	271,132	30.6750	276,077	30.6495
Zelda		253,841	31.5579	138,904	33.9124	284,866	31.5299	288,527	31.4777
Average		221,128	29.9278	124,474	31.9481	250,207	29.9158	264,892	30.1266

In Table 1, Bai and Chang's PSNR (=33.0656 dB) is measured as higher than that (=32.6911 dB) of our proposed method. Here, the EC of Bai and Chang [30] is 55,407 bits, and the EC of our method is 151,173 bits. In the end, our proposed method shows the capability to conceal about 95,000 bits more than that of Bai and Chang.

If the threshold T and EC are given for a faithful measurement, the PSNR of our proposed method may be the highest. This is because the size of the hidden bits affects the PSNR. Apparently, a relative good method has high values both for the PSNR and EC. When $T = 10$, we can see that our method's EC ($-219,847$ bits) is the largest. The method of W Hong [33] and our proposed method both have the same PSNR (31.7 dB), which is 0.1 dB lower than that of Ou and Sun's method [29]. However, in this case as well, when considering the amount of EC, our method outperforms the other two methods.

When $T = 20$, the PSNR of the proposed method is higher than the previous two methods (Ou and Sun [29] and W Hong [33]), and the EC of our method has the highest performance. It can be seen from the simulation results in Table 1 that the proposed method has 140,000 bits more than that of Bai and Chang [30] in terms of EC.

Figure 8 shows the performance comparison between our proposed method and the existing methods, where we measured the PSNR while increasing the capacity of the secret bits from 20,000 bits to 310,000 bits in four images ((a) Lena, (b) Boat, (c) Pepper, and (d) Zelda) by using the proposed and existing methods.

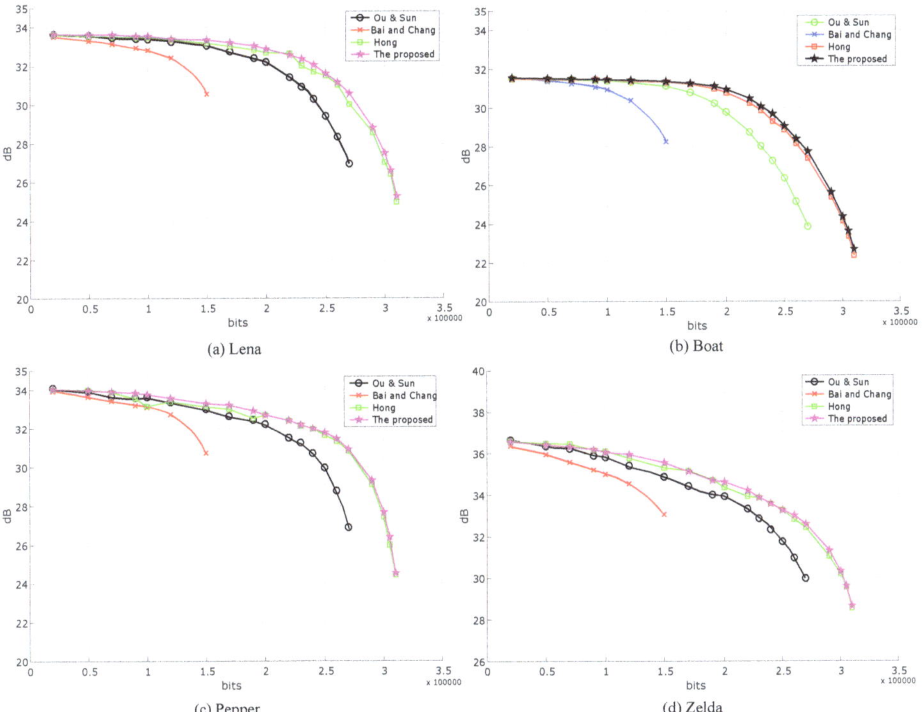

Figure 8. Performance comparisons of the proposed method and other related methods (i.e., Ou and Sun, Bai and Chang, Hong) based on four images: (**a**) Lena, (**b**) Boat, (**c**) Pepper, and (**d**) Zelda.

We propose a way to improve the performance of Bai and Chang's method [30], and as shown in Figure 8, it is confirmed that our proposed method is superior to existing methods. On the other hand, our proposed method shows almost the same performance as W Hong's method [33], but it can be confirmed that the performance of our proposed scheme is slightly better. Ou and Sun's method [29] is superior to Bai and Chang's method [20], but the performance is not as high as that of our proposed method.

AMBTC has difficulty hiding enough data, because it is a compressed code, and unlike conventional grayscale images, it is not easy to exploit high embedding capacity by the constraint of compressed pixels. It is difficult to improve the DH performance for images with many complex blocks, and if we exploit many pixels for high data concealment, the image quality may deteriorate.

In Figure 8, we can see that the EC of Bai and Chang's method [30] is very low. That is because this method can hide only six bits of data while inverting up to two pixels in each bitmap. Thus, there is a limit to embedding enough data in the *trio*'s bitmaps. Since this method cannot conceal many secret bits for the threshold T of the same condition, it shows a relatively high PSNR. After all, that is why this method is inferior to other methods. If we would like to increase the number of secret bits even at the expense of the PSNR, it is possible to increase the size of the threshold T. However, it may often be the case that the PSNR becomes worse than expected without increasing the number of hidden bits. For example, when $T \leq 4$, the three methods except Bai and Chang's method can hide about 130,000 bits, while the PSNRs are slightly reduced. For such a large amount of data to embed, they exploited the DBS method with respect to *BM* equally.

Bai and Chang's method must increase the T value in order to conceal 130,000 bits of data, and as a result, the errors accumulate rapidly. Since Bai and Chang's method [30] uses up to four LSB for data concealment, the size of the error inevitably increases. Since our proposed method uses up to three LSB and the frequent count of three LSB is also not very high, the negative effect on image quality is less than that of Bai and Chang's method [30]. In the end, we prove that the proposed method has a better optimization performance than Bai and Chang's method [30].

Figure 9 shows the evaluation by comparing the histograms of stego images generated from the proposed method and existing methods, i.e., W Hong, Ou and Sun, and Bai and Chang. Here, stego images are generated after concealing 150,000 bits in the cover Lena image by the existing and proposed methods. The pixel value range on the x-axis is [95, 115]. In Figure 9, the curves of our proposed method and the two existing methods (i.e., W Hong and Ou and Sun) are similar, while Bai and Chang's histogram curve has a larger amplitude than the other methods. The reason is that the maximum EC of Bai and Chang's method is up to 150,000 bits. In other words, we can see that the quality of the image reaches the lower limit because it exhausts all possible resources. The histogram does not show much difference because our proposed method and the two existing methods keep more than 33 dB in common when concealing 150,000 bits. As shown in Figure 8, as the EC increases, the histogram of the stego image also is far from the histogram of the original cover image.

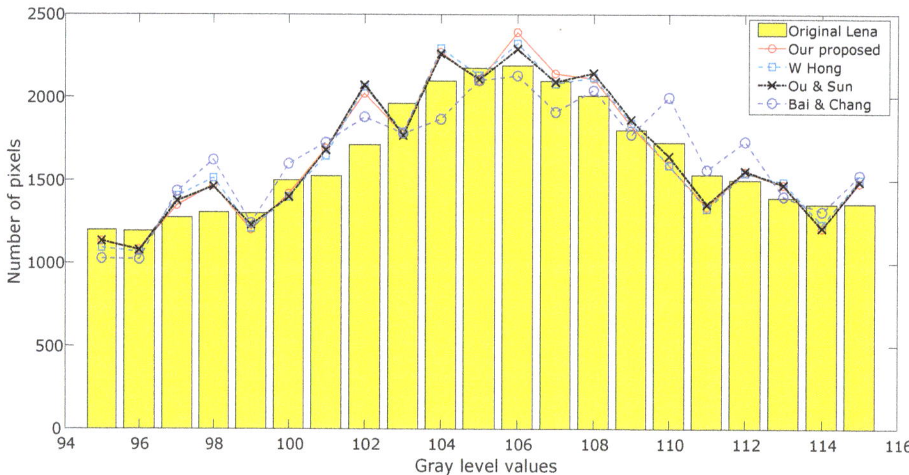

Figure 9. Compared histograms among the proposed method and other related methods with the Lena image when the number of hidden bits is 150,000.

Table 2 shows an experiment to compare the PSNR and SSIM after concealing the same amount of data (120,000 bits) in the cover image for a more objective performance check and reliable comparison. The SSIM of the proposed method shows the highest value. On the other hand, in the case of the

PSNR, W Hong's method [33] shows a high average. In fact, the PSNR only quantifies the quality of reconstructed or damaged images in relation to the facts. For this reason, we introduce SSIM as a criterion for the secondary evaluation. SSIM evaluates the structure of the image. The SSIM of the reconstructed image for the ground image is always one, and if the value is close to one, you can see that the image quality is excellent. Therefore, we can see that our proposed method is superior to the existing methods in terms of SSIM.

Table 2. Performance comparison of the PSNR and SSIM among the proposed and previous schemes (using 120,000 bits).

Images	Ou and Sun [29]		Bai and Chang [30]		W Hong [33]		The Proposed	
	PSNR	SSIM	PSNR	SSIM	PSNR	SSIM	PSNR	SSIM
Boats	31.3506	0.6433	30.3823	0.6675	31.3846	0.6828	31.4158	0.7298
Goldhill	31.7499	0.6942	31.1779	0.7345	32.2203	0.7279	31.4158	0.7642
Airplane	31.7282	0.6614	31.1754	0.6526	31.9034	0.7042	31.3737	0.7305
Lena	33.2362	0.6614	32.4231	0.6870	33.3540	0.7094	33.4090	0.7566
Peppers	33.3636	0.6556	32.7389	0.6966	33.3905	0.7081	33.5822	0.7316
Zelda	35.4041	0.6936	34.5442	0.7177	35.7839	0.7335	35.9442	0.7778
Average	32.8054	0.6683	32.0736	0.6943	33.0061	0.7110	32.8568	0.7484

Table 3 shows the MSE and NCC simulation results for the existing and proposed methods for the four images. The average MSE value of the proposed method is lower than those of the three other methods. The MSE value of the Airplane image in our proposed method is slightly higher than those of Ou and Sun [29] and W Hong [33]. However, from the NCC scores, there is no difference, so it is objectively proven that there is no problem with the performance of our proposed method. Furthermore, when the maximum EC of Ou and Sun reaches 270,000 bits, the PSNR drops to 23 dB. On the other hand, our proposed method can maintain the PSNR higher than 30 dB, so the DH performance of our proposed method is useful. Our proposed method can obtain better performance by creating a lookup table to obtain more optimal values than W Hong's method.

Table 3. Performance comparison of the MSE and Normalized Cross-Correlation (NCC) between the proposed and previous schemes (using 150,000 bits).

Images	Ou and Sun [29]		Bai and Chang [30]		W Hong [33]		The Proposed	
	MSE	NC	MSE	NC	MSE	NC	MSE	NC
Boats	49.9795	0.9946	97.2898	0.9932	51.0112	0.9945	47.6810	0.9950
Goldhill	50.2434	0.9939	70.2292	0.9934	53.1714	0.9938	52.0115	0.9940
Airplane	43.4923	0.9960	88.157	0.9952	44.2237	0.9960	48.1400	0.9961
Lena	32.3098	0.9954	57.5453	0.9948	33.3644	0.9953	31.3258	0.9957
Peppers	32.3199	0.9955	55.0679	0.9948	34.0308	0.9943	30.4184	0.9958
Zelda	21.1771	0.9943	32.3215	0.9938	21.4966	0.9943	18.0991	0.9948
Average	38.2537	0.9943	66.7685	0.9942	39.5497	0.9949	37.9460	0.9952

Table 4 shows the comparison of the CPU execution time between the proposed and the existing methods. The computer for the experiment is a YOGA 730, and the CPU processor is Intel(R) Core(TM) i5-8250U CPU 1.6 GHz. The software is MATLAB R2019a. Here, we measure the CPU time to conceal a random number from 20,000 bits to 200,000 bits in the Lena image by using the four methods (i.e., Ou and Sun, W Hong, Bai and Chang, our proposed method). The process of the measurement includes AMBTC compression, data embedding, and AMBTC decompression. The most time-consuming method is that of Bai and Chang, and the least time-consuming method is that of Ou and Sun. The method we propose is faster than Bai and Chang's, but it is time consuming compared to the other two. However, if we code using the C language, the required time would be less than 1 s.

Table 4. Comparing the CPU time between the proposed and the existing methods (measurement: seconds).

Methods	Hidden Bits									
	20,000	50,000	70,000	90,000	100,000	120,000	150,000	170,000	190,000	200,000
Ou and Sun	1.4531	1.5313	1.6094	1.6563	1.7188	1.7969	1.8281	1.9063	1.9219	1.9531
W Hong	1.4688	1.5938	1.6406	1.7813	1.8281	1.8594	1.875	1.9063	1.9531	2.0313
Bai and Chang	2.4688	3.7344	5.25	5.9688	6.5625	7.3594	8.4219	-	-	-
The proposed	1.6875	1.8281	2.125	2.2656	2.5625	2.8594	3.1563	3.5313	3.5938	3.6094

5. Conclusions

In this paper, we introduced a DH method that applies DBS and optimized HC(7,4) to AMBTC compressed grayscale images. The basic unit of AMBTC is the *trio*, which consists of two quantization levels and one bitmap and is represented by $trio(a, b, bitmap)$. Therefore, AMBTC is a trioset, and the proposed DH method is applied to each block of an image. The proposed method may have different final performance results depending on the characteristics of each block. Therefore, we divided every block into two groups (smooth blocks and complex blocks) and applied the proposed method. The distinction of whether a block is a smooth block or a complex block is determined by the difference between the two quantization levels of the block. That is, if the difference $(|a - b|)$ between two quantization levels is smaller than or equal to the threshold T, it is categorized as a smooth block. When hiding data in a complex block with a difference higher than threshold T, the MSE errors increase compared to a smooth block. Therefore, it is important in terms of DH to distinguish the blocks. In other words, the smoother the blocks are, the more they help to maintain the image quality while concealing more data. In this paper, our proposed method achieved the optimized level by HC(7,4) based on the lookup table. As a result, it was shown through experiments that our proposed scheme surpasses the performance of Hong's method. Experimental results show that the proposed scheme provides a high EC while suppressing the loss of quality of the cover image. In the future, we will devise a method to calculate a more optimal distance when applying HC(7,4) to two quantization levels and conduct research to find a way to minimize data concealment errors.

Author Contributions: Conceptualization, C.-N.Y. and C.K.; methodology, D.S. and C.K.; validation, C.-N.Y. and C.K.; formal analysis, C.-N.Y.; writing, original draft preparation, C.K. and L.L.; funding acquisition, D.-K.S., C.-N.Y., C.K., and L.L. All authors read and agreed to the published version of the manuscript.

Funding: This work was supported by the National Natural Science Foundation of China (61866028), the Key Program Project of Research and Development (Jiangxi Provincial Department of Science and Technology) (20171ACE50024), the Foundation of China Scholarship Council (CSC201908360075), and the Open Foundation of Key Laboratory of Jiangxi Province for Image Processing and Pattern Recognition (ET201680245, TX201604002). This research was supported in part by the Ministry of Science and Technology (MOST), under grants MOST 108-2221-E-259-009-MY2 and 109-2221-E-259-010. This research was supported by the Basic Science Research Program through the National Research Foundation of Korea (NRF) funded by 2015R1D1A1A01059253 and 2018R1D1A1B07047395 and was supported under the framework of the international cooperation program managed by NRF (2016K2A9A2A05005255).

Conflicts of Interest: The authors declare no conflict of interest.

References

1. Chang, C.C.; Li, C.T.; Shi, Y.Q. Privacy-aware reversible watermarking in cloud computing environments. *IEEE Access* **2018**, *6*, 70720–70733. [CrossRef]

2. Byun, S.W.; Son, H.S.; Lee, S.P. Fast and robust watermarking method based on DCT specific location. *IEEE Access* **2019**, *7*, 100706–100718. [CrossRef]

3. Kim, C.; Yang, C.N. Watermark with DSA signature using predictive coding. *Multimed. Tools Appl.* **2015**, *74*, 5189–5203. [CrossRef]

4. Kim, H.J.; Kim, C.; Choi, Y.; Wang, S.; Zhang, X. Improved modification direction methods. *Comput. Math. Appl.* **2010**, *60*, 319–325. [CrossRef]

5. Kim, C. Data hiding by an improved exploiting modification direction, *Multimed. Tools Appl.* **2010**, *69*, 569–584. [CrossRef]

6. Petitcolas, F.A.P.; Anderson, R.J.; Kuhn, M.G. Information hiding—A survey. *Proc. IEEE* **1999**, *87*, 1062–1078. [CrossRef]

7. Kim, C.; Shin, D.; Yang, C.N.; Chen, Y.C.; Wu, S.Y. Data hiding using sequential hamming + k with m overlapped pixels. *KSII Trans. Internet Inf.* **2019**, *13*, 6159–6174.

8. Mielikainen, J. LSB matching revisited. *IEEE Signal Proc. Lett.* **2006**, *13*, 285–287. [CrossRef]

9. Chan, C.K.; Cheng, L.M. Hiding data in images by simple LSB substitution. *Pattern Recognit.* **2004**, *37*, 469–474. [CrossRef]

10. Kim, C.; Shin, D.; Kim, B.G.; Yang, C.N. Secure medical images based on data hiding using a hybrid scheme with the Hamming code. *J. Real-Time Image Process.* **2018**, *14*, 115–126. [CrossRef]

11. Tian, J. Reversible data embedding using a difference expansion. *IEEE Trans. Circuits Syst. Video Technol.* **2003**, *13*, 890–896. [CrossRef]

12. Hu, Y.; Lee, H.-K.; Li, J. DE-based reversible data hiding with improved overflow location map. *IEEE Trans. Circuits Syst. Video Technol.* **2008**, *19*, 250–260.

13. Celik, M.U.; Sharma, G.; Tekalp, A.M.; Saber, E. Lossless generalized-LSB data embedding. *IEEE Trans. Image Process.* **2005**, *14*, 253–266. [CrossRef] [PubMed]

14. Ni, Z.; Shi, Y.-Q.; Ansari, N.; Su, W. Reversible data hiding. *IEEE Trans. Circuits Syst. Video Technol.* **2006**, *16*, 354–362.

15. Kim, C.; Baek, J.; Fisher, P.S. Lossless Data Hiding for Binary Document Images Using n-Pairs Pattern. In Proceedings of the Information Security and Cryptology—ICISC 2014, Lecture Notes in Computer Science (LNCS), Seoul, Korea, 3–5 December 2014; Volume 8949, pp. 317–327.

16. Hong, W.; Chen, T.-S.; Chang, Y.-P.; Shiu, C.W. A high capacity reversible data hiding scheme using orthogonal projection and prediction error modification. *Signal Process.* **2010**, *90*, 2911–2922. [CrossRef]

17. Carpentieri, B.; Castiglione, A.; Santis, A.D.; Palmieri, F.; Pizzolante, R. One-pass lossless data hiding and compression of remote sensing data. *Future Gener. Comput. Syst.* **2019**, *90*, 222–239. [CrossRef]

18. Zhang, X. Reversible data hiding in encrypted image. *IEEE Signal Process. Lett.* **2011**, *18*, 255–258. [CrossRef]

19. Puteaux, P.; Puech, W. An efficient MSB prediction-based method for high-ca- pacity reversible data hiding in encrypted images. *IEEE Trans. Inf. Forensics Secur.* **2018**, *13*, 1670–1681. [CrossRef]

20. Zhang, F.; Lu, W.; Liu, H.; Yeung, Y.; Xue, Y. Reversible data hiding in binary images based on image magnification. *Multimed. Tools Appl.* **2019**, *78*, 21891–21915. [CrossRef]

21. Leng, L.; Li, M.; Kim, C.; Bi, X. Dual-source discrimination power analysis for multi-instance contactless palmprint recognition. *Multimed. Tools Appl.* **2017**, *76*, 333–354. [CrossRef]

22. Leng, L.; Zhang, J.S.; Khan, M.K.; Chen, X.; Alghathbar, K. Dynamic weighted discrimination power analysis: A novel approach for face and palmprint recognition in DCT domain. *Int. J. Phys. Sci.* **2010**, *5*, 2543–2554.

23. Deeba, F.; Kun, S.; Dharejo, F.A.; Zhou, Y. Wavelet-Based Enhanced Medical Image Super Resolution. *IEEE Access* **2020**, *8*, 37035–37044. [CrossRef]

24. Delp, E.; Mitchell, O. Image compression using block truncation coding. *IEEE Trans. Commun.* **1979**, *27*, 1335–1342. [CrossRef]

25. Lema, M.D.; Mitchell, O.R. Absolute moment block truncation coding and its application to color images. *IEEE Trans. Commun.* **1984**, *COM-32*, 1148–1157. [CrossRef]

26. Kumar, R.; Singh, S.; Jung, K.H. Human visual system based enhanced AMBTC for color image compression using interpolation. In Proceedings of the 2019 6th International Conference on Signal Processing and Integrated Networks (SPIN), Noida, India, 7–8 March 2019; Volume 385, pp. 903–907.

27. Hong, W.; Chen, T.S.; Shiu, C.W. Lossless steganography for AMBTC compressed images. In Proceedings of the 2008 Congress on Image and Signal Processing, Sanya, China, 27–30 May 2008; pp. 13–17.

28. Chuang, J.C.; Chang, C.C. Using a simple and fast image compression algorithm to hide secret information. *Int. J. Comput. Appl.* **2006**, *28*, 329–333.

29. Ou, D.; Sun, W. High payload image steganography with minimum distortion based on absolute moment block truncation coding. *Multimed. Tools Appl.* **2015**, *74*, 9117–9139. [CrossRef]

30. Bai, J.; Chang, C.C. High payload steganographic scheme for compressed images with Hamming code. *Int. J. Netw. Secur.* **2016**, *18*, 1122–1129.

31. Kumar, R.; Kim, D.S.; Jung, K.H. Enhanced AMBTC based data hiding method using hamming distance and pixel value differencing. *J. Inf. Secur. Appl.* **2019**, *47*, 94–103. [CrossRef]

32. Chen, J.; Hong, W.; Chen, T.S.; Shiu, C.W. Steganography for BTC compressed images using no distortion technique. *Imaging Sci. J.* **2013**, *58*, 177–185. [CrossRef]

33. Hong, W. Efficient data hiding based on block truncation coding using pixel pair matching technique. *Symmetry* **2018**, *10*, 36. [CrossRef]

34. Hong, W.; Chen, T.S. A novel data embedding method using adaptive pixel pair matching. *IEEE Trans. Inf. Forensics Secur.* **2012**, *7*, 176–184. [CrossRef]

35. Huang, Y.H.; Chang, C.C.; Chen, Y.H. Hybrid secret hiding schemes based on absolute moment block truncation coding. *Multimed. Tools Appl.* **2017**, *76*, 6159–6174. [CrossRef]

36. Chen, Y.Y.; Chi, K.Y. Cloud image watermarking: High quality data hiding, and blind decoding scheme based on block truncation coding. *Multimed. Syst.* **2019**, *25*, 551–563. [CrossRef]

37. Malik, A.; Sikka, G.; Verma, H.K. An AMBTC compression-based data hiding scheme using pixel value adjusting strategy. *Multidimens. Syst. Signal Process.* **2018**, *29*, 1801–1818. [CrossRef]

38. Lin, J.; Weng, S.; Zhang, T.; Ou, B.; Chang, C.C. Two-Layer Reversible Data Hiding Based on AMBTC Image With (7, 4) Hamming Code. *IEEE Access* **2019**, *8*, 21534–21548. [CrossRef]

39. Sampat, M.P.; Wang, Z.; Gupta, S.; Bovik, A.C.; Markey, M.K. Complex wavelet structural similarity: A new image similarity index. *IEEE Trans. Image Process.* **2009**, *18*, 2385–2401. [CrossRef]

Article

Practical Inner Product Encryption with Constant Private Key †

Yi-Fan Tseng, Zi-Yuan Liu * and Raylin Tso

Department of Computer Science, National Chengchi University, Taipei 11605, Taiwan;
yftseng@cs.nccu.edu.tw (Y.-F.T.); raylin@cs.nccu.edu.tw (R.T.)
* Correspondence: zyliu@cs.nccu.edu.tw; Tel.: +886-2-29393091 (ext. 62329)

† Proceedings of the 17th International Joint Conference on e-Business and Telecommunications—Volume 3:
 SECRYPT, INSTICC; SciTePress: Setubal, Portugal, 2020; pp. 553–558, doi:10.5220/0009804605530558.

Received: 7 November 2020; Accepted: 1 December 2020; Published: 3 December 2020

Abstract: Inner product encryption, first introduced by Katz et al., is a type of predicate encryption in which a ciphertext and a private key correspond to an attribute vector and a predicate vector, respectively. Only if the attribute and predicate vectors satisfy the inner product predicate will the decryption in this scheme be correct. In addition, the ability to use inner product encryption as an underlying building block to construct other useful cryptographic primitives has been demonstrated in the context of anonymous identity-based encryption and hidden vector encryption. However, the computing cost and communication cost of performing inner product encryption are very high at present. To resolve this problem, we introduce an efficient inner product encryption approach in this work. Specifically, the size of the private key is only one $\mathbb{G}$ element and one $\mathbb{Z}_p$ element, and decryption requires only one pairing computation. The formal security proof and implementation result are also demonstrated. Compared with other state-of-the-art schemes, our scheme is the most efficient in terms of the number of pairing computations for decryption and the private key length.

Keywords: predicate encryption; inner product encryption; constant-size private key; efficient decryption; constant pairing computations

---

1. Introduction

Inner product encryption (IPE), first introduced by Katz et al. [1], is a type of predicate encryption [2] in which a ciphertext and a private key correspond to an attribute vector $\mathbf{x}$ and a predicate vector $\mathbf{y}$, respectively. In particular, the decryption will be correct if and only if the attribute vector and the predicate vector satisfy the inner product predicate, meaning that the inner product operation of $\mathbf{x}$ and $\mathbf{y}$ equals zero ($\langle \mathbf{x}, \mathbf{y} \rangle = 0$). Over the past decade, many IPE schemes have been proposed, such as those based on pairing [3–7] and lattice [8–11]. The security definition of an IPE scheme [1] can be naturally extended from the IND–CPA security of identity-based encryption [12–14]. More precisely, under the security approach of IPE, an adversary learns nothing about the encrypted message from a ciphertext associated with an attribute vector $\mathbf{x}$ if they do not own the private key associated with a predicate vector $\mathbf{y}$ such that $\langle \mathbf{x}, \mathbf{y} \rangle = 0$. Such a definition is also called the IND–CPA security for IPE scheme in some papers [15] and is defined as the payload-hiding property in [1]. Alternatively, the security definition defined in [1], called the attribute-hiding property, states that a ciphertext reveals nothing about the corresponding ciphertext attribute $\mathbf{x}$. However, we emphasize that the attribute-hiding property is not an absolutely necessary property for IPE. Many IPE schemes proposed in the literature achieve only IND–CPA security/payload hiding, such as that in [15–17].

In addition to their usefulness in fine-grained access control, IPE schemes can be used to construct various cryptographic primitives or can be converted to more complex primitives, such as identity-based encryption [12–14], hidden vector encryption [2,18] and subset predicate encryption [19,20]. We refer readers to the work presented in [1,19] for details.

Although many IPE schemes have been introduced, the computing cost and communication cost of these schemes are high. In particular, the pairing operation required by existing pairing-based IPE schemes is typically linearly related to the vector length; therefore, the computational efficiency of these schemes is low. Moreover, the size of the private key of most schemes is linearly related to vector lengths. However, although the existing lattice-based IPE schemes are considered quantum-resistant, the key size of almost all schemes is too large or the message space is too small. In addition, Internet of Things devices are gradually becoming common in daily life; however, the problems mentioned in the preceding discussion make the application of an IPE scheme impractical for these resource-constrained devices. Thus, an unresolved question remains: can we obtain an efficient IPE scheme by reducing the cost of decryption and optimizing the length of the private key?

1.1. Our Contributions

Herein, we resolve the aforementioned problem by introducing an effective IPE scheme. In particular, in the proposed scheme, the length of a private key is independent of the length of the predicate vector. In addition, the decryption only requires one pairing operation; thus, the decryption is also independent of the length of the predicate vector. Rigorous proofs are provided to demonstrate that, under a modified decisional Diffie–Hellman assumption, our proposed scheme is coselective IND–CPA secure. Moreover, our proposed scheme is more efficient than other advanced schemes, as listed in Tables 1 and 3.

1.2. Related Works

1.2.1. Pairing-Based IPE Schemes

The first IPE scheme, introduced by Katz et al. [1], entails the evaluation of predicates over $\mathbb{Z}_N$ using the inner product, where N is a composite number. After this pioneering work, many studies followed. For example, Okamoto and Takashima [3] proposed the first hierarchical predicate encryption method (or delegable predicate encryption) for inner product predicates; this provides a user with functionality to delegate more restrictive functionality to another user. Attrapadung and Libert [16] constructed an IPE scheme that solves the inefficiency problem of the previous scheme. More precisely, provided that the description of the ciphertext attribute vector is not included in the ciphertext, the ciphertext overhead of the scheme is reduced to $O(1)$. By combining dual system encryption [21] and dual pairing vector spaces [3] carefully, Lewko et al. [22] obtained the first fully secure IPE scheme and hierarchical predicate encryption under the n-extended decisional Diffie–Hellman assumption. However, the security of all these previous studies was based on nonstandard assumptions. To resolve this issue, Park [23] developed the first IPE scheme under the standard assumptions (i.e., decisional bilinear Diffie–Hellman and decisional linear (DLIN) assumptions). Okamoto and Takashima [24] then introduced two nonzero inner product encryption schemes that support constant-size ciphertexts and a constant-size secret key, respectively, which are adaptively secure under the DLIN assumption in the standard model. The authors also proposed the first IPE scheme that is fully secure and fully attribute-hiding [25] as well as the first unbounded IPE scheme that is also fully secure and fully attribute-hiding in the standard model under the DLIN assumption [26]. Kawiai and Takashima [27] introduced a new notion, called IPE with ciphertext conversion, which considers the security of predicate-hiding. Zhenlin and Wei [28] then introduced another concept, called multiparty cloud computation IPE with multiplicative homomorphic property, which enables an IPE scheme to support multiparty cloud computation. Kim et al. [29] proposed a new efficient IPE scheme that only requires n exponentiation and three pairing computations for decryption. Huang et al. [30] proposed the first enabled–disabled IPE, which supports timed-release services and data self-destruction. Ramanna [15] constructed two IPE schemes using tag-based quasi-adaptive noninteractive zero knowledge, where the first and second both have the property of constant-size ciphertext but only the second has the property of attribute-hiding. Zhang et al. [7] recently proposed a new IPE scheme based on a double encryption system; it has been demonstrated to achieve adaptive security under a weak attribute-hiding model.

As discussed subsequently, extensive research has focused on the developed and proposed schemes; however, the private key length of most schemes is linearly dependent on the vector length or requires many pairing operations, making these schemes impractical. Thus, determining how to construct a more practical scheme remains a critical area of research.

1.2.2. Lattice-Based IPE Schemes

To fend off attack from quantum computers in the future, Agrawal et al. [8] proposed the first IPE scheme based on the lattice hard assumption (i.e., the learning with error assumption, which is believed to be able to withstand quantum attacks); to do so, they modified an identity-based encryption approach proposed by Agrawal et al. [31]. Xagawa [9], inspired by the work of Agrawal et al., proposed an improved lattice-based IPE scheme that reduced the size of public parameters and ciphertext. Li et al. [10] proposed a lattice-based IPE scheme that further reduced the size of public parameters and ciphertext. In contrast to [9], their work reduced the size by a factor of $\log n$, where n is the security parameter. Wang et al. [11] recently proposed the first compact IPE scheme that employs an IPE scheme [9], fully homomorphic encryption [32] and vector-encoding schemes [33]. Although these constructions are thought to be able to withstand quantum computer attacks, they are based on the learning with errors assumption, resulting in key lengths that are still too large to be practical.

1.3. Organization

The remainder of this paper is organized as follows. In Section 2, we start by discussing some preliminaries on bilinear maps, complexity assumptions and the definition of IPE. In Section 3, we then propose our IPE scheme and demonstrate its correctness. In Section 4, we subsequently demonstrate security proofs using a modified decisional Diffie–Hellman problem, and then in Section 5, we compare our approach with other state-of-the-art schemes and reveal the implementation results. In Section 6, we finally conclude the paper.

2. Preliminaries

Herein, we present the necessary preliminaries, such as notations, complex assumptions, and the definition of an IPE scheme.

2.1. Notations

Throughout this paper, we use $x \xleftarrow{\$} S$ to denote "choose an element x randomly and uniformly from the set S" and $x \leftarrow A$ to denote "x is the output of the algorithm A". Moreover, we use $\mathbf{a}$ to denote a vector and use $\mathbf{a}_i$ to denote the i-th entry of vector $\mathbf{a}$. The inner product of these two vectors $\mathbf{x}, \mathbf{y}$ is denoted as $\langle \mathbf{x}, \mathbf{y} \rangle$. For a prime p, we use $\mathbb{Z}_p$ to denote the set of integers modulo p. Finally, we use $\mathbb{N}$ and $\mathbb{Z}$ to denote the set of positive integers and integers, respectively.

2.2. Bilinear Maps

Let $\mathbb{G}$ and $\mathbb{G}_T$ be an additive and a multiplicative cyclic group, respectively; here, the order of $\mathbb{G}$ and $\mathbb{G}_T$ is a large prime p (i.e., $|\mathbb{G}| = |\mathbb{G}_T| = p$). Then, let P be a generator of $\mathbb{G}$. A bilinear map (pairing) $e : \mathbb{G} \times \mathbb{G} \to \mathbb{G}_T$ is a mapping with the following properties:

- Bilinearity: For $a, b \in \mathbb{Z}_p, e(aP, bP) = e(P, P)^{ab}$.
- Nondegeneracy: $\exists P \in \mathbb{G}$, such that $e(P, P) \neq 1_{\mathbb{G}_T}$.
- Computability: The mapping e is efficiently computable.

In this work, we take advantage of the generalized decisional Diffie–Hellman exponent (GDDHE) problem, based on [34]. The GDDHE problem is a generic framework within which new complexity assumptions can be created. We first give an overview of the GDDHE problem. Let

- p be a prime;
- s, n be two positive integers;
- $P, Q \in \mathbb{F}_p[X_1, \ldots, X_n]^s$ be two s-tuple of n-variate polynomials over $\mathbb{F}_p$; and
- f be an n-variate polynomial in $\mathbb{F}_p[X_1, \ldots, X_n]$.

Q, Q_T are two ordered sets with multivariate polynomials, and thus, we define $Q = (q_1, q_2, \ldots, q_s)$ and $R = (r_1, r_2, \ldots, r_s)$. As stated in [34], we require $p_1 = q_1 = 1$ to be two constant polynomials. Consider a bilinear map $e : \mathbb{G} \times \mathbb{G} \to \mathbb{G}_T$ with the generator P of $\mathbb{G}$ and $g_T = e(P, P) \in \mathbb{G}_T$. For a vector $(x_1, x_2, \ldots, x_n) \in \mathbb{F}_p^n$, we define

$$Q(x_1, x_2, \ldots, x_n)P = (q_1(x_1, x_2, \ldots, x_n)P, \ldots, q_s(x_1, x_2, \ldots, x_n)P) \in \mathbb{G}^s,$$

and

$$g_T^{R(x_1, x_2, \ldots, x_n)} = (g_T^{r_1(x_1, x_2, \ldots, x_n)}, \ldots, g_T^{r_s(x_1, x_2, \ldots, x_n)}) \in \mathbb{G}_T^s.$$

By "f depends on (Q, R)" we mean that there are $s^2 + s$ constants $\{a_{i,j}\}_{i,j=1}^s$ and $\{b_k\}_{k=1}^s$ such that

$$f = \sum_{i,j=1}^s a_{i,j} q_i q_j + \sum_{k=1}^s b_k r_k.$$

We say that f is independent of (Q, R) if f does not depend on (Q, R).

Definition 1 (The (Q, R, f)-GDDHE Problem). *Given* $(Q(x_1, \ldots, x_n)P, g_T^{R(x_1, \ldots, x_n)}, Z) \in \mathbb{G}^s \times \mathbb{G}_T^s \times \mathbb{G}_T$, *decide if* $Z \stackrel{?}{=} g_T^{f(x_1, \ldots, x_n)}$.

Then, for an algorithm $\mathcal{A}$, the advantage of $\mathcal{A}$ in solving the (Q, R, f)-GDDHE problem is defined as

$$Adv^{(Q,R,f)\text{-GDDHE}}(\mathcal{A}) = \left| \mathcal{A}\left(Q(x_1, \ldots, x_n)P, g_T^{R(x_1, \ldots, x_n)}, g_T^{f(x_1, \ldots, x_n)}\right) - \mathcal{A}\left(Q(x_1, \ldots, x_n)P, g_T^{R(x_1, \ldots, x_n)}, Z \stackrel{\$}{\leftarrow} \mathbb{G}_T\right) \right|.$$

Boneh et al. propose that the (Q, R, f)-GDDHE problem is difficult if f is independent of (Q, R) and demonstrate that a large class of hard problems can be fit into the framework of the GDDHE problem; for instance, the DDH problem over $\mathbb{G}_T$.

Definition 2 (The decisional Diffie–Hellman problem over $\mathbb{G}_T$ ($\text{DDH}_{\mathbb{G}_T}$ problem)). *Let* $g_T = e(P, P)$ *be a generator of* $\mathbb{G}_T$. *Given* $(P, g_T, A = g_T^a, B = g_T^b, C) \in \mathbb{G} \times \mathbb{G}_T^4$, *where* $a, b \stackrel{\$}{\leftarrow} \mathbb{Z}_p$, *decide whether* $C = g_T^{ab}$ *or an random element from* $\mathbb{G}_T$.

By setting $Q = (1), R = (1, a, b), f = ab$, the DDH problem over $\mathbb{G}_T$ is equivalent to the (Q, R, f)-GDDHE problem. Observe that no constants exist such that the linear combination of $1, a, b$ equals ab; therefore, f is independent of (Q, R). Given the result of Boneh et al., we conclude that no algorithm is available with which to solve the $\text{DDH}_{\mathbb{G}_T}$ problem with a nonnegligible advantage. See [34] for additional details.

Next, we present a modified version of the $\text{DDH}_{\mathbb{G}_T}$ problem, which will be used in the security proof.

Definition 3 (The modified decisional Diffie–Hellman problem over $\mathbb{G}_T$ (M-DDH$_{\mathbb{G}_T}$ problem)). *Let* $g_T = e(P, P)$ *be a generator of* $\mathbb{G}_T$. *Given* $(P, A' = aP, g_T, A = g_T^a, B = g_T^b, C) \in \mathbb{G}^2 \times \mathbb{G}_T^4$, *where* $a, b \xleftarrow{\$} \mathbb{Z}_p$, *decide whether* $C = g_T^{ab}$ *or a random element from* $\mathbb{G}_T$.

Theorem 1 (The modified decisional Diffie–Hellman assumption over $\mathbb{G}_T$ (M-DDH$_{\mathbb{G}_T}$ assumption)). *We say that the M-DDH$_{\mathbb{G}_T}$ assumption holds if there is no algorithm $\mathcal{D}$ for solving the M-DDH$_{\mathbb{G}_T}$ problem with a nonnegligible advantage.*

Proof. Compared with the DDH$_{\mathbb{G}_T}$ problem, the instance of the M-DDH$_{\mathbb{G}_T}$ problem contains an additional element $A' = aP$. The M-DDH$_{\mathbb{G}_T}$ problem is equivalent to the (Q, R, f)-GDDHE problem with

$$Q = (1, a), R = (1, a, b), f = ab.$$

No constants exist such that the linear combination of the monomials $(1 \cdot a), 1, a, b$ equals the polynomial ab. Therefore, considering the the results of Boneh et al., we conclude that the M-DDH$_{\mathbb{G}_T}$ problem is hard. Moreover, we define the advantage for an algorithm $\mathcal{D}$ in solving the M-DDH$_{\mathbb{G}_T}$ problem as

$$Adv^{\text{M-DDH}_{\mathbb{G}_T}}(\mathcal{D}) = \left| \Pr[\mathcal{D}(P, A', g_T, A, B, C = g_T^{ab}) = 1] - \Pr[\mathcal{D}(P, A', g_T, A, B, C \xleftarrow{\$} \mathbb{G}_T) = 1] \right|.$$

□

2.3. Definition of Inner Product Encryption

An IPE scheme consists of four algorithms: **Setup, KeyGen, Encrypt** and **Decrypt**. The details of the algorithms are as follows:

- **Setup**$(1^\lambda, 1^\ell)$. Take as inputs the security parameters $(1^\lambda, 1^\ell)$, where $\lambda, \ell \in \mathbb{N}$, and the algorithm outputs the system parameter params and the master secret key msk. The descriptions of the attribute vector space $\mathfrak{A}$ and the predicate vector space $\mathfrak{P}$ are implicitly included in params. Moreover, the inner product operation over $\mathfrak{A}$ and $\mathfrak{P}$ must be well defined.
- **Encrypt**$(\text{params}, \mathbf{x}, M)$. Given the system parameter params, an attribute vector $\mathbf{x} \in \mathfrak{A}$ and a message M, the algorithm outputs a ciphertext $C_\mathbf{x}$ for the attribute vector $\mathbf{x}$.
- **KeyGen**$(\text{params}, \text{msk}, \mathbf{y})$. Given the system parameter params and a predicate vector $\mathbf{y} \in \mathfrak{P}$, the algorithm outputs the private key $K_\mathbf{y}$ for the predicate vector $\mathbf{y}$.
- **Decrypt**$(\text{params}, C_\mathbf{x}, K_\mathbf{y})$. Given the system parameter params, a ciphertext $C_\mathbf{x}$ and the private key $K_\mathbf{y}$, the algorithm outputs a message M or a error symbol $\perp$.

The correctness is defined as follows. For all $\lambda, \ell \in \mathbb{N}$, let $C_\mathbf{x} \leftarrow$ **Encrypt**$(\text{params}, \mathbf{x} \in \mathfrak{A}, M)$ and let $K_\mathbf{y} \leftarrow$ **KeyGen**$(\text{params}, \text{msk}, \mathbf{y} \in \mathfrak{P})$; thus, we have

$$\begin{aligned} M &\leftarrow \textbf{Decrypt}(\text{params}, C_\mathbf{x}, K_\mathbf{y}) && \text{if } \langle \mathbf{x}, \mathbf{y} \rangle = 0; \\ \perp &\leftarrow \textbf{Decrypt}(\text{params}, C_\mathbf{x}, K_\mathbf{y}) && \text{if } \langle \mathbf{x}, \mathbf{y} \rangle \neq 0, \end{aligned}$$

where $(\text{params}, \text{msk}) \leftarrow$ **Setup**$(1^\lambda, 1^\ell)$.

2.4. Security Model

Here, we first introduce IND–CPA security for IPE. The IND–CPA game of IPE for the attribute vector space $\mathfrak{A}$ and predicate vector space $\mathfrak{P}$ is defined as an interactive game between a challenger $\mathcal{C}$ and an adversary $\mathcal{A}$.

- **Setup**. The challenger $\mathcal{C}$ runs **Setup**$(1^\lambda, 1^\ell)$ and sends the system parameter params to the adversary $\mathcal{A}$.

- **Query Phase 1**. The challenger polynomially answers many private key queries for $\mathbf{y} \in \mathfrak{P}$ for the adversary $\mathcal{A}$ by returning $K_{\mathbf{y}} \leftarrow \mathbf{KeyGen}(\texttt{params}, \texttt{msk}, \mathbf{y})$.
- **Challenge**. The adversary $\mathcal{A}$ submits an attribute vector $\mathbf{x}^* \in \mathfrak{A}$ such that $\langle \mathbf{x}^*, \mathbf{y} \rangle \neq 0$ for all $\mathbf{y}$ that have been queried in **Query Phase 1** and two messages M_0, M_1 with the same length to challenger $\mathcal{C}$. Then, $\mathcal{C}$ randomly chooses $\beta \in \{0,1\}$ and returns a challenge ciphertext $C_{\mathbf{x}^*} \leftarrow \mathbf{Encrypt}(\texttt{params}, \mathbf{x}^*, M_\beta)$.
- **Query Phase 2**. This phase is the same as **Query Phase 1**, except that the adversary is not allowed to make a query with $\mathbf{y} \in \mathfrak{P}$ such that $\langle \mathbf{x}^*, \mathbf{y} \rangle \neq 0$.
- **Guess**. The adversary $\mathcal{A}$ outputs a bit β' and wins the game if $\beta' = \beta$.

The advantage of an adversary for winning the IND–CPA game is defined as

$$Adv^{\mathsf{IND\text{-}CPA}}(\mathcal{A}) = \left| \Pr[\beta' = \beta] - \frac{1}{2} \right|.$$

Definition 4 (IND–CPA Security for IPE). *We say that an IPE is IND–CPA secure if there is no probabilistic polynomial-time adversary $\mathcal{A}$ who wins the IND–CPA game with a nonnegligible advantage.*

As we mentioned in Section 1, in some literature [1,23], the security notions for an IPE are defined with the notions "payload hiding" and "attribute hiding". Informally, payload-hiding (or attribute-hiding) is defined to argue that a ciphertext leaks no information about the encrypted message (or attribute vector). The IND–CPA security shown in this section is equivalent to payload-hiding. We emphasize that attribute-hiding is unnecessary for an IPE scheme; in [15–17], schemes have been proposed satisfying only payload hiding.

We next present the selective security and the coselective security [16,35] for IPE. The selective IND-CPA (sIND-CPA) game is defined the same as the IND-CPA game, except that the adversary $\mathcal{A}$ is forced to commit before the **Setup** phase to an attribute vector $\mathbf{x}^*$, and $\mathcal{A}$ is not allowed to make private key queries with $\mathbf{y}$ such that $\langle \mathbf{x}^*, \mathbf{y} \rangle \neq 0$ in both **Query Phase 1** and **Query Phase 2**.

Definition 5 (sIND-CPA Security for IPE). *An IPE scheme is said to be sIND–CPA secure if no probabilistic polynomial-time adversary wins the sIND–CPA game with a nonnegligible advantage.*

The coselective IND–CPA (csIND–CPA) game is defined as equal to the IND–CPA game, except that the adversary $\mathcal{A}$ is forced to commit before the **Setup** phase q to predicate vectors $\mathbf{y}^{(1)}, \ldots, \mathbf{y}^{(q)}$ for the private key queries, where q is a polynomial in the security parameter λ and $\mathcal{A}$ is required to invoke the **Challenge** phase with an attribute vector $\mathbf{x}^*$ such that $\langle \mathbf{x}^*, \mathbf{y}^{(j)} \rangle \neq 0$ for $j = 1, \ldots, q$.

Definition 6 (csIND–CPA Security for IPE). *An IPE scheme is said to be csIND–CPA secure if no probabilistic polynomial-time adversary wins the csIND–CPA game with a nonnegligible advantage.*

Coselective security can be understood as a complementary notion to selective security. In the selective security game, the adversary can learn the private key in accordance with its previous choices, whereas in the coselective security game, the adversary can choose its target after seeing the public parameter and learning the private keys of its choice. Although selective security and coselective security are weaker than full security, both notions are, by definition, incomparable in general by definition.

3. Proposed Inner Product Encryption Scheme

Our IPE scheme consists of four algorithms: **Setup**, **KeyGen**, **Encrypt** and **Decrypt**. The details of the proposed scheme are explained in the following.

- **Setup**$(1^\lambda, 1^\ell)$. Given the security parameters $(1^\lambda, 1^\ell)$, where $\lambda, \ell \in \mathbb{N}$, the algorithm performs as follows.

  1. Choose bilinear groups $\mathbb{G}, \mathbb{G}_T$ of prime order $p > 2^\lambda$. Let P and $g_T = e(P, P)$ be the generator of $\mathbb{G}$ and $\mathbb{G}_T$, respectively.
  2. Set the predicate vector space and the attribute vector space to $\mathbb{Z}_p^\ell$.
  3. Choose $\mathbf{s} = (s_1, s_2, \ldots, s_\ell) \xleftarrow{\$} \mathbb{Z}_p^\ell$.
  4. Compute $\widehat{\mathbf{h}} = (g_T^{s_i})_{i=1}^\ell = (\widehat{h}_1, \ldots, \widehat{h}_\ell)$.
  5. Output the system parameter $\mathtt{params} = (P, g_T, \widehat{\mathbf{h}})$, and the master secret key $\mathtt{msk} = \mathbf{s}$.

- **Encrypt**$(\mathtt{params}, \mathbf{x}, M)$. Given the system parameter $\mathtt{params}$, a vector $\mathbf{x} = (x_1, x_2, \ldots, x_\ell) \in \mathbb{Z}_p^\ell$, and a message $M \in \mathbb{G}_T$, the algorithm performs as follows.

  1. Choose $r, \delta \xleftarrow{\$} \mathbb{Z}_p$.
  2. Compute $C_0 = rP$, and $\widehat{C}_0 = g_T^r$.
  3. Compute $C_i = \widehat{h}_i^r \cdot g_T^{\delta x_i} \cdot M$ for $i = 1$ to ℓ.
  4. Output the ciphertext $C_\mathbf{x} = (C_0, \widehat{C}_0, C_1, C_2, \ldots, C_\ell)$.

- **KeyGen**$(\mathtt{params}, \mathtt{msk}, \mathbf{y})$. Given the system parameter $\mathtt{params}$, a master secret key $\mathtt{msk}$, and a vector $\mathbf{y} = (y_1, y_2, \ldots, y_\ell) \in \mathbb{Z}_p^\ell$, where $\sum_{i=1}^\ell y_i \neq 0$, the algorithm performs as follows.

  1. Choose $k \xleftarrow{\$} \mathbb{Z}_p$.
  2. Compute $K_0 = kP$, and $K_1 = \langle \mathbf{s}, \mathbf{y} \rangle + k \mod p$.
  3. Output the private key $K_\mathbf{y} = (K_0, K_1)$.

- **Decrypt**$(\mathtt{params}, C_\mathbf{x}, K_\mathbf{y})$. Given the system parameter $\mathtt{params}$, a ciphertext $C_\mathbf{x}$, and the private key $K_\mathbf{y}$, where $\mathbf{y} = (y_1, y_2, \ldots, y_\ell)$ the algorithm performs as follows.

  1. Compute $D_0 = e(K_0, C_0)$.
  2. Compute $D_1 = \prod_{i=1}^\ell C_i^{y_i}$.
  3. Compute $D = \dfrac{D_0 \cdot D_1}{\widehat{C}_0^{K_1}}$.
  4. Compute $d = (\sum_{i=1}^\ell y_i)^{-1} \mod p$.
  5. Compute $M = D^d$.

Correctness

The correctness of the proposed scheme is shown as follows.

- $D_0 = e(K_0, C_0) = e(kP, rP) = g_T^{kr}$.
- $D_1 = \prod_{i=1}^\ell C_i^{y_i} = \prod_{i=1}^\ell (\widehat{h}_i^r \cdot g_T^{\delta x_i} \cdot M)^{y_i} = \prod_{i=1}^\ell (\widehat{h}_i^{y_i})^r \cdot (g_T^{\delta x_i y_i}) \cdot (M^{y_i}) = \prod_{i=1}^\ell ((g_T^{s_i})^{y_i})^r \prod_{i=1}^\ell (g_T^{\delta x_i y_i}) \prod_{i=1}^\ell (M^{y_i}) = g_T^{r\langle \mathbf{s}, \mathbf{y} \rangle} \cdot g_T^{\delta \langle \mathbf{x}, \mathbf{y} \rangle} \cdot M^{\sum_{i=1}^\ell y_i}$.
- $\widehat{C}_0^{K_1} = g_T^{rK_1} = g_T^{r\langle \mathbf{s}, \mathbf{y} \rangle + rk}$.
- $D = \dfrac{D_0 \cdot D_1}{\widehat{C}_0^{K_1}} = \dfrac{g_T^{r\langle \mathbf{s}, \mathbf{y} \rangle} \cdot g_T^{\delta \langle \mathbf{x}, \mathbf{y} \rangle} \cdot M^{\sum_{i=1}^\ell y_i} \cdot g_T^{kr}}{g_T^{r\langle \mathbf{s}, \mathbf{y} \rangle + rk}} = g_T^{\delta \langle \mathbf{x}, \mathbf{y} \rangle} \cdot M^{\sum_{i=1}^\ell y_i}$.
- We have $D = M^{\sum_{i=1}^\ell y_i}$ iff $\langle \mathbf{x}, \mathbf{y} \rangle = 0$.
- Thus $D^d = M^{\sum_{i=1}^\ell y_i \cdot ((\sum_{i=1}^\ell y_i)^{-1} \mod p)} = M$.

4. Security Analysis of the Proposed Scheme

We now provide the security proof for the coselective security of the proposed IPE scheme. In the subsequent proof, we view a vector as a row vector.

Theorem 2. *The proposed scheme is csIND–CPA secure for q private key queries, where q is a polynomial in the security parameter λ, under the M-DDH$_{\mathbb{G}_T}$ assumption.*

Proof. Given $(P, A' = aP, g_T, A = g_T^a, B = g_T^b, C)$, we build an algorithm $\mathcal{C}$ using the adversary $\mathcal{A}$ to solve the M-DDH$_{\mathbb{G}_T}$ problem as follows.

- **Init.** The adversary $\mathcal{A}$ commits q predicate vectors $\mathbf{y}^{(1)}, \ldots, \mathbf{y}^{(q)}$.
- **Setup.** $\mathcal{C}$ first finds a vector $\mathbf{u} = (u_1, u_2, \ldots, u_\ell)$ such that

$$
\begin{bmatrix} \mathbf{y}_1 \\ \mathbf{y}_2 \\ \vdots \\ \mathbf{y}_q \end{bmatrix} \mathbf{u}^\top = \mathbf{0}_\ell^\top,
$$

where $\mathbf{0}_\ell = \underbrace{(0, 0, \ldots, 0)}_{\ell}$. Such $\mathbf{u}$ exists when $q > \ell$. The operation is to find a vector $\mathbf{u}$ such that $\langle \mathbf{u}, \mathbf{y}_j \rangle = 0$ for $j = 1$ to q. $\mathcal{C}$ then chooses $\mathbf{v} = (v_1, v_2, \ldots, v_\ell) \xleftarrow{\$} \mathbb{Z}_p^\ell$. Next, $\mathcal{C}$ computes $\widehat{\mathbf{h}} = (B^{u_i} \cdot g_T^{v_i})_{i=1}^\ell = (\widehat{h}_1, \ldots, \widehat{h}_\ell)$. Finally, $\mathcal{C}$ sets params $= (P, g_T, \widehat{\mathbf{h}})$ and sends params to $\mathcal{A}$. Note that $\mathcal{C}$ implicitly sets $\mathtt{msk} = \mathbf{s} = (s_i = u_i \cdot b + v_i)_{i=1}^\ell$.

- **Query Phase 1.** After receiving $\mathbf{y}^{(i)} = (y_1^{(i)}, \ldots, y_\ell^{(i)})$ from $\mathcal{A}$, where $i \in [1, 2, \ldots, q]$, $\mathcal{C}$ first chooses $k \xleftarrow{\$} \mathbb{Z}_p$ and then computes $\mathsf{K}_{\mathbf{y}^{(i)}} = (\mathsf{K}_0, \mathsf{K}_1) = (kP, \langle \mathbf{v}, \mathbf{y}^{(i)} \rangle + k \mod p)$. The correctness of the private key $\mathsf{K}_{\mathbf{y}^{(i)}}$ is demonstrated as follows.

$$
\begin{aligned}
\mathsf{K}_1 \\
&= \langle \mathbf{s}, \mathbf{y}^{(i)} \rangle + k \mod p \\
&= \textstyle\sum_{j=1}^\ell s_j y_j^{(i)} + k \mod p \\
&= \textstyle\sum_{j=1}^\ell (u_j \cdot b + v_j) \cdot y_j^{(i)} + k \mod p \\
&= b \textstyle\sum_{j=1}^\ell u_j y_j^{(i)} + \textstyle\sum_{j=1}^\ell v_j y_j^{(i)} + k \mod p \\
&= b \langle \mathbf{u}, \mathbf{y}^{(i)} \rangle + \langle \mathbf{v}, \mathbf{y}^{(i)} \rangle + k \mod p \\
&= \langle \mathbf{v}, \mathbf{y}^{(i)} \rangle + k \mod p.
\end{aligned}
$$

- **Challenge.** Upon receiving $\mathbf{x}^*$, where $\langle \mathbf{x}^*, \mathbf{y}^{(i)} \rangle \neq 0$ for $i = 1, \ldots, q$, and two equal-length messages M_0, M_1 from $\mathcal{A}$, the challenger $\mathcal{C}$ performs the following.

  1. Choose $\beta \in \{0, 1\}$.
  2. Choose $\delta \xleftarrow{\$} \mathbb{Z}_p$.
  3. Set $\mathsf{C}_0' = A'$ and $\widehat{\mathsf{C}}_0' = A$.
  4. For $i = 1$ to ℓ, compute $\mathsf{C}_i' = (C^{u_i} \cdot A^{v_i} \cdot g_T^{\delta x_i^*}) \cdot M_\beta$.
  5. Set the challenge ciphertext $\mathsf{C}^* = (\mathsf{C}_0', \widehat{\mathsf{C}}_0', \mathsf{C}_1', \mathsf{C}_2', \ldots, \mathsf{C}_\ell')$.
  6. Return C^* to $\mathcal{A}$.

Here, we implicitly set the randomness of the encryption procedure to a. Therefore, if $C = g_T^{ab}$, then we have $C_0' = aP, \widehat{C}_0' = g_T^a$ for $i = 1, \ldots, \ell$,

$$
\begin{aligned}
C_i' &= (C^{u_i} \cdot A^{v_i} \cdot g_T^{\delta x_i^*}) \cdot M_\beta \\
&= (g_T^{abu_i} \cdot g_T^{av_i} \cdot g_T^{\delta x_i^*}) \cdot M_\beta \\
&= (g_T^{a(bu_i + v_i)}) \cdot (g_T^{\delta x_i^*}) \cdot M_\beta \\
&= h_i^a \cdot g_T^{\delta x_i^*} \cdot M_\beta.
\end{aligned}
$$

Thus, the challenge ciphertext C^* is a valid ciphertext.

- **Query Phase 2**. This phase is the same as **Query Phase 1**.
- **Guess**. The adversary $\mathcal{A}$ outputs a bit β'. The challenger $\mathcal{C}$ outputs 1 if $\mathcal{A}$ wins the game and outputs a random bit otherwise.

Assume that the adversary $\mathcal{A}$ wins the game with advantage ϵ:

$$
\left| \Pr[\beta' = \beta] - \frac{1}{2} \right| \geq \epsilon.
$$

If $C = g_T^{ab}$, then the view of the adversary is identical as that in real world. Thus, we have

$$
\begin{aligned}
&\Pr[\mathcal{C}(P, A', g_T, A, B, C = g_T^{ab}) = 1] \\
=\ &\Pr[\beta' = \beta] \\
\geq\ &\tfrac{1}{2} + \epsilon.
\end{aligned}
$$

However, if C is a random element in $\mathbb{G}_T$, then the choice of β is independent from the adversary's view and we have

$$
\begin{aligned}
&\Pr[\mathcal{C}(P, A', g_T, A, B, C \xleftarrow{\$} \mathbb{G}_T) = 1] \\
=\ &\Pr[\beta' = \beta] \\
=\ &\tfrac{1}{2}.
\end{aligned}
$$

Therefore, the advantage of $\mathcal{C}$ in solving the M-DDH$_{\mathbb{G}_T}$ problem is

$$
\begin{aligned}
&\left| \Pr[\mathcal{C}(P, A', g_T, A, B, C = g_T^{ab}) = 1] \right. \\
&\left. -\ \Pr[\mathcal{C}(P, A', g_T, A, B, C \xleftarrow{\$} \mathbb{G}_T) = 1] \right| \\
\geq\ &\left| (\tfrac{1}{2} + \epsilon) - \tfrac{1}{2} \right| \\
\geq\ &\epsilon.
\end{aligned}
$$

This means that if there is an adversary winning the game with nonadvantage ϵ, then there is an algorithm $\mathcal{C}$ solving the M-DDH$_{\mathbb{G}_T}$ problem with a probability greater than ϵ. $\square$

5. Efficiency Analysis and Implementation Results

Herein, we compare the efficiency of the proposed IPE scheme with the schemes proposed in [1,3,5–7,15,16,22–30,36] (Because [4,17] are the complete versions of [16,24], we only compare our work with [16,24]). As shown in Table 1, we compare our scheme to others in two aspects: the size of the private key and the number of pairing operations for decryption. The type of group order is also presented because the efficiency of prime order groups is higher than that of composite order bilinear groups.

As is evident in Table 1, our proposed scheme has the shortest private key length and smallest number of pairings. Moreover, both the private key length and the number of pairings in our proposed scheme are independent of the length of the predicate and attribute vectors. The most efficient existing scheme is [29], where the private key length is three group elements and three pairings are needed for decryption. In our scheme, the private key is only an element of $\mathbb{G}$ and an element of $\mathbb{Z}_p$, and only one pairing is necessary during decryption. Furthermore, in [5], the private key length $(2m|\mathbb{G}|)$ and the number of pairings $(2m)$ are independent of the lengths of the vectors, where m is the leakage-resilience parameter. However, m must be at least equal to or greater than 2. Therefore, the private key length and pairing number are still larger than those obtained with our approach (this is because their scheme degenerates to a conventional IPE scheme without leakage resilience when $m = 1$).

Table 1. Comparison of our scheme's efficiency with that of other schemes. The vector length for an IPE scheme is denoted by ℓ; the bit lengths of the representations for an element in $\mathbb{Z}_p$ and $\mathbb{G}$ are denoted by $|\mathbb{Z}_p|$ and $|\mathbb{G}|$, respectively; the leakage resilience parameter is denoted by m.

Scheme	Private Key Length	Number of Pairings for Decryption	Group Order
[1]	$(2\ell+1)|\mathbb{G}|$	$2\ell+1$	Composite
[3]	$(\ell+3)|\mathbb{G}|$	$\ell+3$	Prime
[16]-1	$(\ell+1)|\mathbb{G}|$	2	Prime
[16]-2	$(\ell+6)|\mathbb{G}| + (\ell-1)|\mathbb{Z}_p|$	9	Prime
[22]	$(2\ell+3)|\mathbb{G}|$	$2\ell+3$	Prime
[24]-1	$(4\ell+1)|\mathbb{G}|$	9	Prime
[24]-2	$9|\mathbb{G}|$	9	Prime
[24]-3	$11|\mathbb{G}|$	11	Prime
[23]	$(4\ell+2)|\mathbb{G}|$	$4\ell+2$	Prime
[25]	$(4\ell+2)|\mathbb{G}|$	$4\ell+2$	Prime
[26]-1	$(15\ell+5)|\mathbb{G}|$	$15\ell+5$	Prime
[26]-2	$(21\ell+9)|\mathbb{G}|$	$21\ell+9$	Prime
[27]	$6\ell|\mathbb{G}|$	6ℓ	Prime
[28]	$\ell|\mathbb{G}|$	ℓ	Composite
[29]	$3|\mathbb{G}|$	3	Prime
[30]	$(4\ell+2)|\mathbb{G}|$	$4\ell+4$	Prime
[15]-1	$(2\ell+1)|\mathbb{G}| + (\ell-1)|\mathbb{Z}_p|$	3	Prime
[15]-2	$5|\mathbb{G}|$	3	Prime
[5]	$2m|\mathbb{G}|$	$2m$	Prime
[36]	$(4\ell+5)|\mathbb{G}|$	$4\ell+5$	Prime
[6]-1	$5|\mathbb{G}|$	5	Prime
[6]-2	$7|\mathbb{G}|$	7	Prime
[7]	$(\ell+1)|\mathbb{G}|$	$\ell+1$	Composite
Ours	$1|\mathbb{G}| + 1|\mathbb{Z}_p|$	1	Prime

We also implemented our scheme and the schemes of [15,17,29] to compare efficiency. We chose these three schemes for the following reasons:

- Among all the existing IPE schemes, the first scheme of [16] requires the smallest number of pairings for decryption (only two pairings required);
- Among the schemes supporting constant private key length, the schemes of [15,29] require the smallest number of pairings for decryption (only three pairings required).

The environment of the implementation is presented in Table 2, and the implementation results are shown in Table 3. We implemented these schemes by using the Charm-Crypto library [37] and Python language. For schemes constructed over symmetric paring groups (the approach in [16] and our method), we selected the pairing group SS512 in [38] (also known as type A groups), and for the schemes constructed over asymmetric pairing groups (in [15,29]), we chose the pairing group BN254 in [39] (also known as type F groups). The SS512 group is a supersingular elliptic curve group where the size of the base field order is 512 bits and the embedding degree is two. For a bilinear map $e : \mathbb{G} \times \mathbb{G} \rightarrow \mathbb{G}_T$ over the SS512 group, the bit lengths of elements in $\mathbb{G}$ and $\mathbb{G}_T$ are 64 and

128 bytes, respectively. In the case of the BN254 group, the size of the base field order is 256 bits and the embedding degree is 12. For a bilinear map $e : \mathbb{G}_1 \times \mathbb{G}_2 \to \mathbb{G}_T$ over the BN254 group, the bit lengths of elements in $\mathbb{G}_1$, $\mathbb{G}_2$, and $\mathbb{G}_T$ are 64, 128, and 384 bytes, respectively. For the length of predicate and attribute vectors, we chose $\ell = 100$. As evident in Table 3, the encryption and decryption algorithms of our scheme were highly efficient. For decryption and encryption, only 10 and 20 ms was required, respectively. Our encryption algorithm was 5, 8.5, and 13 times faster than that in [15,16,29], respectively, and our decryption algorithm was 10, 14, and 14 times faster than that in [15,16,29], respectively. Moreover, our private key length was 86, 2.6, and 4.3 times shorter than that in [15,16,29], respectively. However, as a trade-off, the length of the ciphertext in our scheme was the largest among these schemes.

Table 2. Environment of the implementation.

	Specification
OS	Ubuntu 18.04 LTS
CPU	Intel i7-4790 3.6 GHz
RAM	8 gb
Language	Python 3.6
Library	Charm-Crypto v0.50

Table 3. Implementation results.

Scheme	Encryption Time (ms)	Decryption Time (ms)	Private Key Length (kb)	Ciphertext Length (kb)
[16]	100	100	31.7	0.937
[29]	170	140	0.955	17.5
[15]	260	140	1.59	25.9
Ours	20	10	0.37	31.3

6. Conclusions

In this work, an efficient IPE scheme in which the size of the private keys and the number of pairings for decryption are constant is introduced; moreover, this scheme is coselective IND–CPA secure under the modified decisional Diffie–Hellman assumption. Comparison and experimental results are also provided to illustrate that the size and computing cost of this scheme are small. In future works, we aim to improve the efficiency by reducing the ciphertext length and provide a security proof for stronger security concerns under standard assumptions. Because the proposed scheme is based on bilinear pairing, it cannot resist quantum attacks, unlike lattice-based IPE schemes. In future work, we will explore how to construct an efficient and practical quantum-resistant IPE scheme.

Author Contributions: Conceptualization, Y.-F.T. and Z.-Y.L.; Methodology, Y.-F.T. and Z.-Y.L.; Investigation, Z.-Y.L.; Writing—Original Draft Preparation, Z.-Y.L.; Writing—Review and Editing, Y.-F.T. and R.T.; Supervision, R.T.; Project Administration, R.T.; Funding Acquisition, Y.-F.T. and R.T. All authors have read and agreed to the published version of the manuscript.

Funding: This research was supported by the Ministry of Science and Technology, Taiwan (ROC), under Project Numbers MOST 108-2218-E-004-001-, MOST 108-2218-E-004-002-MY2, MOST 109-2218-E-011-007-, and by Taiwan Information Security Center at National Sun Yat-sen University (TWISC@NSYSU).

Conflicts of Interest: The authors declare no conflict of interest.

References

1. Katz, J.; Sahai, A.; Waters, B. Predicate Encryption Supporting Disjunctions, Polynomial Equations, and Inner Products. In *Advances in Cryptology—EUROCRYPT 2008, LNCS*; Smart, N., Ed.; Springer: Berlin/Heidelberg, Germany, 2008; Volume 4965, pp. 146–162. [CrossRef]
2. Boneh, D.; Waters, B. Conjunctive, Subset, and Range Queries on Encrypted Data. In *Theory of Cryptography, LNCS*; Vadhan, S.P., Ed.; Springer: Berlin/Heidelberg, Germany, 2007; Volume 4392, pp. 535–554. [CrossRef]
3. Okamoto, T.; Takashima, K. Hierarchical Predicate Encryption for Inner-Products. In *Advances in Cryptology—ASIACRYPT 2009, LNCS*; Matsui, M., Ed.; Springer: Berlin/Heidelberg, Germany, 2009; Volume 5192, pp. 214–231. [CrossRef]
4. Okamoto, T.; Takashima, K. Achieving Short Ciphertexts or Short Secret-keys for Adaptively Secure General Inner-product Encryption. *Des. Codes Cryptogr.* **2015**, *77*, 725–771. [CrossRef]
5. Kurosawa, K.; Phong, L.T. Anonymous and Leakage Resilient IBE and IPE. *Des. Codes Cryptogr.* **2017**, *85*, 273–298. [CrossRef]
6. Chen, J.; Gong, J.; Wee, H. Improved Inner-Product Encryption with Adaptive Security and Full Attribute-Hiding. In *Advances in Cryptology—ASIACRYPT 2018, LNCS*; Peyrin, T., Galbraith, S., Eds.; Springer: Berlin/Heidelberg, Germany, 2018; Volume 1127, pp. 673–702. [CrossRef]
7. Zhang, Y.; Li, Y.; Wang, Y. Efficient Inner Product Encryption for Mobile Client with Constrained Capacity. *Int. J. Innov. Comput. I* **2019**, *15*, 209–226. [CrossRef]
8. Agrawal, S.; Freeman, D.M.; Vaikuntanathan, V. Functional Encryption for Inner Product Predicates from Learning with Errors. In *Advances in Cryptology—ASIACRYPT 2011, LNCS*; Lee, D.H., Wang, X., Eds.; Springer: Berlin/Heidelberg, Germany, 2011; Volume 7073, pp. 21–40. [CrossRef]
9. Xagawa, K. Improved (Hierarchical) Inner-Product Encryption from Lattices. In *Public-Key Cryptography—PKC 2013, LNCS*; Kurosawa, K., Hanaoka, G., Eds.; Springer: Berlin/Heidelberg, Germany, 2013; Volume 7778, pp. 235–252. [CrossRef]
10. Li, J.; Zhang, D.; Lu, X.; Wang, K. Compact (Targeted Homomorphic) Inner Product Encryption from LWE. In *Information and Communications Security, LNCS*; Qing, S., Mitchell, C., Chen, L., Liu, D., Eds.; Springer: Cham, Switzerland, 2017; Volume 10631, pp. 132–140. [CrossRef]
11. Wang, Z.; Fan, X.; Wang, M. Compact Inner Product Encryption from LWE. In *Information and Communications Security, LNCS*; Qing, S., Mitchell, C., Chen, L., Liu, D., Eds.; Springer: Cham, Switzerland, 2018; Volume 10631, pp. 141–153. [CrossRef]
12. Shamir, A. Identity-Based Cryptosystems and Signature Schemes. In *Advances in Cryptology—CRYPTO 1984, LNCS*; Springer: Berlin/Heidelberg, Germany, 1985; Volume 196, pp. 47–53. [CrossRef]
13. Boneh, D.; Franklin, M. Identity-Based Encryption from the Weil Pairing. In *Advances in Cryptology—CRYPTO 2001, LNCS*; Kilian, J., Ed.; Springer: Berlin/Heidelberg, Germany, 2001; Volume 2139, pp. 213–229. [CrossRef]
14. Boneh, D.; Boyen, X. Efficient Selective-ID Secure Identity Based Encryption without Random Oracles. In *Advances in Cryptology—EUROCRYPT 2004, LNCS*; Cachin, C., Camenisch, J., Eds.; Springer: Berlin/Heidelberg, Germany, 2004; Volume 3027, pp. 223–238. [CrossRef]
15. Ramanna, S.C. More Efficient Constructions for Inner-Product Encryption. In *Applied Cryptography and Network Security, LNCS*; Manulis, M., Sadeghi, A.R., Schneider, S., Eds., Springer: Cham, Switzerland, 2016; Volume 9696, pp. 231–248. [CrossRef]
16. Attrapadung, N.; Libert, B. Functional Encryption for Inner Product: Achieving Constant-Size Ciphertexts with Adaptive Security or Support for Negation. In *Public Key Cryptography—PKC 2010, LNCS*; Nguyen, P.Q., Pointcheval, D., Eds.; Springer: Berlin/Heidelberg, Germany, 2010; Volume 6056, pp. 384–402. [CrossRef]
17. Attrapadung, N.; Libert, B. Functional Encryption for Public-attribute Inner Products: Achieving Constant-size Ciphertexts with Adaptive Security or Support for Negation. *J. Math. Cryptol.* **2012**, *5*, 115–158. [CrossRef]
18. Lee, K. Efficient Hidden Vector Encryptions and Its Applications. *arXiv* **2017**, arXiv:1702.07456.
19. Katz, J.; Maffei, M.; Malavolta, G.; Schröder, D. Subset Predicate Encryption and Its Applications. In *Cryptology and Network Security, LNCS*; Capkun, S., Chow, S.S.M., Eds.; Springer: Cham, Switzerland, 2018; Volume 11261, pp. 115–134. [CrossRef]

20. Chatterjee, S.; Mukherjee, S. Large Universe Subset Predicate Encryption based on Static Assumption (without Random Oracle). In *Topics in Cryptology—CT-RSA 2019, LNCS*; Matsui, M., Ed.; Springer: Cham, Switzerland, 2019; Volume 11405, pp. 62–82. [CrossRef]

21. Waters, B. Dual System Encryption: Realizing Fully Secure IBE and HIBE under Simple Assumptions. In *Advances in Cryptology—CRYPTO 2009, LNCS*; Springer: Berlin/Heidelberg, Germany, 2009; Volume 5677, pp. 619–636. [CrossRef]

22. Lewko, A.; Okamoto, T.; Sahai, A.; Takashima, K.; Waters, B. Fully Secure Functional Encryption: Attribute-Based Encryption and (Hierarchical) Inner Product Encryption. In *Advances in Cryptology—EUROCRYPT 2010, LNCS*; Gilbert, H., Ed.; Springer: Berlin/Heidelberg, Germany, 2010; Volume 6110, pp. 62–91. [CrossRef]

23. Park, J.H. Inner-Product Encryption under Standard Assumptions. *Des. Codes Cryptogr.* **2011**, *58*, 235–257. [CrossRef]

24. Okamoto, T.; Takashima, K. Achieving Short Ciphertexts or Short Secret-Keys for Adaptively Secure General Inner-Product Encryption. In *Cryptology and Network Security, LNCS*; Lin, D., Tsudik, G., Wang, X., Eds.; Springer: Berlin/Heidelberg, Germany, 2011; Volume 7092, pp. 138–159. [CrossRef]

25. Okamoto, T.; Takashima, K. Adaptively Attribute-Hiding (Hierarchical) Inner Product Encryption. In *Advances in Cryptology—EUROCRYPT 2012, LNCS*; Pointcheval, D., Johansson, T., Eds.; Springer: Berlin/Heidelberg, Germany, 2012; Volume 7237, pp. 591–608. [CrossRef]

26. Okamoto, T.; Takashima, K. Fully Secure Unbounded Inner-Product and Attribute-Based Encryption. In *Advances in Cryptology—ASIACRYPT 2012, LNCS*; Wang, X., Sako, K., Eds.; Springer: Berlin/Heidelberg, Germany, 2012; Volume 7658, pp. 349–366. [CrossRef]

27. Kawai, Y.; Takashima, K. Predicate- and Attribute-Hiding Inner Product Encryption in a Public Key Setting. In *Pairing-Based Cryptography—Pairing 2013, LNCS*; Cao, Z., Zhang, F., Eds.; Springer: Cham, Switzerland, 2014; Volume 836, pp. 113–130. [CrossRef]

28. Zhenlin, T.; Wei, Z. A Predicate Encryption Scheme Supporting Multiparty Cloud Computation. In Proceedings of the 2015 International Conference on Intelligent Networking and Collaborative Systems, Taipei, Taiwan, 2–4 September 2015; pp. 252–256. [CrossRef]

29. Kim, I.; Hwang, S.O.; Park, J.H.; Park, C. An Efficient Predicate Encryption with Constant Pairing Computations and Minimum Costs. *IEEE Trans. Comput.* **2016**, *65*, 2947–2958. [CrossRef]

30. Huang, S.Y.; Fan, C.I.; Tseng, Y.F. Enabled/Disabled Predicate Encryption in Clouds. *Future Gener. Comput. Syst.* **2016**, *62*, 148–160. [CrossRef]

31. Agrawal, S.; Boneh, D.; Boyen, X. Efficient Lattice (H)IBE in the Standard Model. In *Advances in Cryptology—EUROCRYPT 2010, LNCS*; Gilbert, H., Ed.; Springer: Berlin/Heidelberg, Germany, 2010; Volume 6110, pp. 553–572. [CrossRef]

32. Gentry, C.; Sahai, A.; Waters, B. Homomorphic Encryption from Learning with Errors: Conceptually-simpler, Asymptotically-faster, Attribute-based. In *Advances in Cryptology—CRYPTO 2013, LNCS*; Springer: Berlin/Heidelberg, Germany, 2013; Volume 8042, pp. 75–92. [CrossRef]

33. Apon, D.; Fan, X.; Liu, F.H. Vector Encoding over Lattices and Its Applications. *IACR Cryptol. EPrint Arch.* **2017**, *2017*, 455. Available online: https://eprint.iacr.org/2017/155 (accessed on 14 January 2020).

34. Boneh, D.; Boyen, X.; Goh, E.J. Hierarchical Identity Based Encryption with Constant Size Ciphertext. In *Advances in Cryptology—EUROCRYPT 2005, LNCS*; Cramer, R., Ed.; Springer: Berlin/Heidelberg, Germany, 2005; Volume 3494, pp. 440–456. [CrossRef]

35. Attrapadung, N. Dual System Encryption via Doubly Selective Security: Framework, Fully Secure Functional Encryption for Regular Languages, and More. In *Advances in Cryptology—EUROCRYPT 2014, LNCS*; Nguyen, P.Q., Oswald, E., Eds.; Springer: Berlin/Heidelberg, Germany, 2014; Volume 8441, pp. 557–577. [CrossRef]

36. Xiao, S.; Ge, A.; Zhang, J.; Ma, C.; Wang, X. Asymmetric Searchable Encryption from Inner Product Encryption. In *Advances on P2P, Parallel, Grid, Cloud and Internet Computing*; Xhafa, F., Barolli, L., Amato, F., Eds.; Springer: Cham, Switzerland, 2017; pp. 123–132. [CrossRef]

37. Akinyele, J.A.; Garman, C.; Miers, I.; Pagano, M.W.; Rushanan, M.; Green, M.; Rubin, A.D. Charm: A Framework for Rapidly Prototyping Cryptosystems. *J. Cryptogr. Eng.* **2013**, *3*, 111–128. [CrossRef]

38. Lee, K.; Park, J.H. Identity-Based Revocation from Subset Difference Methods under Simple Assumptions. *IEEE Access* **2019**, *7*, 60333–60347. [CrossRef]
39. Barreto, P.S.L.M.; Naehrig, M. Pairing-Friendly Elliptic Curves of Prime Order. In *Selected Areas in Cryptography, LNCS*; Preneel, B., Tavares, S., Eds.; Springer: Berlin/Heidelberg, Germany, 2006; Volume 3897, pp. 319–331. [CrossRef]

Publisher's Note: MDPI stays neutral with regard to jurisdictional claims in published maps and institutional affiliations.

Article

You Only Look Once, But Compute Twice: Service Function Chaining for Low-Latency Object Detection in Softwarized Networks †

Zuo Xiang [1,*,‡], Patrick Seeling [2,‡] and Frank H. P. Fitzek [1,‡]

1 Centre for Tactile Internet with Human-in-the-Loop, Technische Universität Dresden, 01187 Dresden, Germany; frank.fitzek@tu-dresden.de

2 Department of Computer Science, Central Michigan University, Mount Pleasant, MI 48859, USA; patrick.seeling@cmich.edu

* Correspondence: zuo.xiang@tu-dresden.de

† Extended version of Xiang, Z.; Zhang, R.; Seeling, P. Machine learning for object detection. In *Computing in Communication Networks*; Fitzek, F.H., Granelli, F., Seeling, P., Eds.; Elsevier/Academic Press: Cambridge, MA, USA, 2020.

‡ The authors contributed equally to this work.

Featured Application: Splitting of formerly only integrated inference from object recognition and other trained (and potentially untrained) machine learning approaches has broad applicability in all application scenarios that rely on these types of models, with connected autonomous cars, smart city applications, and video surveillance being prominent examples.

Abstract: With increasing numbers of computer vision and object detection application scenarios, those requiring ultra-low service latency times have become increasingly prominent; e.g., those for autonomous and connected vehicles or smart city applications. The incorporation of machine learning through the applications of trained models in these scenarios can pose a computational challenge. The softwarization of networks provides opportunities to incorporate computing into the network, increasing flexibility by distributing workloads through offloading from client and edge nodes over in-network nodes to servers. In this article, we present an example for splitting the inference component of the YOLOv2 trained machine learning model between client, network, and service side processing to reduce the overall service latency. Assuming a client has 20% of the server computational resources, we observe a more than 12-fold reduction of service latency when incorporating our service split compared to on-client processing and and an increase in speed of more than 25% compared to performing everything on the server. Our approach is not only applicable to object detection, but can also be applied in a broad variety of machine learning-based applications and services.

Keywords: object detection; latency optimization; mobile edge cloud; connected autonomous cars; smart city; video surveillance

Citation: Xiang, Z.; Seeling, P.; Fitzek, F.H.P. You Only Look Once, But Compute Twice: Service Function Chaining for Low-Latency Object Detection in Softwarized Networks. *Appl. Sci.* **2019**, *11*, 2177. https://doi.org/10.3390/app11052177

Academic Editor: Cheonshik Kim

Received: 22 December 2020
Accepted: 25 February 2021
Published: 2 March 2021

Publisher's Note: MDPI stays neutral with regard to jurisdictional claims in published maps and institutional affiliations.

1. Introduction

Multimedia network traffic has permeated all types of networks, and its dominance continues with increased adoptions of new connected services. Within the range of multimedia network traffic types, video is typically the most dominant form, especially with respect to bandwidth requirements. For example, Cisco forecasts in [1] that 82% of Internet Protocol (IP) traffic will be comprised of video by the year 2022. Within the video domain, specifically the object detection sub-category has an additional significant latency requirement, especially when applied in certain scenarios, see, e.g., [2]. The object identification and understanding within an ongoing video stream is based on the Computer Vision (CV) domain of real-time video analysis. Prominent examples for real-time object detection and analysis include Google Lens or smart city applications that perform video surveillance [3–5] or for connected autonomous cars, as illustrated in Figure 1. Especially for the latter, incorporating new sensor data such as from LIDAR and other on-board sensors that goes beyond image data alone is also attracting interest [6–8].

(a)

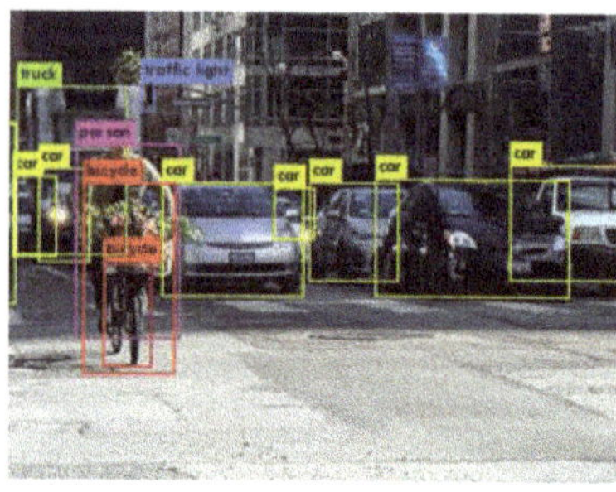

(b)

Figure 1. Object detection use cases including pedestrians and vehicles detection. (**a**) Pedestrian data set detection by YOLOv2 (image from [9]). (**b**) Object detection on the street (image from [10]).

Significant challenges exist to reliably perform real-time video analysis on resource-limited devices, such as mobile phones or ad-hoc deployed video monitoring, when considering higher frame rates of live video captures. The requirements are typically high when locally processing data, as captured image analysis and machine vision tasks that comprise visual understanding commonly encompass involved Artificial Intelligence (AI) approaches. The AI component of these types of systems has undergone steady improvements in recent years as well, with increasing precision and recall, especially for Deep Learning (DL) approaches [11]. As these approaches exceed traditional methods, deep learning-based mechanisms have become increasingly popular, themselves commonly based on Convolutional Neural Networks (CNN) [12]. This enables CV systems to more reliably detect objects even in complicated scenes. The training of these models is typically highly resource-intensive; however, continuous improvements in hardware alleviate some of these problems and make a focus on the inference from these models more important. Example approaches include R-CNN [13], Faster R-CNN [14], and YOLO [15] combine precision with improved detection speed (also referred to as the inference speed).

The focus on latency optimization in a mobile context has to combine several requirements, such as resource usage and low latency of detection. Common resources considered include memory, CPU, and bandwidth on the computing side, however, overall system costs commonly need to be factored into solutions as well. For example, future intelligent transport system and connected autonomous vehicle applications of object detection are highly latency sensitive and mission-critical at the same time. Current approaches commonly are limited in realizing the full potential that upcoming network softwarization provides:

- Object detection as outlined above is resource demanding and commonly not suitable for prolonged execution on mobile (i.e., battery-limited) devices and can overwhelm the computational resources of embedded solutions.
- Instead, cloud computing typically offers flexible resource management for computationally intensive tasks through computational offloading, see, e.g., [16,17]. The need to communicate with far-away cloud computing resources in traditional network infrastructures, however, increases the overall service latency significantly.
- One approach to overcome the limitations of mobile processing while providing low latency services is to combine local processing and geographically close cloud services for more computationally expensive processing. While current communication networks infrastructure does not typically allow for in-network computing, new softwarized networks provide this flexibility.

- In this article, we focus on the latency optimization aspects of mobile object detection by combining on-device and in-network computing. Our approach can be applied in 5G and beyond networks (as well as any network that has in-situ computing enabled).

In this article, we describe the implementation and performance analysis for a real-time object detection method that incorporates this network softwarization and computing resource provisioning.

The current trend to edge computing [18,19] and network softwarization in general enables the flexible service and application deployment under tight latency constraints, such as the one we consider here. Typically, deployments in softwarized networks include a combination of technologies to fulfill the requirements of real-time use cases: Software-Defined Networking (SDN) [20], Network Function Virtualization (NFV) [21], and Service Function Chaining (SFC) [22]. As the network becomes softwarized, Computing in the Network (COIN) and the Mobile Edge Cloud (MEC) [23] become powerful concepts to combine mobile, local, and far computing resources in a flexible fashion per use-case. Computing in the network will significantly reduce latency and issues that stem from extended packet switching across multiple networks, such as congestion. Virtualized resources can be flexibly deployed at various locations closer to the user, follow the user, and be reallocated in a dynamic fashion. In such a setup, initial pre-processing could be performed at edge nodes and reduce the subsequent nodes' latency requirements for real-time services. This split of overall service processing needs is enabled by the layer-based approach used in object detection neural networks and the ability to split the location of processing by connecting the different layers flexibly over the network.

We describe the overall approach in the following Section 2, which contains information about the general on-device or on-server object recognition approach. Additionally, we describe the implementation of a single service function split between an initial service client and the server, noting that multiple splits could be performed as well. We follow with the description of results for a latency-focused performance evaluation in Section 3 and discussion in Section 4 before concluding in Section 5.

2. Materials and Methods

In this manuscript, we employ the You Only Look Once (YOLO) object detection library as a concrete example, noting that similarities with other neural networks can be exploited to modify our described approach with those models and mechanisms as well. In this section, initially discuss the general approach before describing YOLO and our setup in greater detail.

2.1. CNN Object Detection Model Split

CNN approaches for object detection generally feature several types of interconnected layers: convolutional layers, pooling layers, fully-connected layers, and batch normalization layers. These layers are typically stacked in a pattern of convolutional layers and activation functions followed by pooling layers, which (in multiple iterations) reduces the overall size of the image to a smaller size. Once a desired small size has been reached, fully connected layers are used, whereby the final layer contains the output. The output of each convolutional or pooling layer is an intermediate representation of the original image data relying on convolutional filters, their parameters derived via CNNs. The parameters (or weights) are dynamic while the feature maps representing different features of an image remain static and the overall outcome depends on the image input. Typically, the weights and resulting output data types are floating-point numbers. After a convolution layer, activation functions such as ReLU [24] are applied. To simplify the overall process, it is also common that the overall image will be initially pre-processed, as multi-layer models typically were trained for and assume a specific image size.

The limitations of computing resources (here, processing and memory) of edge nodes motivates a split of the overall processing to take place via different levels of offloading. For example, should traditional cloud computing approaches be involved, the entire sequence of images (or video frames) generated at the client on the network edge would have to be forwarded to centralized cloud servers. In compute-and-forward networks, on the other hand, computing resources are available inside the network which enables intermediate processing. In turn, reduced amounts of data alleviate network congestion and can improve overall service latency. We assume that deep learning frameworks such as Tensorflow [25] can be deployed as VNFs inside the network as well as on the centralized server. We additionally note that here, we consider a general CPU-based baseline evaluation, which can greatly be enhanced with additional accelerators, such as GPUs or FPGAs.

A significant initial consideration is how and where to perform a potential split between the on-device, edge, and centralized server processing in this overall architecture. Table 1 provides the initial layers for YOLOv2 [26], SSD [27], VGG16 [28], and Faster R-CNN [14].

Table 1. Initial 10-layer designs for example object detection models.

Model	Structure of first 10 layers
YOLOv2	Conv. + Pool. + Conv. + Pool. + 3 Conv. + Pool. + 2 Conv.
SSD	2 Conv. + Pool. + 2 Conv. + Pool. + 3 Conv. + Pool.
VGG16	2 Conv. + Pool. + 2 Conv. + Pool. + 3 Conv. + Pool.
Faster R-CNN	2 Conv. + Pool. + 2 Conv. + Pool. + 3 Conv. + Pool.

Comparing these entries, all feature different combinations of similar layers that can be evaluated to determine a favorable point to split the original model such that the part before a split can be executed on a network device and bandwidth savings result. This requires limiting the number of layers prior to a split. Consequently, the number of layers before the split point should not be too high and the output data of the front part should be smaller than the original input image size in order to realize bandwidth savings.

Given a particular split to enable the offloading of processing parts, the structure of the pipeline for evaluating the performance of deploying object detection services in edge computing such as MEC is presented in Figure 2 with a detailed visualization of basic components.

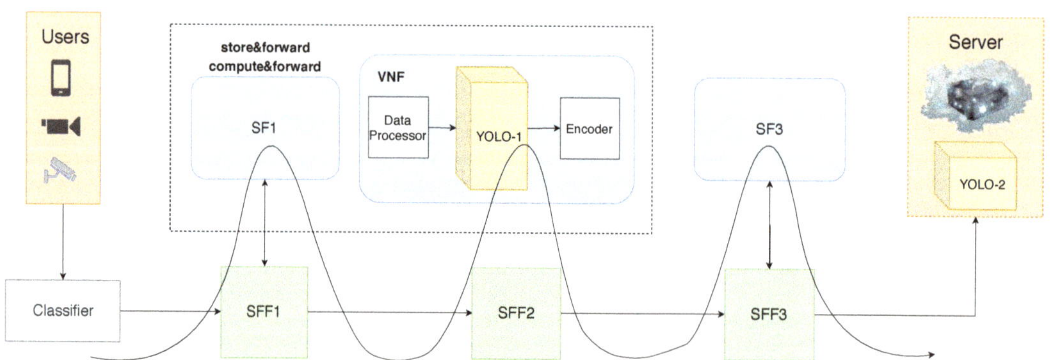

Figure 2. Overview of the distributed architecture, here for the example employing YOLO [29].

The implementation of this example is focused on the VNF, which supports both store-and-forward and compute-and-forward to adapt to the network state. The outer Service Function Path is not modified during computation, i.e., the VNF will not affect other protocols or the SFC architecture.

The VNF is employed to offload part of the overall computational burden of the CNN related computations in the object detection from centralized servers to the network edge.

We employ YOLOv2 as example for such object detection methods. YOLOv2 is deployed in the VNF at the edge and the server. As described, we follow the outlined approach of splitting the CNN model into two parts. The first part is deployed in the VNF and the second part is deployed on the server. Following the overall desire to reduce the overall service latency under the computational constraints, the complexity of the first part is lower than that of the second part, where in our case, the first part will be the pre-processor for video frames.

2.2. You Only Look Once (YOLO), But Twice

We now focus on the concrete implementation employed in the remainder of this article. YOLOv2 is mainly constructed of convolutional layers and max-pooling layers [26], similar to several other approaches highlighted in Table 1 and illustrated in Figure 3.

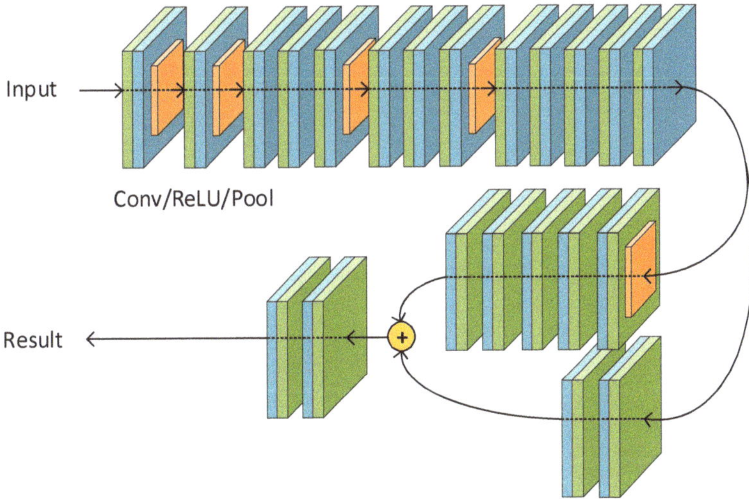

Figure 3. Combined Model structure of YOLOv2 as executed as a single instance.

Following our assumption of computational resource availability at clients, edge nodes, and centralized cloud computing servers, increasing distance from the network edge corresponds to higher computational resources. Subsequently, splitting workloads should focus on the initial layers, provided that the split takes place at an advantageous processing step in the neural network. Similarly, not too many layers should have been processed at the initial nodes to improve the overall service latency and adhere to computing resource restrictions. Figure 4 illustrates the different layer outputs in relation to the initial input image for YOLOv2. Figure 4 additionally contains the reference input size (i.e., $1 \times 608 \times 608 \times 3$).

While some initial layers clearly outsize the original input, the outputs of the latter layers are very small. For example, the final convolutional layer has only 13% of the original input size. In the first 10 layers, the output size of *max_8* and *conv_10* are both 66% of the input size, which are both candidates for a potential early split. To expedite the processing, we here consider the first candidate max-pooling layer's output as a split point. This provides a possibility to compress the resulting feature maps (which should result in smaller sizes than the input images). The resulting model's split is illustrated in Figure 5, showcasing how the outputs are communicated further into the network.

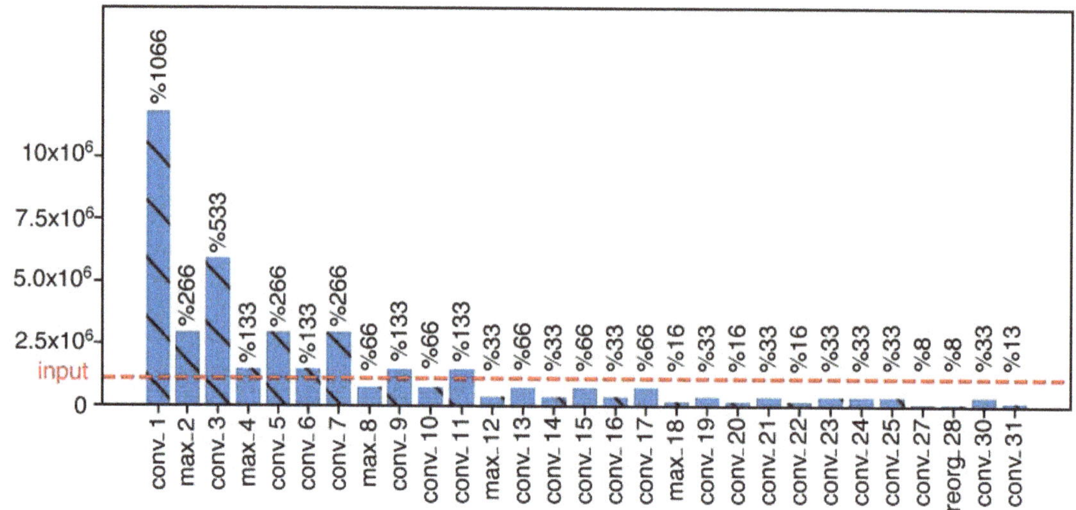

Figure 4. Output size of each layer in YOLOv2 for *conv*-olutional and *max*-pooling layers [29].

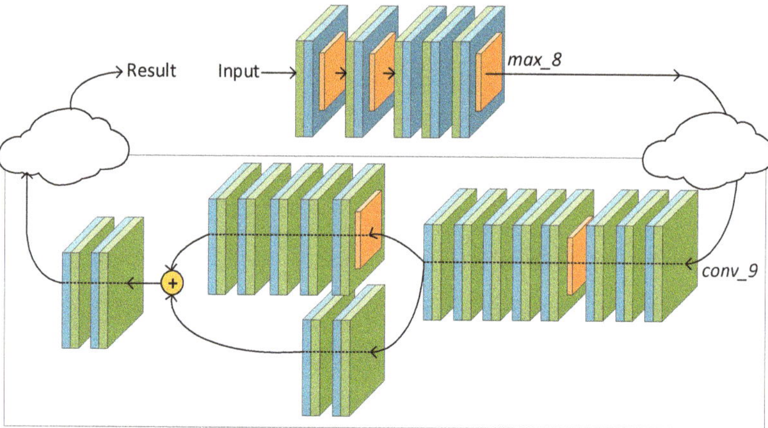

Figure 5. YOLOv2 split into two separate instances with the output of the eighth layer communicated over the network.

In our particular example, the VNF consists of the following three components packaged as container:

Data Processor The data processor collects the incoming video packets and performs relevant pre-processing tasks. These tasks could encompass video decoding, image manipulations (especially reshaping to proper input dimensions), or pixel representation changes.

YOLO Part 1 The initial part of YOLO as VNF provides initial detection model processing as outlined in this section. The resulting feature maps contain the extracted information from the original image.

Encoder The encoder ecodes (compresses) the resulting feature maps before sending them to the server to reduce bandwidth requirements even further. As the feature maps themselves are representable as image data data, we consider several image compression approaches.

The alternative approach to the YOLO service function split is the monolithic deployment on the central cloud server. A significant benefit is that cloud servers are generally assumed to have an abundance of computing resources at their disposal. In our example implementation, the server deploys the regular (full as in Figure 3) YOLOv2. Additionally, the server also deploys the remaining layers of the split YOLOv2 service (as in Figure 5). To enable separation of the server-side service to use, the VNF adds a small header indicating which approach to use. Should the received data be pre-processed by the VNF, the potentially compressed feature maps are decoded and entered in the remaining chain of layers. Alternatively, should the received data be simply forwarded data from user equipment, the traditional YOLOv2 pre-processing chain commences (employing the same mechanisms as in the VNF). In either case, the object detection result is obtained on the server and sent back to the user equipment after processing is completed.

2.3. Testbed Input Data Performance Metrics

Our example evaluation is based on the COCO data set [30], employing YOLOv2 [26] as described in this section. We consider three different object detection scenarios, namely (*i.*) on-device, (*ii.*) server-based, and (*iii.*) service function split. In addition to pre-processing and subsequent YOLOv2 object detection fully deployed on the client/server, we also perform a split with only layers after the *max_8* on the server, and the layers and processing before being implemented as VNF. In our example, the input images are normalized to the range [0–1], i.e., the data type of all feature maps will be 32-bit float. For the overall testbed, we employ a generic computer system with an i7-6700T CPU with 16GB RAM using Ubuntu 18.04 LTS and implement the system in the Communication Networks Emulator (ComNetsEmu), see [31]. The Tensorflow library v1.13 is used to implement the object detection function of VNF and server. All programs and measurement scripts are implemented in Python 3.6 and are publicly accessible in the repository of ComNetsEmu [32]. All source code can be found in the folder: `app/machine_learning_for_object_detection`. Detailed descriptions (for reproducible measurements) of all the libraries and environments used can be found in the `Dockerfile` included in the repository. Provided the nature of non-accelerated performance evaluation here, our results provide an upper first limit to attainable latency, which can be improved upon, e.g., with GPU accelerations. The client, VNF and server are running on different physical CPUs (using Linux `cpuset_cpus`) to minimize interference. For the latency measurements, a multi-hop topology is used connecting client to in-network service function (processing or forwarding as illustrated in Figure 2) and server. All links in the topology have the same homogeneous bandwidth is limits of to 10 Mbit/s with a fixed propagation delay of 150 ms. The same source image data is sent by the client (pedestrian.jpg with a original size of 48 kB) for 30 repeated measurements. All measurements were performed utilizing JPEG compression for the original and in-network computation's intermediate result forwarding.

As not only latency performance, but also the actual prediction outcome performance are important for object detection services, a careful trade-off between the two should be made. For the 8th layer (split point) of YOLOv2, the output data shape is $1 \times 78 \times 78 \times 128$, which results in approximately 92% as a baseline average precision for the entire COCO data set without YOLOv2 modifications. We select compression format and working point following our reasoning in [29], with results illustrated in Figure 6 for multiple compression approaches' compression factor versus the average attainable precision.

As illustrated, only JPEG and WebP result in higher attainable average precision beyond the 92% baseline. Subsequently, we select JPEG compression of about 50% to be applied to the 7th layer output; the compressed data is about half of the original image data (also in JPEG format).

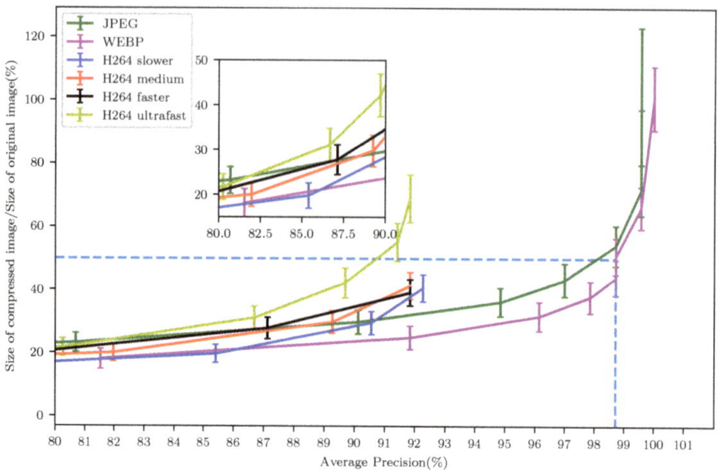

Figure 6. Image-based compression methods for JPEG input assumption, from [29].

We initially assume that the client features a limited processing capacity that is 20% that of the server/service function in a common scenario. We base this split on the CoreMark Benchmark [33] values per MHz for the Samsung Exynos 5422 (15.077 for four cores at 2.1 GHz) and the Intel Core i5-8500 (57.207 for four cores at 3 GHz). The Samsung Exynos as a popular mobile device CPU and representative for a low-power fixed smart city device or smartphone at just below 20% performance of the i5-8500. Similar comparisons for other benchmarks confirm this general approach, e.g., the Passmark Average CPU Mark [34] results for the entire CPU of current Android phones are around 6000 while current dual CPU server systems are rated around 90,000. Based on single thread ratings, it would require 1/10th of a modern server's threads to replicate the entire available CPU performance of a smartphone. Similarly, multi-core benchmarks from Geekbench v5 for a Google Pixel 5 smartphone range around 1500 while the AMD Threadripper 3990X is rated at around 27,000. Again, the idea of providing fractional resources for NFV would allow us to serve 18 phones at full virtualized CPU performance in this foundational comparison. In turn, we reason that our split is representative of the common performance differences between mobile and short-term available edge computing resources. As we perform our evaluations in the ComNetsEmu environment with the above settings, we note that during the experimentation, the server is always allocated with 100% CPU time while the client is allocated a dynamic portion of the server's CPU time, denoted as α. With the overall service latency T as the main focus of this article, we determine it as

$$T = t_{CPU}^{Client} + t_{CPU}^{Server} + 2 \cdot t_{prop} + t_{tran}^{up} + t_{tran}^{down}. \tag{1}$$

where intuitively $t_{CPU}^{Client|Server}$ denotes the required CPU times for client and server, respectively. Similarly, we denote the fixed propagation delay as t_{prop} and the up- or downstream transmission delays as $t_{tran}^{up|down}$.

3. Results

In this section, we describe the obtained service latency results for the three evaluated scenarios of on-device, server-based, and service function split object detection service with YOLOv2 as described in prior sections. We initially present our overarching results in Table 2.

Table 2. Overview of obtained service latency T results for YOLOv2 performed on-device (with varying degrees α of server computation resource), store-and-forward networking with server-side processing, and compute-and-forward with $\alpha = 0.2$ client-side processing up to layer 8 of YOLOv2 and remainder processing server-side. All results are in seconds.

| | Client, α | | | | | | | | | | Store | Compute |
	0.1	0.2	0.3	0.4	0.5	0.6	0.7	0.8	0.9	1		($\alpha = 0.2$)
Min	116.498	74.990	50.103	28.381	16.132	11.883	9.517	8.229	7.368	6.575	8.937	6.012
Median	150.701	81.901	54.140	31.276	16.693	12.359	9.984	8.735	7.678	6.882	9.170	6.581
Average	147.884	81.656	53.902	31.275	16.746	12.406	9.964	8.695	7.712	6.904	9.242	6.643
Max	155.296	85.397	57.722	34.629	17.996	13.049	10.623	9.319	8.228	7.370	9.772	7.496
StdDev	8.952	2.565	1.844	1.639	0.450	0.320	0.313	0.296	0.253	0.226	0.237	0.408

We first observe that for the two scenarios of fully on-device ($\alpha = 1$) and fully on-server (Store), the server-side processing incurs a delay of just over 2 seconds. For the client-only service latency, we notice an exponential increase as the performance of the client in relation to the server diminishes. At $\alpha = 0.2$, the client requires almost a 12-fold increase to process the image. As outlined in the motivation in Section 2.3, we employ this as a comparison point to the server for the compute-and-forward scenario. The compute-and-forward case provides a total service latency that is just below that of the client having the full server resources itself. We additionally notice from the table that the median and average are fairly close to another, with generally less than one percent difference. A visual comparison of these three service approaches is illustrated in Figure 7.

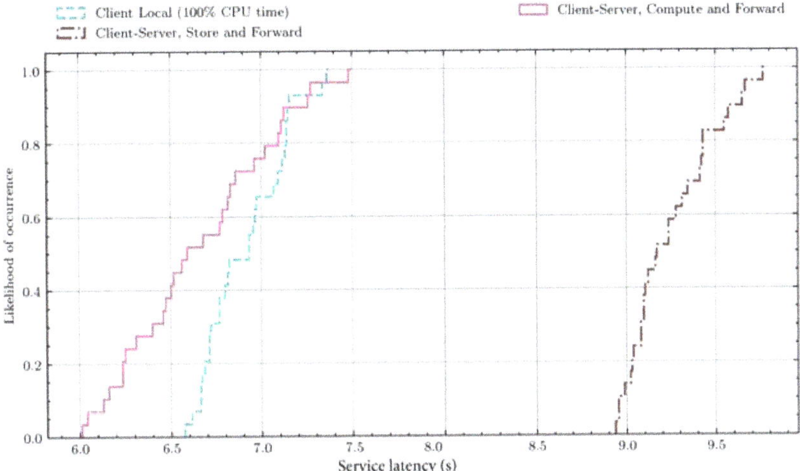

Figure 7. Service latency likelihood for YOLOv2 performed on-device only (with device computational resources equal to server-side resources, $\alpha = 1$), store-and-forward networking with server-side processing only, and compute-and-forward with $\alpha = 0.2$ client-side processing up to layer 8 of YOLOv2 and remainder processing server-side.

We observe that the store-and-forward approach is in this comparison not desirable at all, as it exhibits the highest service latency. The comparison of an assumed full server-level CPU performance on the client side with the compute-and-forward approach with only 20% server-side equivalent resources on the client side showcases a significant overlap in service time distributions. Particularly, we notice that 50% of the compute-and-forward latency times observed are lower than any local processing, while the remaining 50% are spread over the entire client-side processing range. In comparison, the store-and-forward approach yields a lower spread of latency values and is more comparable to the on-client processing in this regard.

We now consider the impact of different local processing capabilities of the client in comparison to the server. We illustrate the outcomes for the overall service latency for different client computational resources in Figure 8. We initially note the increase in service latency as the evaluation moves from compute-and-forward over store-and-forward to the scenario of $\alpha = 0.5$ in Table 2, assuming the client's processing resources are 50% of the server resources. We observe that the visual difference to the other two server-side approaches is significant. We additionally observe the continuous increase of service latency to the $\alpha = 0.2$ case, which is the alternative to the compute-and-forward case and showcases the immense benefit that can be obtained from our described approach visually. Overall, we derive that the split between in-network processing and server-side processing heavily favors the service function split, especially for scenarios where clients have low computational resources when compared to available server-side resources.

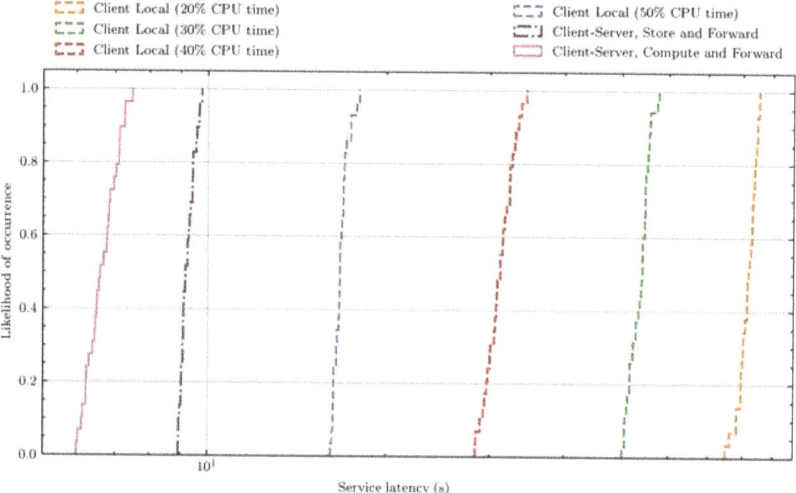

Figure 8. Overall service latency times for YOLOv2 object detection for on-client (with client computation resources equal to 20–50% server resources), traditional store-and-forward of image data to the server for object detection, and service split between in-network computing and forwarding to server.

4. Discussion

Overall, our results are indicative of significant service latency reductions that can be attained through splitting the inference workload in the multi-layer YOLOv2 object recognition model. Some of our results have show an increasing spread across service latency values, especially in scenarios where the client has only smaller fractional CPU times. This spread can be attributed to the increased burden on the CPU of performing multiple operations and the overhead, especially when considering the computational burden of the various layers in the YOLOv2 model. It is particularly noteworthy that the emulation framework employed (ComNetsEmu) was not designed for ultra-low latency usage and is originally a prototyping and teaching tool and we expect additional gains can be realized when implementing our approach on production-level systems.

We note that our assumptions were based around similar architectures employed on client and server implementations here, which could be even further abstracted across different platforms and, most importantly, through the utilization of GPUs on the server side rather than the CPU-driven approach we are evaluating here. Indeed, the comparison between server and client is based on a generic viewpoint and does not account for potential additional gains due to parallel processing and multi-threading. Significant increases in server core density also will increase the potential for the server side having significantly more computational resources available for bursty operations such as individual image operations even without GPUs.

Indeed, moving into ultra-low latency application scenarios will require changes to the current approach to networked services, such as with a ChAin-based Low latency VNF ImplemeNtation (CALVIN) [35], which significantly reduced processing times at the network's MEC. While negative effects can result [36], we showcased that in the generic scenario we considered this was not the case. While commonly, specific hardware is required to provide speed-up factors for learning, not inference, recent research has also evaluated the possibility to employ commodity hardware for these scenarios [37,38]. Specifically, in [39], the authors were able to achieve a throughput of 19 decisions per second for autonomous line following on a smart network interface. While the task at hand is different, the overall concept of offloading potrtions or all of the computer vision tasks into the network is similar.

Ongoing research takes place that continues on the various facets of object detection mechanisms as well – in our context with continuous upgrades of the YOLO model. In [40], the authors describe and improve upon YOLOv3 for the outlined significant ITS scenario. They derive processing times of just below 10 ms, which reaches service latency levels that are suitable for real-time object detection. Indeed, the interest for improvement and implementation for YOLO at the network edge is continuously attracting research interest [41–43] to improve upon the continuously developed YOLO, including hardware implementations [44]. Comparing these optimized approaches to our evaluation base don CPU processing alone is limited, as mostly, GPU or specialized hardware is employed for this type of task. In turn, our results can be seen as a ceiling evaluation of the resulting service latency for cases where no specialized hardware is available and processing needs to be performed on the CPU.

5. Conclusions

There will be an increased need for object detection as well as other machine learning-based approaches that are performed in a low-latency fashion in future application scenarios. For example, future Intelligent Transport Systems (ITS) will rely on pedestrian and car detection mechanisms to avoid loss of life and damage to property. Similarly, in connected autonomous driving, an object detection service is helpful for decision-making, such as for braking and obstacle avoidance. In the driver view, for example, object detection services can help the car to protect vulnerable road users (VRUs) such pedestrians and bicycles as we originally illustrated in Figure 1b.

Approaches that rely on machine learning commonly require significant processing, which is not always available on device, but becomes available in the softwarized 5G and beyond cellular networks. We present an approach to implement a service that splits the traditional YOLOv2 model between an on-device client and centralized server component by performing only the initial layers' processing on the client and the remainder on the server. Comparing our approach with traditional on-client and on-server processing with varying degrees of client computational resources, we find that a 12-fold reduction of the service latency can be achieved when the client has 20% of the server's resources—a scenario we deem likely in future connected device scenarios, especially for battery-limited devices.

The approach to split the intermediate results in systems incorporating neural network layers is not limited to object recognition tasks alone, but can be applied for all such systems. The increased embedding of AI approaches in modern networked systems provides broad opportunities to employ approaches such as ours to improve service levels and decrease their latency times. A particularly interesting future avenue here would be the reliance on partially pre-determined outcomes from prior cached results for distributed edge systems.

Another venue currently under consideration is the combination of the service function split we showcased here together with network coding.

Author Contributions: Conceptualization, Z.X., P.S., and F.H.P.F.; methodology, Z.X., P.S., and F.H.P.F.; software, Z.X.; validation, Z.X. and P.S.; formal analysis, Z.X.; investigation, Z.X.; resources, F.H.P.F.; data curation, Z.X.; writing—original draft preparation, Z.X.; writing—review and editing, Z.X., P.S., and F.H.P.F.; visualization, Z.X.; supervision, F.H.P.F.; project administration, F.H.P.F.; funding acquisition, F.H.P.F. All authors have read and agreed to the published version of the manuscript.

Funding: Partially funded by the German Research Foundation (DFG, Deutsche Forschungsgemeinschaft) as part of Germany's Excellence Strategy–EXC 2050/1–Project ID 390696704 –Cluster of Excellence "Centre for Tactile Internet with Human-in-the-Loop" (CeTI) of Technische Universität Dresden.

Conflicts of Interest: The authors declare no conflict of interest. The funders had no role in the design of the study; in the collection, analyses, or interpretation of data; in the writing of the manuscript, or in the decision to publish the results.

Abbreviations

The following abbreviations are used in this manuscript:

5G	Fifth-Generation Cellular Networks
AI	Artificial Intelligence
CNN	Convolutional Neural Network
COIN	COmputing In the Network
CPU	Central Processing Unit
CV	Computer Vision
DL	Deep Learning
FPGA	Field-Programmable Gate Array
GPU	Graphics Processing Unit
IP	Internet Protocol
ITS	Intelligent Transport System
JPEG	Joint Photographic Experts Group
LIDAR	Light Detection and Ranging
MEC	Mobile Edge Cloud
NFV	Network Function Virtualization
RAM	Random Access Memory
ReLU	Rectified Linear Unit
SDN	Software-Defined Network
SFC	Service Function Chaining
UDP	User Datagram Protocol
VNF	Virtual Network Function
VRU	Vulnerable Road User
YOLO	You Look Only Once
WebP	Web Picture

References

1. CISCO. VNI Global Fixed and Mobile Internet Traffic Forecasts. Available online: https://www.cisco.com/c/en/us/solutions/service-provider/visual-networking-index-vni/index.html (accessed on 28 February 2021).
2. Kim, J.; Cho, J. Exploring a Multimodal Mixture-Of-YOLOs Framework for Advanced Real-Time Object Detection. *Appl. Sci.* **2020**, *10*, 612. [CrossRef]
3. Yoon, C.S.; Jung, H.S.; Park, J.W.; Lee, H.G.; Yun, C.H.; Lee, Y.W. A Cloud-Based UTOPIA Smart Video Surveillance System for Smart Cities. *Appl. Sci.* **2020**, *10*, 6572. [CrossRef]
4. Mandal, V.; Mussah, A.R.; Jin, P.; Adu-Gyamfi, Y. Artificial Intelligence-Enabled Traffic Monitoring System. *Sustainability* **2020**, *12*, 9177. [CrossRef]

5. Wei, P.; Shi, H.; Yang, J.; Qian, J.; Ji, Y.; Jiang, X. City-Scale Vehicle Tracking and Traffic Flow Estimation Using Low Frame-Rate Traffic Cameras. In *Adjunct Proceedings of the 2019 ACM International Joint Conference on Pervasive and Ubiquitous Computing and Proceedings of the 2019 ACM International Symposium on Wearable Computers*; UbiComp/ISWC '19 Adjunct; Association for Computing Machinery: New York, NY, USA, 2019; pp. 602–610. [CrossRef]

6. Yang, W.; Zhang, X.; Lei, Q.; Shen, D.; Xiao, P.; Huang, Y. Lane Position Detection Based on Long Short-Term Memory (LSTM). *Sensors* **2020**, *20*, 3115. [CrossRef]

7. Kim, W.; Cho, H.; Kim, J.; Kim, B.; Lee, S. YOLO-Based Simultaneous Target Detection and Classification in Automotive FMCW Radar Systems. *Sensors* **2020**, *20*, 2897. [CrossRef]

8. Castelló, V.O.; del Tejo Catalá, O.; Igual, I.S.; Perez-Cortes, J.C. Real-time on-board pedestrian detection using generic single-stage algorithms and on-road databases. *Int. J. Adv. Robot. Syst.* **2020**, *17*, 1729881420929175. [CrossRef]

9. Dominguez-Sanchez, A.; Cazorla, M.; Orts-Escolano, S. Pedestrian Movement Direction Recognition Using Convolutional Neural Networks. *IEEE Trans. Intell. Transp. Syst.* **2017**, *18*, 3540–3548. [CrossRef]

10. Hui, J. Real-time Object Detection with YOLO, YOLOv2 and now YOLOv3. Available online: https://medium.com/@jonathan_hui/real-time-object-detection-with-yolo-yolov2-28b1b93e2088 (accessed on 28 February 2021).

11. Sze, V.; Chen, Y.; Yang, T.; Emer, J.S. Efficient Processing of Deep Neural Networks: A Tutorial and Survey. *Proc. IEEE* **2017**, *105*, 2295–2329. [CrossRef]

12. Gu, J.; Wang, Z.; Kuen, J.; Ma, L.; Shahroudy, A.; Shuai, B.; Liu, T.; Wang, X.; Wang, G.; Cai, J.; et al. Recent advances in convolutional neural networks. *Pattern Recognit.* **2018**, *77*, 354–377. doi:10.1016/j.patcog.2017.10.013. [CrossRef]

13. Girshick, R.B.; Donahue, J.; Darrell, T.; Malik, J. Rich feature hierarchies for accurate object detection and semantic segmentation. In Proceedings of the IEEE Conference on Computer Vision and Pattern Recognition (CVPR), Columbus, OH, USA, 23–28 June 2014; pp. 580–587.

14. Ren, S.; He, K.; Girshick, R.B.; Sun, J. Faster R-CNN: Towards Real-Time Object Detection with Region Proposal Networks. *arXiv* **2015**, arXiv:1506.01497.

15. Redmon, J.; Divvala, S.K.; Girshick, R.B.; Farhadi, A. You Only Look Once: Unified, Real-Time Object Detection. *arXiv* **2015**, arXiv:1506.02640.

16. Lin, L.; Liao, X.; Jin, H.; Li, P. Computation Offloading Toward Edge Computing. *Proc. IEEE* **2019**, *107*, 1584–1607. [CrossRef]

17. Melendez, S.; McGarry, M.P. Computation offloading decisions for reducing completion time. In Proceedings of the 2017 14th IEEE Annual Consumer Communications Networking Conference (CCNC), Las Vegas, NV, USA, 8–11 January 2017; pp. 160–164. [CrossRef]

18. Abbas, N.; Zhang, Y.; Taherkordi, A.; Skeie, T. Mobile Edge Computing: A Survey. *IEEE Internet Things J.* **2018**, *5*, 450–465. [CrossRef]

19. Taleb, T.; Samdanis, K.; Mada, B.; Flinck, H.; Dutta, S.; Sabella, D. On Multi-Access Edge Computing: A Survey of the Emerging 5G Network Edge Cloud Architecture and Orchestration. *IEEE Commun. Surv. Tutor.* **2017**, *19*, 1657–1681. [CrossRef]

20. Haleplidis, E.; Pentikousis, K.; Denazis, S.; Salim, J.H.; Meyer, D.; Koufopavlou, O. Software-Defined Networking (SDN): Layers and Architecture Terminology. RFC 7426, RFC Editor. 2015. Available online: http://www.rfc-editor.org/rfc/rfc7426.txt (accessed on 28 February 2021).

21. Duan, Q.; Ansari, N.; Toy, M. Software-defined network virtualization: An architectural framework for integrating SDN and NFV for service provisioning in future networks. *IEEE Netw.* **2016**, *30*, 10–16. [CrossRef]

22. Intel. Internet Engineering Task Force (IETF). Available online: https://tools.ietf.org/html/rfc7665 (accessed on 28 February 2021).

23. Doan, T.V.; Fan, Z.; Nguyen, G.T.; You, D.; Kropp, A.; Salah, H.; Fitzek, F.H.P. Seamless Service Migration Framework for Autonomous Driving in Mobile Edge Cloud. In Proceedings of the 2020 IEEE 17th Annual Consumer Communications Networking Conference (CCNC), Las Vegas, NV, USA, 10–13 January 2020; pp. 1–2. [CrossRef]

24. Krizhevsky, A.; Sutskever, I.; Hinton, G.E. ImageNet Classification with Deep Convolutional Neural Networks. In *Advances in Neural Information Processing Systems 25*; Pereira, F., Burges, C.J.C., Bottou, L., Weinberger, K.Q., Eds.; Curran Associates, Inc.: Red Hook, NY, USA, 2012; pp. 1097–1105.

25. Tensorflow Official Website. Available online: https://www.tensorflow.org (accessed on 15 December 2020).

26. Redmon, J.; Farhadi, A. YOLO9000: Better, Faster, Stronger. In Proceedings of the 2017 IEEE Conference on Computer Vision and Pattern Recognition (CVPR), Honolulu, HI, USA, 21–27 July 2017; pp. 6517–6525. [CrossRef]

27. Liu, W.; Anguelov, D.; Erhan, D.; Szegedy, C.; Reed, S.; Fu, C.Y.; Berg, A.C. SSD: Single Shot MultiBox Detector. In *Computer Vision—ECCV 2016*; Leibe, B., Matas, J., Sebe, N., Welling, M., Eds.; Springer International Publishing: Cham, Switzerland, 2016; pp. 21–37.

28. Liu, S.; Deng, W. Very deep convolutional neural network based image classification using small training sample size. In Proceedings of the 2015 3rd IAPR Asian Conference on Pattern Recognition (ACPR), Kuala Lumpur, Malaysia, 3–6 November 2015; pp. 730–734. [CrossRef]

29. Xiang, Z.; Zhang, R.; Seeling, P. Chapter 19—Machine learning for object detection. In *Computing in Communication Networks*; Fitzek, F.H., Granelli, F., Seeling, P., Eds.; Elsevier/Academic Press: Cambridge, MA, USA, 2020; pp. 325–338. [CrossRef]

30. Lin, T.; Maire, M.; Belongie, S.J.; Bourdev, L.D.; Girshick, R.B.; Hays, J.; Perona, P.; Ramanan, D.; Dollár, P.; Zitnick, C.L. Microsoft COCO: Common Objects in Context. In *European Conference on Computer Vision*; Springer: Cham, Switzerland, 2014.

31. Xiang, Z.; Pandi, S.; Cabrera, J.; Granelli, F.; Seeling, P.; Fitzek, F.H.P. An Open Source Testbed for Virtualized Communication Networks. *IEEE Commun. Mag.* **2021**, 1–7. in print
32. ComNetsEmu Public Repository. 2020. Available online: https://git.comnets.net/public-repo/comnetsemu (accessed on 28 February 2021).
33. (EMBC), E.M.B.C. CoreMark CPU Benchmark Scores. Available online: https://www.eembc.org/coremark/ (accessed on 28 February 2021).
34. Software, P. PassMark CPU Benchmark Datasets. Available online: https://www.cpubenchmark.net/ (accessed on 28 February 2021).
35. Xiang, Z.; Gabriel, F.; Urbano, E.; Nguyen, G.T.; Reisslein, M.; Fitzek, F.H.P. Reducing Latency in Virtual Machines: Enabling Tactile Internet for Human-Machine Co-Working. *IEEE J. Sel. Areas Commun.* **2019**, *37*, 1098–1116. [CrossRef]
36. Yang, F.; Wang, Z.; Ma, X.; Yuan, G.; An, X. Understanding the Performance of In-Network Computing: A Case Study. In Proceedings of the 2019 IEEE Intl Conf on Parallel Distributed Processing with Applications, Big Data Cloud Computing, Sustainable Computing Communications, Social Computing Networking (ISPA/BDCloud/SocialCom/SustainCom), Xiamen, China, 16–18 December 2019; pp. 26–35. [CrossRef]
37. Xiong, Z.; Zilberman, N. Do Switches Dream of Machine Learning? Toward In-Network Classification. In *Proceedings of the 18th ACM Workshop on Hot Topics in Networks*; HotNets '19; Association for Computing Machinery: New York, NY, USA, 2019; pp. 25–33. [CrossRef]
38. Sanvito, D.; Siracusano, G.; Bifulco, R. Can the Network Be the AI Accelerator? In *Proceedings of the 2018 Morning Workshop on In-Network Computing*; NetCompute '18; Association for Computing Machinery: New York, NY, USA, 2018; pp. 20–25. [CrossRef]
39. Glebke, R.; Krude, J.; Kunze, I.; Rüth, J.; Senger, F.; Wehrle, K. Towards Executing Computer Vision Functionality on Programmable Network Devices. In *Proceedings of the 1st ACM CoNEXT Workshop on Emerging In-Network Computing Paradigms*; ENCP '19; Association for Computing Machinery: New York, NY, USA, 2019; pp. 15–20. [CrossRef]
40. Cao, J.; Song, C.; Peng, S.; Song, S.; Zhang, X.; Shao, Y.; Xiao, F. Pedestrian Detection Algorithm for Intelligent Vehicles in Complex Scenarios. *Sensors* **2020**, *20*, 3646. [CrossRef] [PubMed]
41. Han, B.G.; Lee, J.G.; Lim, K.T.; Choi, D.H. Design of a Scalable and Fast YOLO for Edge-Computing Devices. *Sensors* **2020**, *20*, 6779. [CrossRef]
42. Zhao, H.; Zhou, Y.; Zhang, L.; Peng, Y.; Hu, X.; Peng, H.; Cai, X. Mixed YOLOv3-LITE: A Lightweight Real-Time Object Detection Method. *Sensors* **2020**, *20*, 1861. [CrossRef]
43. Yang, Y.; Deng, H. GC-YOLOv3: You Only Look Once with Global Context Block. *Electronics* **2020**, *9*, 1235. 081235. [CrossRef]
44. Wang, Z.; Xu, K.; Wu, S.; Liu, L.; Liu, L.; Wang, D. Sparse-YOLO: Hardware/Software Co-Design of an FPGA Accelerator for YOLOv2. *IEEE Access* **2020**, *8*, 116569–116585. [CrossRef]

Article

Power Allocation for Secrecy-Capacity-Optimization-Artificial-Noise Secure MIMO Precoding Systems under Perfect and Imperfect Channel State Information

Yebo Gu [1], Bowen Huang [2] and Zhilu Wu [1,*]

[1] School of Electronics and Information Engineering, Harbin Institute of Technology, Harbin 150001, China; 16B305002@hit.edu.cn
[2] JushriTechnologies Inc., Shanghai 200335, China; bowen.huang@jushri.com
* Correspondence: wuzhilu@hit.edu.cn

Citation: Gu, Y.; Huang, B.; Wu, Z. Power Allocation for Secrecy-Capacity-Optimization-Artificial-Noise Secure MIMO Precoding Systems under Perfect and Imperfect Channel State Information. *Appl. Sci.* **2021**, *11*, 4558. https://doi.org/10.3390/app11104558

Academic Editor: Cheonshik Kim

Received: 2 April 2021
Accepted: 13 May 2021
Published: 17 May 2021

Publisher's Note: MDPI stays neutral with regard to jurisdictional claims in published maps and institutional affiliations.

Abstract: In this paper, we consider the physical layer security problem of the wireless communication system. For the multiple-input, multiple-output (MIMO) wireless communication system, secrecy capacity optimization artificial noise (SCO−AN) is introduced and studied. Unlike its traditional counterpart, SCO−AN is an artificial noise located in the range space of the channel state information space and thus results in a significant increase in the secrecy capacity. Due to the limitation of transmission power, making rational use of this power is crucial to effectively increase the secrecy capacity. Hence, in this paper, the objective function of transmission power allocation is constructed. We also consider the imperfect channel estimation in the power allocation problems. In traditional AN research conducted in the past, the expression of the imperfect channel estimation effect was left unknown. Still, the extent to which the channel estimation error impacts the accuracy of secrecy capacity computation is not negligible. We derive the expression of channel estimation error for least square (LS) and minimum mean squared error (MMSE) channel estimation. The objective function for transmission power allocation is non-convex. That is, the traditional gradient method cannot be used to solve this non-convex optimization problem of power allocation. An improved sequence quadratic program (ISQP) is therefore applied to solve this optimization problem. The numerical result shows that the ISQP is better than other algorithms, and the power allocation as derived from ISQP significantly increases secrecy capacity.

Keywords: physical layer security; secure transmission; secrecy capacity; secrecy capacity optimization artificial noise; power allocation; channel estimation error

1. Introduction

Secure transmission is a fundamental problem in wireless communications due to the broadcast nature of the wireless medium. Along with the rapid advancement of information technology, the higher information transmission rate has called for a stricter standard of information transmission security. For a long time, the primary method of guaranteeing the secure transmission of information has been via encryption technology. Encryption technology utilizes the limitation in computing speed to prevent the eavesdropper from deciphering all encrypted information in a limited time. However, as computer technology advances with faster computation, the decryption of information becomes more straightforward. In theory, no encrypted information is indecipherable if the computer's calculation speed is fast enough. This indeed is the inherent flaw in the current information encryption technology. Therefore, the physical layer security technology has been proposed to solve the problems of secure information transmission.

The physical layer security technology differs substantially from the information encryption technology. Unlike encryption technology, which relies on the limitation in computation speed, the physical layer security technology has its basis in the randomness of the wireless communication channel. The physical layer security technology tries to prevent eavesdroppers from decoding information, regardless of the amount of time or the computing speed. One of the most innovative physical layer security technologies is artificial noise (AN). AN adds extra noise to the information. This noise solely impacts the eavesdropper's channel but does not affect the legitimate receiver channel. That is, only the signal received by the eavesdropper is reduced in this method. The effectiveness of the physical layer technology is then evaluated by secrecy capacity.

The study of physical layer security begins from [1]. This paper proposes unconditional secure transmission as the ultimate goal of physical layer security technology study.

After [1,2] is the first paper to study the secure transmission of information from the perspective of information theory. In [2], wiretap communication model with the eavesdropping channel is proposed, and the aforementioned secrecy capacity was also first proposed in this paper. Paper [3] studies the physical layer security technology based on [2]. In [3], a broadcast channel model with confidential messages is proposed to extend Wyner's work.

Currently, the physical layer security technology has not been at the center of public attention, primarily due to a strict restriction that the eavesdropper's channel must be strictly worse than the legitimate channel. Considering the following cases: the eavesdropper is closer to the transmitter, or the eavesdropper has more antennas than the transmitter. These mentioned conditions will make the eavesdropper's channel better than the legitimate channel and thus reduces the effectiveness of the physical layer security technology.

To help with the issue above, AN technology is introduced. The proposal of AN technology reduces the difficulty of applying the physical layer security technology in the multiple-input, multiple-output (MIMO) communication system [4]. AN is in the null space of the legitimate channel, which mean the legitimate channel is not affected. There is no need to employ any additional signal processing device to the legitimate receiver. Meanwhile, the eavesdropper's channel capacity is reduced significantly. To show this result quantitatively, let A denote the channel capacity of the legitimate receiver and B denote the channel capacity of the eavesdropper. The principle of AN is to increase the difference $A - B$ by reducing B and keeping A constant.

There have been many outstanding works in the realm of AN technology. In [5,6], AN and the interference alignment technology are creatively merged to introduce AN featuring interference alignment. In [7], the lower bound on the secrecy capacity of artificial noise wireless communication systems subject to transmit power is proposed. Ref. [8] proposes the secrecy capacity expression with imperfect channel estimation. This expression is non-convex, so the gradient descent method cannot be used for this optimization problem. Therefore, it is impossible to get the optimal solution of the secrecy capacity expression. The study in [9–12] consider the effects of active eavesdropper. The active eavesdropper can interfere with pilot to reduce the secrecy capacity of the wire-tap system. This is something that hasn't been explored in previous studies.

The past research on AN is summarized into two main aspects:

(1)   Research on AN noise technology under different communication modes [13–21]: examples include the AN power allocation problem in OFDM, GSM, and other communication modes [22] and the application of AN under intelligent reflecting surface [23]. The simplified communication model is $Y = HX + e$, where Y denotes the received signal, H denotes the channel, X denotes the transmitted signal, and e is the noise. The above researches focus on "H".

(2)   Reshaping certain features of AN. For example, Ref. [24] designs an artificial noise that has interference alignment characteristics. The research focused on "X" from the equation above [25–28].

Still, there has been little to no research attention on redesigning the core of AN. Therefore, our research focus on creating a new kind of AN. Our research shows that our new artificial noise has a better performance compared to its traditional counterpart.

In [29], the secrecy capacity optimization artificial noise (SCO−AN) is proposed. The core of AN technology is to design a noise in the null space of the channel state information space. Unlike the traditional AN, which ignores the range space of the channel state information space, SCO−AN is located in that range space. While SCO−AN may slightly impact the channel capacity of the legitimate receiver, SCO−AN significantly reduces the channel capacity of the eavesdropper. Therefore, this method still increases the difference between the legitimate channel capacity and the eavesdropping channel capacity. SCO−AN is a tool to convert the noise immunity of communication systems into secrecy capacity.

As there is a limitation in the transmission power, it is critical to draw an optimization problem to maximize the secrecy capacity under that limitation. The power allocation problem becomes essential. Therefore, in this paper, we study the power allocation problem of SCO−AN. The Hessian matrix of the SCO−AN power allocation objective function is not positive definite, which means the objective function is non-convex. The maximum value of the SCO−AN power allocation function cannot be obtained by the gradient descent method. An improved sequential quadratic programming (ISQP) is proposed to solve this problem. With the effects of imperfect channel estimation considered, the objective power allocation function containing imperfect channel estimation parameters is constructed.

The main contributions of this paper are summarized as follows:

(1) In reality, the secrecy capacity of a wireless communication system using SCO−AN is limited by transmission power. Considering this limitation, this paper constructs a power distribution function for SCO−AN and the information-bearing signal.

(2) Since the power allocation objective function is non-convex, it is difficult to optimize the power distribution function using a power optimization scheme based on gradient descent. ISQP is then proposed to allocate power between SCO−AN and the information-bearing signal. ISQP improves the traditional iterative algorithm and reduces the computational complexity by simplifying the initial iterative matrix and improving computational efficiency.

(3) Due to the influence of Gaussian white noise in the channel, there is an error in the channel estimation, resulting in an error in the SCO−AN design. The channel estimation error affects the accuracy of the power allocation optimization. This paper considers the imperfect channel state information for power allocation. The power allocation objective function of SCO−AN and the information-bearing signal containing channel estimation errors is constructed. The expression for the channel estimation errors is derived for the first time. This expression can then be applied to future physical layer security research examining imperfect channel estimation.The power allocation function is then converted to a function with only one variable–the SCO−AN–simplifying the function's overall computational complexity.

This paper is structured as follows:

- In Section 2, the system model and the framework are introduced.
- In Section 3, the objective function for the power allocation between SCO−AN and the information-bearing signal, with and without considering imperfect channel estimation, is proposed. ISQP is then applied to optimize the power allocation. The algorithm flow of ISQP algorithm is constructed.
- In Section 4, simulation results are shown and discussed.
- In Section 5, the conclusion is drawn, and the suggestions for future work are presented.

In this paper, the following notations are used: Boldface upper case denotes matrices, boldface lower case denotes vectors, italics case denotes numbers; $[\cdot]^T$ denotes the matrix transpose operation; $[\cdot]^*$ denotes the complex conjugate operation; $[\cdot]^{\dagger}$ denotes the conjugate transpose operation (conjugate complex number) for the matrix (number) "$\cdot$"; $E\{\cdot\}$ denotes the mathematical expectation; $\|\cdot\|$ denotes the norm of a vector; and $|\cdot|$ denotes the determinant of a matrix.

2. Related Work and System Model

2.1. Related Work–Wireless Communication Model with Eavesdroppers

In this section, we review the artificial noise technology and the method of SCO−AN. Moreover, the effects of imperfect channel estimation are analyzed in detail.

Figure 1 shows a wireless communication system model with an eavesdropper. In this model, Alice is the transmitter of the message, Bob is the legitimate receiver, and Eve is the eavesdropper. Alice has N_A antennas, Bob has N_B antennas and Eve has N_E antennas. **H** represents the channel state information (CSI) of the legitimate channel (Alice to Bob), while **G** represents the CSI of the eavesdropper channel (Alice to Eve). $\mathbf{H}_k$ and $\mathbf{G}_k$ represent the CSI of **H** and **G** at time k respectively. The element $h_{i,j}$ (or $g_{i,j}$) in **H** (or **G**) is the channel gain coefficient between the i_{th} transmitter antenna and the j_{th} receiver's (or eavesdropper's) antenna. $\mathbf{x}_k \in \mathbb{C}^{N_A}$ represents the signal transmitted by Alice at time k; $\mathbf{y}_k \in \mathbb{C}^{N_E}$ represents the signal received by Bob at time k; and $\mathbf{z}_k \in \mathbb{C}^{N_B}$ represents the signal received by Eve at time k.

$$\mathbf{z}_k = \mathbf{H}_k\mathbf{x}_k + \mathbf{n}_k, \tag{1}$$

$$\mathbf{y}_k = \mathbf{G}_k\mathbf{x}_k + \mathbf{e}_k, \tag{2}$$

where $\mathbf{n}_k$ and $\mathbf{e}_k$ are independent and identically distributed (i.i.d) additive Gaussian white noise (AGWN) with the variance of σ_n^2 and σ_e^2 respectively. For the convenience of discussion, we assume that the CSI of **G** and **H** can be obtained by Alice without delay. The maximum transmitting power is assumed to be P, where $E[\mathbf{x}_k^\dagger\mathbf{x}_k] \leq P$.

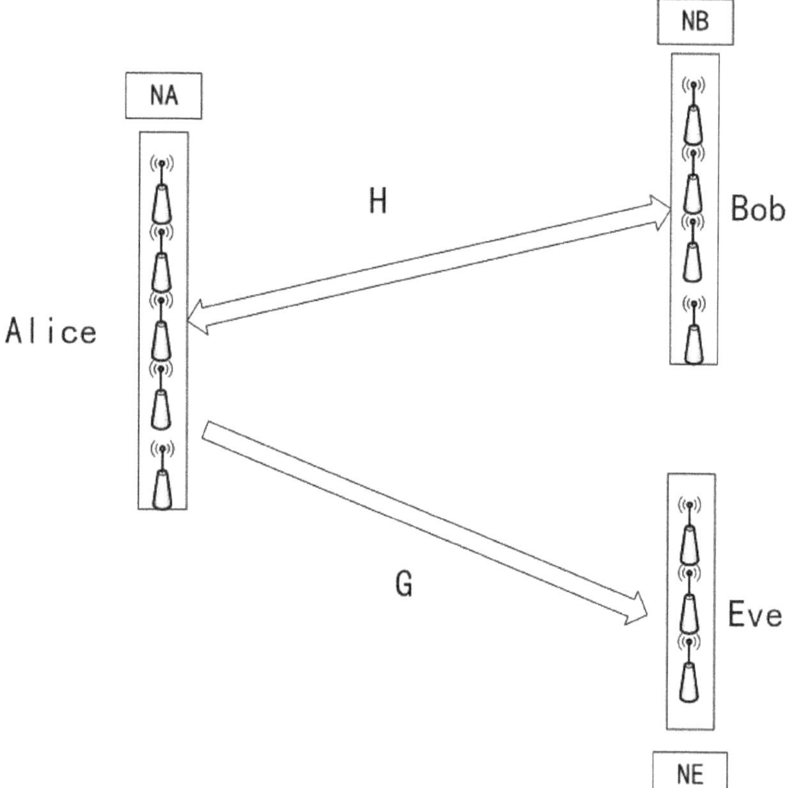

Figure 1. Wireless communication model with eavesdropper.

2.2. Related Work–The Artificial Noise

Located in the null space of legitimate channel (i.e., Bob's channel), AN does not affect Bob's reception of information. For Eve, however, AN reduces Eve's channel capacity significantly. Alice sends AN simultaneously while sending the information-bearing signal; that is,

$$\mathbf{x}_k = \mathbf{w}_k + \mathbf{s}_k, \tag{3}$$

In (3), $\mathbf{w}_k \in \mathbb{C}^{N_A}$ denotes AN; $\mathbf{s}_k \in \mathbb{C}^{N_A}$ denotes the information-bearing signal; and $\mathbf{w}_k$ is artificial noise, which is located in the null space of $\mathbf{H}_k$, such that $\mathbf{H}_k \mathbf{w}_k = 0$. Let $\mathbf{Z}_k$ be a standard orthonormal basis for $\mathbf{H}_k$ and $\mathbf{v}_k$ be a complex random variable with the variance of σ_v^2 such that $\mathbf{w}_k = \mathbf{Z}_k \mathbf{v}_k$ and $\mathbf{Z}_k^\dagger \mathbf{Z}_k = \mathbf{I}$. Then, the signals received by Bob and Eve are:

$$\begin{aligned} \mathbf{z}_k &= \mathbf{H}_k \mathbf{x}_k + \mathbf{n}_k \\ &= \mathbf{H}_k \mathbf{w}_k + \mathbf{H}_k \mathbf{s}_k + \mathbf{n}_k \\ &= \mathbf{H}_k \mathbf{s}_k + \mathbf{n}_k, \end{aligned} \tag{4}$$

$$\mathbf{y}_k = \mathbf{G}_k \mathbf{s}_k + \mathbf{G}_k \mathbf{w}_k + \mathbf{e}_k, \tag{5}$$

where $\mathbf{y}_k$ is the signal received by Eve, and $\mathbf{z}_k$ is the signal received by Bob. $\mathbf{y}_k$ and $\mathbf{z}_k$ are Gaussian vectors. As $\mathbf{w}_k$ is in the null space of $\mathbf{H}_k$, we have $\mathbf{H}_k \mathbf{w}_k = 0$ and the term with $\mathbf{w}_k$ vanishes in (4). That is, the artificial noise does not impact Bob, while Eve is affected.

In [4], the transmitted signal is chosen as $\mathbf{s}_k = \mathbf{p}_k \mathbf{u}_k$, where $\mathbf{u}_k$ is the information signal with the variance of σ_u^2 and $\mathbf{p}_k$ obeys the independent Gaussian distribution. Here, $\mathbf{p}_k$ is chosen such that: (a) $\mathbf{H}_k \mathbf{p}_k \neq 1$, and (b) $\|\mathbf{p}_k\| = 1$.

In [4], Goel considers two scenarios:

(a)   A single-input, single-output (SISO) wireless communication system where the transmitter, the receiver, and the eavesdropper equip one antenna each, i.e., $N_A = N_B = N_R = 1$; and

(b)   A MIMO wireless communication system where the the the transmitter, the receiver, and the eavesdropper each equip multiple antennas, i.e., $N_A = N_B = N_R > 1$.

For scenario a, the variables in (4)–(6) are Gaussian complex variables. $\log_e(*)$ is used to calculate entropy, so the lower bound on secrecy capacity after adding artificial noise is given by:

$$\begin{aligned} C_{sec}^a &= I(Z;S) - I(Y;S) \\ &= \log\left(1 + \frac{|H_k p_k|^2 \sigma_u^2}{\sigma_n^2}\right) - \log\left(1 + \frac{|G_k p_k|^2 \sigma_u^2}{E|G_k w_k|^2 + \sigma_e^2}\right), \end{aligned} \tag{6}$$

where $E|G_k w_k|^2 = (G_k Z_k Z_k^\dagger G_k^\dagger)\sigma_k^2$. C_{sec}^a denotes the secrecy capacity after adding artificial noise, and $I(A;B)$ denotes mutual information entropy of A and B.

For scenario b, $\mathbf{G}_k$ and $\mathbf{H}_k$ are Gaussian complex matrices. The elements in $\mathbf{G}_k$ and $\mathbf{H}_k$ are Gaussian complex variables. The other variables in (4)–(6) are Gaussian vectors. It then follows that the lower bound on secrecy capacity after adding artificial noise is given by:

$$\begin{aligned} C_{sec}^a &= I(Z;S) - I(Y;S) \\ &= \log\left|I\sigma_n^2 + \mathbf{H}_k E[\mathbf{s}_k \mathbf{s}_k^\dagger] \mathbf{H}_k^\dagger\right| - \log\left(\frac{|\mathbf{G}_k \mathbf{Z}_k \mathbf{Z}_k^\dagger \mathbf{G}_k^\dagger \sigma_v^2 + \mathbf{G}_k E[\mathbf{s}_k \mathbf{s}_k^\dagger] \mathbf{G}_k^\dagger|}{|\mathbf{G}_k \mathbf{Z}_k \mathbf{Z}_k^\dagger \mathbf{G}_k^\dagger \sigma_v^2 + I\sigma_e^2|}\right). \end{aligned} \tag{7}$$

2.3. Related Work–SCO−AN: Perfect Channel Estimation

SCO−AN is proposed in [29]. In this section, SCO−AN is introduced in detail.

The goal of physical layer security is to maximize the secrecy capacity of a communication system. In the wireless wiretap communication model, it is not possible to increase the channel capacity of the legitimate receiver. AN is then proposed to reduce the eavesdropper's channel capacity while the legitimate receiver's channel capacity remains intact. Inspired by AN, the secrecy capacity optimization artificial noise (SCO−AN) is proposed in [29]. Unlike the traditional AN, SCO−AN has a slight impact on the legitimate receiver's channel capacity but reduces the capacity of eavesdropping channels much more significantly. Hence, the system's overall secrecy capacity increases.

Next, we compute the analytical expression of using SCO−AN, in a manner parallel to our computations of AN above. Alice adds SCO−AN to the transmission signal:

$$\mathbf{x}_k = \mathbf{w}_g + \mathbf{s}_k, \tag{8}$$

In [29], the transmitted signal is $\mathbf{s}_k = \mathbf{p}_k \mathbf{u}_k$, where $\mathbf{u}_k$ is the information-bearing signal with variance of σ_u^2 and $\mathbf{p}_k$ obeys the Gaussian distribution. $\mathbf{p}_k$ satisfies the following conditions: (a) $\mathbf{H}_k \mathbf{p}_k \neq 1$; and (b) $\|\mathbf{p}_k\| = 1$. $\mathbf{w}_g \in \mathbb{C}^{N_A}$ denotes the SCO−AN. To facilitate calculations, we assume that $\mathbf{w}_g = \mathbf{Z}_k \mathbf{v}_g$, where $\mathbf{Z}_k$ is a standard orthonormal basis of $\mathbf{H}_k$ and $\mathbf{v}_g$ is a complex random variables with variance σ_g^2. The signals received by the Bob and Eve are:

$$\mathbf{z}_k = \mathbf{H}_k \mathbf{s}_k + \mathbf{H}_k \mathbf{w}_g + \mathbf{n}_k, \tag{9}$$

$$\mathbf{y}_k = \mathbf{G}_k \mathbf{s}_k + \mathbf{G}_k \mathbf{w}_g + \mathbf{e}_k, \tag{10}$$

where $\mathbf{z}_k$ denotes the signal received by Bob and $\mathbf{y}_k$ denotes the signal received by Eve.

For the SISO wireless communication system, all the elements in (8)–(10) are complex variables. So the lower bound on secrecy capacity after adding SCO−AN is:

$$C_{\text{sec}}^g = I(Z;S) - I(Y;S)$$
$$= \log\left(1 + \frac{|H_k p_k|^2 \sigma_u^2}{E|H_k w_g|^2 + \sigma_n^2}\right) - \log\left(1 + \frac{|G_k p_k|^2 \sigma_u^2}{E|G_k w_g|^2 + \sigma_e^2}\right), \tag{11}$$

where $E|H_k w_g|^2 = (H_k Z_k Z_k^\dagger H_k^\dagger)\sigma_g^2$, and $E|G_k w_g|^2 = (G_k Z_k Z_k^\dagger G_k^\dagger)\sigma_g^2$. C_{sec}^g denotes the secrecy capacity after adding SCO−AN. In (11), C_{sec}^g is a non-convex function about σ_g^2.

For the MIMO wireless communication system, $\mathbf{H}_k$ and $\mathbf{G}_k$ are gaussian complex martixs, $\mathbf{x}_k$, $\mathbf{w}_g$, $\mathbf{s}_k$, $\mathbf{n}_k$ and $\mathbf{e}_k$ are gaussian vectors. So the lower bound on secrecy capacity after adding SCO−AN is:

$$C_{\text{sec}}^g = I(Z;S) - I(Y;S)$$
$$= \log\left(\frac{|(\mathbf{H}_k \mathbf{Z}_k \mathbf{Z}_k^\dagger \mathbf{H}_k^\dagger)\sigma_g^2 + I\sigma_n^2 + \mathbf{H}_k \mathbf{Z}_k \mathbf{Z}_k^\dagger \mathbf{H}_k^\dagger \sigma_u^2|}{|(\mathbf{H}_k \mathbf{Z}_k \mathbf{Z}_k^\dagger \mathbf{H}_k^\dagger)\sigma_g^2 + I\sigma_n^2|}\right)$$
$$- \log\left(\frac{|(\mathbf{G}_k \mathbf{Z}_k \mathbf{Z}_k^\dagger \mathbf{G}_k^\dagger)\sigma_g^2 + I\sigma_e^2 + \mathbf{G}_k \mathbf{Z}_k \mathbf{Z}_k^\dagger \mathbf{G}_k^\dagger \sigma_u^2|}{|(\mathbf{G}_k \mathbf{Z}_k \mathbf{Z}_k^\dagger \mathbf{G}_k^\dagger)\sigma_g^2 + I\sigma_e^2|}\right) \tag{12}$$

(12) is a function of σ_g^2.

For the convenience of discussion, C_{sec}^k represents the change of secrecy capacity after adding the SCO−AN when compared to simply adding traditional AN. For the case of SCO−AN, to ensure the effectiveness of physical security, (13) must be guaranteed.

$$C_{\text{sec}}^k = C_{\text{sec}}^g - C_{\text{sec}}^a > 0, \tag{13}$$

In (13), for the SISO communication system, C_{sec}^a is given by (6) and C_{sec}^g is given by (11). For the MIMO communication system, C_{sec}^a is given by (7) and C_{sec}^g is given by (12).

In Figure 2, the dashed line represents the secrecy capacity of AN calculated by (7), and the solid line is the secrecy capacity of SCO−AN calculated by (12). The legitimate channel $\mathbf{H}$ and the eavesdropper channel $\mathbf{G}$ are Rayleigh fading channels. The signal $\mathbf{x}_k$ is a complex covector. Figure 2 shows that SCO−AN provides more secrecy capacity than AN does. The noise in $\mathbf{H}$ and $\mathbf{G}$ are Gaussian white noise. The secrecy capacity increases with higher SNR.

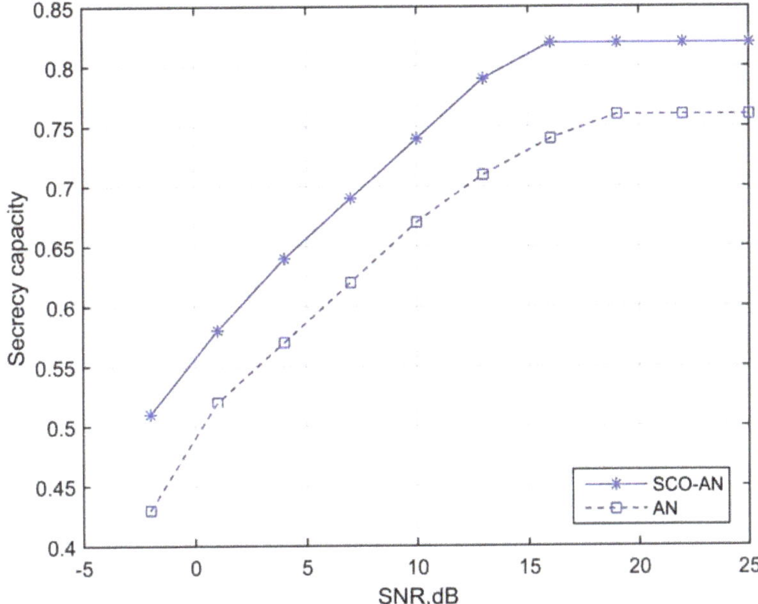

Figure 2. Secrecy capacity comparison of the AN and SCO−AN versus different SNR.

2.4. SCO−AN: Imperfect Channel Estimation

The Gaussian white noise causes the error of channel estimation. The effect of the imperfect channel estimation should be considered.

For the SISO communication system, H_{eo} denotes channel estimation error. The channel state information received by Alice is $\widetilde{H}$:

$$H_k = H_{eo} + \widetilde{H}, \tag{14}$$

The signal received by Bob after adding SCO−AN is:

$$z_k^* = (H_{eo} + \widetilde{H})s_k + (H_{eo} + \widetilde{H})w_g + n_k, \tag{15}$$

For MIMO communication system, $\mathbf{H}_{eo}$ denotes channel estimation error. The channel state information received by Alice is $\widetilde{\mathbf{H}}$:

$$\mathbf{H}_k = \mathbf{H}_{eo} + \widetilde{\mathbf{H}}, \tag{16}$$

The signal received by Bob after adding SCO−AN is:

$$\mathbf{z}_k^* = (\mathbf{H}_{eo} + \widetilde{\mathbf{H}})\mathbf{s}_k + (\mathbf{H}_{eo} + \widetilde{\mathbf{H}})\mathbf{w}_g + \mathbf{n}_k, \tag{17}$$

We assume that the channel estimation of $\mathbf{G}$ is perfect.

For the SISO communication system, H_{eo}, $\widetilde{H}$, and Z_k are independent. Therefore, $|H_{eo}Z_k|^2 = |H_{eo}|^2|Z_k|^2$, $|\tilde{H}Z_k|^2 = |\tilde{H}|^2|Z_k|^2$. The lower bound on secrecy capacity after adding SCO−AN under imperfect channel estimation is:

$$
\begin{aligned}
C^{g}_{sec,eo} &= I(Z;S) - I(Y;S) \\
&= \log\left(1 + \frac{|\tilde{H}p_k|^2\sigma_u^2}{\sigma_n^2 + |H_{eo}p_k|^2\sigma_u^2 + E|\tilde{H}w_g|^2 + E|H_{eo}w_g|^2}\right) - \log\left(1 + \frac{|G_k p_k|^2\sigma_u^2}{E|G_k w_g|^2 + \sigma_e^2}\right) \\
&= \log\left(1 + \frac{|\tilde{H}p_k|^2\sigma_u^2}{\sigma_n^2 + |H_{eo}p_k|^2\sigma_u^2 + |\tilde{H}Z_k|^2\sigma_g^2 + |H_{eo}Z_k|^2\sigma_g^2}\right) - \log\left(1 + \frac{|G_k p_k|^2\sigma_u^2}{|G_k Z_k|^2\sigma_g^2 + \sigma_e^2}\right) \\
&= \log\left(1 + \frac{|\tilde{H}p_k|^2\sigma_u^2}{\sigma_n^2 + |H_{eo}p_k|^2\sigma_u^2 + |\tilde{H}|^2|Z_k|^2\sigma_g^2 + |H_{eo}|^2|Z_k|^2\sigma_g^2}\right) - \log\left(1 + \frac{|G_k p_k|^2\sigma_u^2}{|G_k|^2|Z_k|^2\sigma_g^2 + \sigma_e^2}\right)
\end{aligned}
\tag{18}
$$

In (18), we see that the channel estimation error will affect the channel capacity of the legitimate channel. Meanwhile, the secrecy capacity of the wireless communication system is reduced.

For the MIMO system, the lower bound on secrecy capacity after adding SCO−AN under imperfect channel estimation is:

$$
\begin{aligned}
C^{g}_{sec,eo} &= I(Z;S) - I(Y;S) \\
&= \log\left(\frac{|\mathbf{K}_H + (\mathbf{H}_{eo} + \tilde{\mathbf{H}})\mathbf{Z}_k\mathbf{Z}_k^{\dagger}(\mathbf{H}_{eo} + \tilde{\mathbf{H}})^{\dagger}\sigma_u^2|}{|\mathbf{K}_H|}\right) - \log\left(\frac{|\mathbf{K}_G + \mathbf{G}_k\mathbf{Z}_k\mathbf{Z}_k^{\dagger}\mathbf{G}_k^{\dagger}\sigma_u^2|}{|\mathbf{K}_G|}\right)
\end{aligned}
\tag{19}
$$

In (19), $\mathbf{K}_H = ((\mathbf{H}_{eo} + \tilde{\mathbf{H}})\mathbf{Z}_k\mathbf{Z}_k^{\dagger}(\mathbf{H}_{eo} + \tilde{\mathbf{H}})^{\dagger})\sigma_g^2 + \mathbf{I}\sigma_n^2$ and $\mathbf{K}_G = (\mathbf{G}_k\mathbf{Z}_k\mathbf{Z}_k^{\dagger}\mathbf{G}_k^{\dagger})\sigma_g^2 + \mathbf{I}\sigma_e^2$.

2.5. Comprison of AN and SCO−AN

The artificial noise must be in the null space of the CSI matrix, this condition makes the artificial noise design very challenging. Artificial noise is the solution of homogeneous linear equations $\mathbf{H}_k\mathbf{w}_k = 0$. If the rank of the matrix $\mathbf{H}_k$ is r and the dimension is $n \times m(n \geq m)$, only when $r < m$, the homogeneous linear equation system $\mathbf{H}_k\mathbf{w}_k = 0$ has no solutions, when $r = m$, the homogeneous linear equations have only zero solutions. In the environment of natural communication, the probability of occurrence of $r = m$ is almost zero, that is to say, in the conditions of natural communication, the design of artificial noise is almost impossible.

For example, in MIMO, when the number of transmitting antennas is less than the number of eavesdropping antennas, artificial noise cannot be designed; when the number of transmitting antennas is equal to the number of eavesdropping antennas, artificial noise can be designed under the condition $|\mathbf{H}| = 0$. When the number of transmitting antennas is greater than the number of eavesdropping antennas, artificial noise cannot be designed. This is exactly the opposite of the original intention of AN. AN is designed to solve the condition that the eavesdropping channel must be a weaken version of the legitimate channel.

Therefore, the previous researches discussed some of the characteristics of AN theoretically and ignored its applicability.

For SISO, H_k is a constant and w_k is a constant as well. If $H_k W_k = 0$ has a non-zero solution, H = 0 must be guaranteed. Therefore, AN is not applicable in SISO wireless communication system.

SCO−AN is located in the range space of the legitimate CSI space, so, $\mathbf{H}_k\mathbf{w}_g \neq 0$.

There are countless non-zero solutions to $\mathbf{w}_g$, so we don't worry about to design $\mathbf{w}_g$.

We try to design AN under the condition of Rayleigh fading channels, and carry out a total of 1000 experiments, and all experiments fail. When we try to design SCO−AN, all experiments are successful.

In Table 1, we compare SCO−AN and AN in detail, and briefly summarize the characteristics and applicability of SCO−AN and AN. It can be seen that SCO−AN is better than AN in every aspect.

Table 1. Comprison of AN and SCO−AN.

Method	Design Difficulty	Secrecy Capacity Improvement	Application	Connection with H	Connection with G
AN	tough	normal	normal	in null space	in range space
SCO−AN	easy	good	good	in range space	in range space

3. Power Allocation of SCO−AN

The transmission power of the wireless communication system is limited. It is essential to allocate secrecy capacity under limited transmission power.

3.1. Objective Function of Power Allocation

3.1.1. Objective Function of Power Allocation for SISO Communication System

$\|\mathbf{p}_k\|^2=1$ and $\mathbf{Z}_k$ is a standard orthonormal basis for $\mathbf{H}_k$, which means $\mathbf{Z}_k\mathbf{Z}_k^\dagger = \mathbf{I}$ for MIMO and $Z_kZ_k^\dagger = 1$ for SISO. We assume that the transmission power is P.

$$\sigma_g^2 + \sigma_u^2 \leq P. \tag{20}$$

We use x instead of σ_u^2 and y instead of σ_g^2. The initial states of x and y are x_0 and y_0 respectively. For the SISO communication system, the secrecy capacity before power allocation is C_{sec}^0. Therefore:

$$\log\left(1 + \frac{|H_k|^2 x_0}{\sigma_n^2 + |H_k|^2 y_0}\right) - \log\left(1 + \frac{|G_k|^2 x_0}{|G_k|^2 y_0 + \sigma_e^2}\right) = C_{sec}^0. \tag{21}$$

There are no variables except x_0 and y_0 in (21). The power allocation problem of SCO−AN for SISO is written as:

$$
\begin{aligned}
\min \quad & \log\left(1 + \frac{|G_k|^2 x}{|G_k|^2 y + \sigma_e^2}\right) - \log\left(1 + \frac{|H_k|^2 x}{\sigma_n^2 + |H_k|^2 y}\right) \\
\text{s.t.} \quad & \log\left(1 + \frac{|H_k|^2 x}{\sigma_n^2 + |H_k|^2 y}\right) - \log\left(1 + \frac{|G_k|^2 x}{|G_k|^2 y + \sigma_e^2}\right) > C_{sec}^0 \\
& \log\left(\frac{|H_k|^2 x}{\sigma_n^2 + |H_k|^2 y}\right) \geq K \\
& x + y \leq P \\
& x > 0 \\
& y > 0
\end{aligned}
\tag{22}
$$

In (22), a restricted condition $\log\left(1 + \frac{|H_k|^2 x}{\sigma_n^2 + |H_k|^2 y}\right) - \log\left(1 + \frac{|G_k|^2 x}{|G_k|^2 y + \sigma_e^2}\right) > C_{sec}^0$ is added to make sure that the optimal direction is correct. SCO−AN is an extra noise for Bob the receiver as well. K is the minimum signal-to-noise ratio (SNR) for normal communication. We add another restricted condition $\log\left(1 + \frac{|H_k|^2 x}{\sigma_n^2 + |H_k|^2 y}\right) \geq K$ to ensure normal communication. The value of K varies among different communication systems.

The objective function in (22) is $\log\left(1 + \frac{|G_k|^2 x}{|G_k|^2 y + \sigma_e^2}\right) - \log\left(1 + \frac{|H_k|^2 x}{\sigma_n^2 + |H_k|^2 y}\right)$. The Hessian matrix of the objective function in (22) is not positive definite, so the extremum of the objective function cannot be obtained by the partial derivative method. An improved sequence quadratic program (ISQP) is adopted to optimize power allocation. The basic idea of ISQP is that, at each iterative step, a quadratic programming problem is solved to establish a descent direction, which reduces the value function to obtain compensation. The iterative steps are repeated until the solution of the original problem is obtained.

The Lagrange function of (22) is:

$$L(x, y, \mu, \lambda) = f(x, y) - \mu_1 h_1(x, y) - \sum_{j=1,2,3,4} \lambda_j g_j(x, y), \tag{23}$$

where

$$
\begin{aligned}
f(x, y) &= -\log\left(1 + \frac{|H_k|^2 x}{\sigma_n^2 + |H_k|^2 y}\right) + \log\left(1 + \frac{|G_k|^2 x}{|G_k|^2 y + \sigma_e^2}\right) \\
g_1(x, y) &= \log\left(1 + \frac{|H_k|^2 x}{\sigma_n^2 + |H_k|^2 y}\right) - \log\left(1 + \frac{|G_k|^2 x}{|G_k|^2 y + \sigma_e^2}\right) - C_{sec}0 \\
g_2(x, y) &= x \\
g_3(x, y) &= y \\
g_4(x, y) &= \log\left(\frac{|H_k|^2 x}{\sigma_n^2 + |H_k|^2 y}\right) - K \\
h_1(x, y) &= x + y - P.
\end{aligned}
\tag{24}
$$

For the case of imperfect channel estimation, the initial states of x and y are x_0 and y_0 respectively. The initial secrecy capacity is C_{sec}^{eo}.

$$\log\left(1 + \frac{|\tilde{H}|^2 x}{\sigma_n^2 + |H_{eo}|^2 x + |\tilde{H}|^2 y + |H_{eo}|^2 y}\right) - \log\left(1 + \frac{|G_k|^2 x_0}{|G_k|^2 y_0 + \sigma_e^2}\right) = C_{sec}^{eo} \tag{25}$$

The power allocation problem of SCO−AN for SISO under imperfect channel estimation is written as:

$$
\begin{aligned}
\min \quad & -\log\left(1 + \frac{|\tilde{H}|^2 x}{\sigma_n^2 + |H_{eo}|^2 x + |\tilde{H}|^2 y + |H_{eo}|^2 y}\right) + \log\left(1 + \frac{|G_k|^2 x}{|G_k|^2 y + \sigma_e^2}\right) \\
\text{s.t.} \quad & \log\left(1 + \frac{|\tilde{H}|^2 x}{\sigma_n^2 + |H_{eo}|^2 x + |\tilde{H}|^2 y + |H_{eo}|^2 y}\right) - \log\left(1 + \frac{|G_k|^2 x}{|G_k|^2 y + \sigma_e^2}\right) > C_{sec}^{eo} \\
& \log\left(1 + \frac{|\tilde{H}|^2 x}{\sigma_n^2 + |H_{eo}|^2 x + |\tilde{H}|^2 y + |H_{eo}|^2 y}\right) \geq K \\
& x + y \leq P \\
& x > 0 \\
& y > 0
\end{aligned}
\tag{26}
$$

The Lagrange function of (26) is:

$$L(x, y, \mu, \lambda) = f(x, y) - \mu_1 h_1(x, y) - \sum_{j=1,2,3,4} \lambda_j g_j(x, y), \tag{27}$$

where

$$
\begin{aligned}
f(x, y) &= -\log\left(1 + \frac{|\tilde{H}|^2 x}{\sigma_n^2 + |H_{eo}|^2 x + |\tilde{H}|^2 y + |H_{eo}|^2 y}\right) + \log\left(1 + \frac{|G_k|^2 x}{|G_k|^2 y + \sigma_e^2}\right) \\
g_1(x, y) &= \log\left(1 + \frac{|\tilde{H}|^2 x}{\sigma_n^2 + |H_{eo}|^2 x + |\tilde{H}|^2 y + |H_{eo}|^2 y}\right) - \log\left(1 + \frac{|G_k|^2 x}{|G_k|^2 y + \sigma_e^2}\right) - C_{sec}^{eo} \\
g_2(x, y) &= x \\
g_3(x, y) &= y \\
g_4(x, y) &= \log\left(\frac{|\tilde{H}|^2 x}{\sigma_n^2 + |H_{eo}|^2 x + |\tilde{H}|^2 y + |H_{eo}|^2 y}\right) - K \\
h_1(x, y) &= x + y - P,
\end{aligned}
\tag{28}
$$

The most frequently used methods for channel estimation are least square (LS) channel estimation and minimum mean square error (MMSE) channel estimation.

LS channel estimation is a classic algorithm for non-blind channel estimation. The pilot symbols are used to estimate the channel.

The LS channel estimation is given as:

$$\tilde{\mathbf{H}}_{LS} = \mathbf{z}_k(\mathbf{x}_k)^{-1}. \tag{29}$$

For the MIMO communication system, the LS channel estimation is:

$$\begin{aligned}
\left\|\mathbf{H}_{eo}^{LS}\right\|^2 &= \left\|\mathbf{H} - \tilde{\mathbf{H}}_{LS}\right\|^2 \\
&= \left\|(\mathbf{z}_k - \mathbf{n})\mathbf{x}_k^{-1} - \mathbf{x}_k\mathbf{x}_k^{-1}\right\|^2 \\
&= \left\|\mathbf{n}\mathbf{x}_k^{-1}\right\|^2.
\end{aligned} \tag{30}$$

In (30), $\mathbf{H}_{eo}^{LS}$ denotes the error of LS channel estimation and $\left\|\mathbf{H}_{eo}^{LS}\right\|^2$ denotes the second norm of the LS channel estimation error. $\left\|\mathbf{H}_{eo}^{LS}\right\|^2$ is in proportion to the SNR of the legitimate channel.

For the SISO communication system, the LS channel estimation is:

$$\left|H_{eo}^{LS}\right|^2 = \left|H - \tilde{H}_{LS}\right|^2 = \left|nx_k^{-1}\right|^2. \tag{31}$$

For the MIMO communication system, similar to LS estimation, it is easy to obtain (32):

$$\begin{aligned}
\left\|\mathbf{H}_{eo}^{MMSE}\right\|^2 &= \left\|\mathbf{H} - \tilde{\mathbf{H}}_{\mathbf{MMSE}}\right\|^2 \\
&= \left\|\mathbf{H} - R_{\mathbf{H}\hat{\mathbf{H}}}\left(R_{\mathbf{HH}} + \tfrac{\sigma_n^2}{\sigma_x^2}I\right)^{-1}\tilde{\mathbf{H}}_{LS}\right\|^2 \\
&= \left\|\mathbf{H} - R_{\mathbf{H}\hat{\mathbf{H}}}\left(R_{\mathbf{HH}} + \tfrac{\sigma_n^2}{\sigma_x^2}I\right)^{-1}\mathbf{y}_k\mathbf{x}_k^{-1}\right\|^2,
\end{aligned} \tag{32}$$

where $\mathbf{H}_{eo}^{MMSE}$ denotes the error of MMSE channel estimation and $R_{\mathbf{AB}}$ denotes the cross-correlation matrix of $\mathbf{A}$ and $\mathbf{B}$.

For SISO communication system, R_{AB} denotes the cross-correlation coefficient of A and B.

$$\begin{aligned}
\left|H_{eo}^{MMSE}\right|^2 &= \left|H - \tilde{H}_{MMSE}\right|^2 \\
&= \left|H - R_{H\hat{H}}\left(R_{HH} + \tfrac{\sigma_e^2}{\sigma_x^2}\right)^{-1}\tilde{H}_{LS}\right|^2 \\
&= \left|H - R_{H\hat{H}}\left(R_{HH} + \tfrac{\sigma_n^2}{\sigma_x^2}\right)^{-1}y_k x_k^{-1}\right|^2.
\end{aligned} \tag{33}$$

For the SISO communication system, according to the analysis above, every parameter in (18) except σ_g^2 is available. (31) and (33) are applicable conclusions. However, for MIMO communication system, the expansion of matrices is too complex, rendering (30) and (32) inapplicable.

3.1.2. Objective Function of Power Allocation for MIMO Communication System

For the MIMO communication system, the power for each transmission is $P_{0,m}$. Therefore,

$$\sigma_g^2 + \sigma_u^2 \leq P_{0,m}. \tag{34}$$

The initial secrecy capacity is $C_{sec}^{0,m}$. For the perfect channel estimation, the initial secrecy capacity is given by:

$$\log\left|I\sigma_n^2 + \mathbf{H}_k\mathbf{Z}_k\mathbf{Z}_k^{\dagger}\mathbf{H}_k^{\dagger}x_0\right| - \log\left(\frac{|\mathbf{G}_k\mathbf{Z}_k\mathbf{Z}_k^{\dagger}\mathbf{G}_k^{\dagger}y_0 + I\sigma_e^2 + \mathbf{G}_k\mathbf{Z}_k\mathbf{Z}_k^{\dagger}\mathbf{G}_k^{\dagger}x_0|}{|\mathbf{G}_k\mathbf{Z}_k\mathbf{Z}_k^{\dagger}\mathbf{G}_k^{\dagger}y_0 + I\sigma_e^2|}\right) = C_{sec}^{0,m}. \tag{35}$$

The power allocation problem of SCO−AN for MIMO is written as:

$$
\begin{aligned}
\min \quad & \log\left(\frac{|(\mathbf{G}_k\mathbf{Z}_k\mathbf{Z}_k^\dagger\mathbf{G}_k^\dagger)y+\mathbf{I}\sigma_e^2+\mathbf{G}_k\mathbf{Z}_k\mathbf{Z}_k^\dagger\mathbf{G}_k^\dagger x|}{|(\mathbf{G}_k\mathbf{Z}_k\mathbf{Z}_k^\dagger\mathbf{G}_k^\dagger)y+\mathbf{I}\sigma_e^2|}\right) - \log\left(\frac{|(\mathbf{H}_k\mathbf{Z}_k\mathbf{Z}_k^\dagger\mathbf{H}_k^\dagger)y+\mathbf{I}\sigma_n^2+\mathbf{H}_k\mathbf{Z}_k\mathbf{Z}_k^\dagger\mathbf{H}_k^\dagger x|}{|(\mathbf{H}_k\mathbf{Z}_k\mathbf{Z}_k^\dagger\mathbf{H}_k^\dagger)y+\mathbf{I}\sigma_n^2|}\right) \\[2mm]
\text{s.t.} \quad & \log\left(\frac{|(\mathbf{G}_k\mathbf{Z}_k\mathbf{Z}_k^\dagger\mathbf{G}_k^\dagger)y+\mathbf{I}\sigma_e^2+\mathbf{G}_k\mathbf{Z}_k\mathbf{Z}_k^\dagger\mathbf{G}_k^\dagger x|}{|(\mathbf{G}_k\mathbf{Z}_k\mathbf{Z}_k^\dagger\mathbf{G}_k^\dagger)y+\mathbf{I}\sigma_e^2|}\right) - \log\left(\frac{|(\mathbf{H}_k\mathbf{Z}_k\mathbf{Z}_k^\dagger\mathbf{H}_k^\dagger)y+\mathbf{I}\sigma_n^2+\mathbf{H}_k\mathbf{Z}_k\mathbf{Z}_k^\dagger\mathbf{H}_k^\dagger x|}{|(\mathbf{H}_k\mathbf{Z}_k\mathbf{Z}_k^\dagger\mathbf{H}_k^\dagger)y+\mathbf{I}\sigma_n^2|}\right) > C_{sec}^{0,m} \\[2mm]
& \log\left(\frac{|(\mathbf{H}_k\mathbf{Z}_k\mathbf{Z}_k^\dagger\mathbf{H}_k^\dagger)y+\mathbf{I}\sigma_n^2+\mathbf{H}_k\mathbf{Z}_k\mathbf{Z}_k^\dagger\mathbf{H}_k^\dagger x|}{|(\mathbf{H}_k\mathbf{Z}_k\mathbf{Z}_k^\dagger\mathbf{H}_k^\dagger)y+\mathbf{I}\sigma_n^2|}\right) > K \\[2mm]
& x + y \le P_{0,m} \\
& x > 0 \\
& y > 0
\end{aligned}
\tag{36}
$$

The Lagrange function of (36) is:

$$
L(x,y,\mu,\lambda) = f(x,y) - \mu_1 h_1(x,y) - \sum_{j=1,2,3,4}\lambda_j g_j(x,y),
\tag{37}
$$

where

$$
\begin{aligned}
f(x,y) &= \log\left(\frac{|(\mathbf{G}_k\mathbf{Z}_k\mathbf{Z}_k^\dagger\mathbf{G}_k^\dagger)y+\mathbf{I}\sigma_e^2+\mathbf{G}_k\mathbf{Z}_k\mathbf{Z}_k^\dagger\mathbf{G}_k^\dagger x|}{|(\mathbf{G}_k\mathbf{Z}_k\mathbf{Z}_k^\dagger\mathbf{G}_k^\dagger)y+\mathbf{I}\sigma_e^2|}\right) - \log\left(\frac{|(\mathbf{H}_k\mathbf{Z}_k\mathbf{Z}_k^\dagger\mathbf{H}_k^\dagger)y+\mathbf{I}\sigma_n^2+\mathbf{H}_k\mathbf{Z}_k\mathbf{Z}_k^\dagger\mathbf{H}_k^\dagger x|}{|(\mathbf{H}_k\mathbf{Z}_k\mathbf{Z}_k^\dagger\mathbf{H}_k^\dagger)y+\mathbf{I}\sigma_n^2|}\right) \\[2mm]
g_1(x,y) &= \log\left(\frac{|(\mathbf{H}_k\mathbf{Z}_k\mathbf{Z}_k^\dagger\mathbf{H}_k^\dagger)y+\mathbf{I}\sigma_n^2+\mathbf{H}_k\mathbf{Z}_k\mathbf{Z}_k^\dagger\mathbf{H}_k^\dagger x|}{|(\mathbf{H}_k\mathbf{Z}_k\mathbf{Z}_k^\dagger\mathbf{H}_k^\dagger)y+\mathbf{I}\sigma_n^2|}\right) - \log\left(\frac{|(\mathbf{G}_k\mathbf{Z}_k\mathbf{Z}_k^\dagger\mathbf{G}_k^\dagger)y+\mathbf{I}\sigma_e^2+\mathbf{G}_k\mathbf{Z}_k\mathbf{Z}_k^\dagger\mathbf{G}_k^\dagger x|}{|(\mathbf{G}_k\mathbf{Z}_k\mathbf{Z}_k^\dagger\mathbf{G}_k^\dagger)y+\mathbf{I}\sigma_e^2|}\right) - C_{sec}^{0,m} \\[2mm]
g_2(x,y) &= \log\left(\frac{|(\mathbf{H}_k\mathbf{Z}_k\mathbf{Z}_k^\dagger\mathbf{H}_k^\dagger)y+\mathbf{I}\sigma_n^2+\mathbf{H}_k\mathbf{Z}_k\mathbf{Z}_k^\dagger\mathbf{H}_k^\dagger x|}{|(\mathbf{H}_k\mathbf{Z}_k\mathbf{Z}_k^\dagger\mathbf{H}_k^\dagger)y+\mathbf{I}\sigma_n^2|}\right) - K \\[2mm]
g_3(x,y) &= x \\
g_4(x,y) &= y \\
h_1(x,y) &= x + y - P_{0,m}\,.
\end{aligned}
\tag{38}
$$

For the imperfect channel estimation, the initial secrecy capacity is $C_{sec,eo}^{0,m}$. For the imperfect channel estimation, the initial secrecy capacity is given by:

$$
\log\left(\frac{|\mathbf{K}_{H,0}^{eo}+(\mathbf{H}_{eo}+\tilde{\mathbf{H}})\mathbf{Z}_k\mathbf{Z}_k^\dagger(\mathbf{H}_{eo}+\tilde{\mathbf{H}})^\dagger x_0|}{|\mathbf{K}_{H,0}^{eo}|}\right) - \log\left(\frac{|\mathbf{K}_G+\mathbf{G}_k\mathbf{Z}_k\mathbf{Z}_k^\dagger\mathbf{G}_k^\dagger x_0|}{|\mathbf{K}_G|}\right) = C_{sec,eo}^{0,m}\,,
\tag{39}
$$

In (39), $\mathbf{K}_{H,0}^{eo} = ((\mathbf{H}_{eo}+\tilde{\mathbf{H}})\mathbf{Z}_k\mathbf{Z}_k^\dagger(\mathbf{H}_{eo}+\tilde{\mathbf{H}})^\dagger)y_0+\mathbf{I}\sigma_n^2$. We use x instead of σ_u^2, y instead of σ_g^2 and the initial states of x and y are x_0 and y_0 respectively.

For the imperfect channel estimation, the power allocation problem of SCO−AN is written as:

$$
\begin{aligned}
\min \quad & \log\left(\frac{\left|\mathbf{K}_H+(\mathbf{H}_{eo}+\tilde{\mathbf{H}})\mathbf{Z}_k\mathbf{Z}_k^\dagger(\mathbf{H}_{eo}+\tilde{\mathbf{H}})^\dagger x\right|}{|\mathbf{K}_H|}\right) - \log\left(\frac{|\mathbf{K}_G+\mathbf{G}_k\mathbf{Z}_k\mathbf{Z}_k^\dagger\mathbf{G}_k^\dagger x|}{|\mathbf{K}_G|}\right) \\[2mm]
\text{s.t.} \quad & \log\left(\frac{\left|\mathbf{K}_H+(\mathbf{H}_{eo}+\tilde{\mathbf{H}})\mathbf{Z}_k\mathbf{Z}_k^\dagger(\mathbf{H}_{eo}+\tilde{\mathbf{H}})^\dagger x\right|}{|\mathbf{K}_H|}\right) - \log\left(\frac{|\mathbf{K}_G+\mathbf{G}_k\mathbf{Z}_k\mathbf{Z}_k^\dagger\mathbf{G}_k^\dagger x|}{|\mathbf{K}_G|}\right)) > C_{sec,eo}^{0,m} \\[2mm]
& \log\left(\frac{\left|((\mathbf{H}_{eo}+\tilde{\mathbf{H}})\mathbf{Z}_k\mathbf{Z}_k^\dagger(\mathbf{H}_{eo}+\tilde{\mathbf{H}})^\dagger)y+(\mathbf{H}_{eo}+\tilde{\mathbf{H}})\mathbf{Z}_k\mathbf{Z}_k^\dagger(\mathbf{H}_{eo}+\tilde{\mathbf{H}})^\dagger x\right|}{|\mathbf{K}_H|}\right) \ge K \\[2mm]
& x + y \le P_{0,m} \\
& x > 0 \\
& y > 0
\end{aligned}
\tag{40}
$$

The Lagrange function of (40) is:

$$L(x,y,\mu,\lambda) = f(x,y) - \mu_1 h_1(x,y) - \sum_{j=1,2,3,4} \lambda_j g_j(x,y), \tag{41}$$

where

$$
\begin{aligned}
f(x,y) &= -\log\left(\frac{\left|\mathbf{K}_H + (\mathbf{H}_{eo}+\tilde{\mathbf{H}})\mathbf{Z}_k\mathbf{Z}_k^{\dagger}(\mathbf{H}_{eo}+\tilde{\mathbf{H}})^{\dagger}x\right|}{|\mathbf{K}_H|}\right) + \log\left(\frac{\left|\mathbf{K}_G + \mathbf{G}_k\mathbf{Z}_k\mathbf{Z}_k^{\dagger}\mathbf{G}_k^{\dagger}x\right|}{|\mathbf{K}_G|}\right) \\
g_1(x,y) &= \log\left(\frac{\left|\mathbf{K}_H + (\mathbf{H}_{eo}+\tilde{\mathbf{H}})\mathbf{Z}_k\mathbf{Z}_k^{\dagger}(\mathbf{H}_{eo}+\tilde{\mathbf{H}})^{\dagger}x\right|}{|\mathbf{K}_H|}\right) - \log\left(\frac{\left|\mathbf{K}_G + \mathbf{G}_k\mathbf{Z}_k\mathbf{Z}_k^{\dagger}\mathbf{G}_k^{\dagger}x\right|}{|\mathbf{K}_G|}\right) - C_{sec,eo}^{0,m} \\
g_2(x,y) &= \log\left(\frac{\left|((\mathbf{H}_{eo}+\tilde{\mathbf{H}})\mathbf{Z}_k\mathbf{Z}_k^{\dagger}(\mathbf{H}_{eo}+\tilde{\mathbf{H}})^{\dagger})y + (\mathbf{H}_{eo}+\tilde{\mathbf{H}})\mathbf{Z}_k\mathbf{Z}_k^{\dagger}(\mathbf{H}_{eo}+\tilde{\mathbf{H}})^{\dagger}x\right|}{|\mathbf{K}_H|}\right) - \mathrm{K} \\
g_3(x,y) &= x \\
g_4(x,y) &= y \\
h_1(x,y) &= x + y - P_{0,m}
\end{aligned}
\tag{42}
$$

3.1.3. Objective Function of Power Allocation for SIMO Communication System with Active Eavesdroppers

In this paper, we discussed the case of the passive eavesdropper. Recently, the proposed pilot spoofing attack technology maked EVE to have the ability to attack Alice. Therefore, we will discuss the influence of active eavesdroppers. As shown in Figure 3, in the SIMO wire-tap communication system, Alice equips with N transmit antenna, N one-antenna receivers (Bobs) equip with N one-antenna Eves. Let P_N denotes the pilot symbol. P_N is known to Eve, which concurrently send the same pilot in the training phase withe average transmission power P_e and Alice has a perfect knowledge of P_N. The relevant knowledge [30] of pilot frequency has been introduced in this paper and will not be repeated.

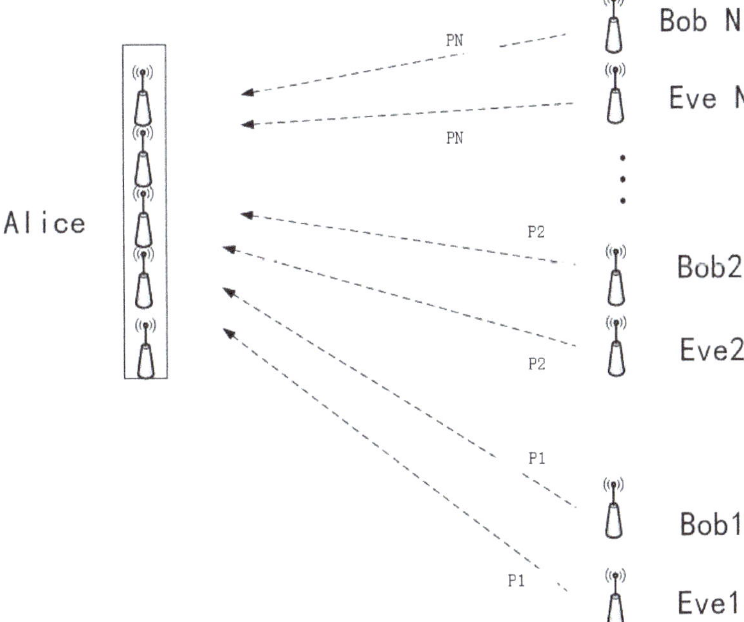

Figure 3. Single-antenna eavesdroppers launch pilot spoofing attack.

According to ([9]), for the LS estimatior,

$$\bar{\mathbf{h}}_{LS} = (\mathbf{K}_B^H \mathbf{K}_B)^{-1} \mathbf{K}_B^H z \tag{43}$$

where $\bar{\mathbf{h}}_{LS}$ denotes the estimate of $\mathbf{h}$ with LS Estimator and $\mathbf{K}_B^H \triangleq \sqrt{P_B} A_B$.

$$\mathbf{A}_B \triangleq \frac{1}{\sqrt{L_B}} \left[a(\theta_{B,1,m}), a(\theta_{B,2,m}), \ldots, a(\theta_{B,L_B,m}) \right] \tag{44}$$

$$\mathbf{A}_B \triangleq \frac{1}{\sqrt{L_B}} \left[a(\theta_{B,1,m}), a(\theta_{B,2,m}), \ldots, a(\theta_{B,L_B,m}) \right] \tag{45}$$

$$a(\theta_{B,l,m}) = \frac{\left[1, e^{-j2\pi \frac{d}{\lambda_c} \cos(\theta_{B,l,m})}, e^{-j4\pi \frac{d}{\lambda_c} \cos(\theta_{B,l,m})}, \ldots, e^{-j2n\pi \frac{d}{\lambda_c} \cos(\theta_{B,l,m})} \right]^T}{\sqrt{N}} \tag{46}$$

L_B is the number of paths between the transmitter and Alice.

For the MMSE estimatior,

$$\bar{\mathbf{h}}_{MMSE} = (\mathbf{I}_{L_B} + \mathbf{K}_B^H R_{dd}^{-1} \mathbf{K}_B)^{-1} \mathbf{K}_B^H R_{dd}^{-1} z, \tag{47}$$

where $\bar{\mathbf{h}}_{MMSE}$ denotes the estimate of $\mathbf{h}$ with MMSE Estimator and $R_{dd} \triangleq K_E K_E^H + \sigma_v^2 I_N$ The power allocation problem of $SCO-AN$ for SIMO with active eavesdropper is written as:

$$\min \quad \log\left(\frac{\left| \mathbf{K}_H + (\mathbf{h}_{eo} + \tilde{\mathbf{h}}) Z_k Z_k^{\dagger} (\mathbf{h}_{eo} + \tilde{\mathbf{h}})^{\dagger} x \right|}{|\mathbf{K}_H|} \right) - \log\left(\frac{\left| \mathbf{K}_G + G_k Z_k Z_k^{\dagger} G_k^{\dagger} x \right|}{|\mathbf{K}_G|} \right)$$

$$\text{s.t.} \quad \log\left(\frac{\left| \mathbf{K}_h + (\mathbf{h}_{eo} + \tilde{\mathbf{h}}) Z_k Z_k^{\dagger} (\mathbf{h}_{eo} + \tilde{\mathbf{h}})^{\dagger} x \right|}{|\mathbf{K}_H|} \right) - \log\left(\frac{\left| \mathbf{K}_G + G_k Z_k Z_k^{\dagger} G_k^{\dagger} x \right|}{|\mathbf{K}_G|} \right)) > C_{sec,eo}^{0,m}$$

$$\log\left(\frac{\left| ((\mathbf{h}_{eo} + \tilde{\mathbf{h}}) Z_k Z_k^{\dagger} (\mathbf{h}_{eo} + \tilde{\mathbf{h}})^{\dagger}) y + (\mathbf{h}_{eo} + \tilde{\mathbf{h}}) Z_k Z_k^{\dagger} (\mathbf{h}_{eo} + \tilde{\mathbf{H}})^{\dagger} x \right|}{|\mathbf{K}_H|} \right) \geq K \tag{48}$$

$$x + y \leq P_{0,m}$$
$$x > 0$$
$$y > 0$$

For SIMO communication system with active eavesdroppers, $\mathbf{h}_{eo} = \mathbf{h} - \tilde{\mathbf{h}}$ in (48). $\tilde{\mathbf{h}}$ denotes $\tilde{\mathbf{h}}_{\mathbf{LS}}$ when LS estimator is used and denotes $\tilde{\mathbf{h}}_{\mathbf{MMSE}}$ when MMSE estimator is used. The Lagrange function of (48) is:

$$L(x, y, \mu, \lambda) = f(x, y) - \mu_1 h_1(x, y) - \sum_{j=1,2,3,4} \lambda_j g_j(x, y), \tag{49}$$

where

$$f(x,y) = -\log\left(\frac{\left|\mathbf{K}_H+(\mathbf{h}_{eo}+\tilde{\mathbf{h}})\mathbf{Z}_k\mathbf{Z}_k^\dagger(\mathbf{h}_{eo}+\tilde{\mathbf{h}})^\dagger x\right|}{|\mathbf{K}_H|}\right)+\log\left(\frac{\left|\mathbf{K}_G+\mathbf{G}_k\mathbf{Z}_k\mathbf{Z}_k^\dagger\mathbf{G}_k^\dagger x\right|}{|\mathbf{K}_G|}\right)$$

$$g_1(x,y) = \log\left(\frac{\left|\mathbf{K}_H+(\mathbf{h}_{eo}+\tilde{\mathbf{h}})\mathbf{Z}_k\mathbf{Z}_k^\dagger(\mathbf{h}_{eo}+\tilde{\mathbf{h}})^\dagger x\right|}{|\mathbf{K}_H|}\right)-\log\left(\frac{\left|\mathbf{K}_G+\mathbf{G}_k\mathbf{Z}_k\mathbf{Z}_k^\dagger\mathbf{G}_k^\dagger x\right|}{|\mathbf{K}_G|}\right)-C_{sec,eo}^{0,m}$$

$$g_2(x,y) = \log\left(\frac{\left|((\mathbf{h}_{eo}+\tilde{\mathbf{h}})\mathbf{Z}_k\mathbf{Z}_k^\dagger(\mathbf{h}_{eo}+\tilde{\mathbf{h}})^\dagger)y+(\mathbf{h}_{eo}+\tilde{\mathbf{h}})\mathbf{Z}_k\mathbf{Z}_k^\dagger(\mathbf{h}_{eo}+\tilde{\mathbf{h}})^\dagger x\right|}{|\mathbf{K}_H|}\right)-K$$

$$g_3(x,y) = x$$
$$g_4(x,y) = y$$
$$h_1(x,y) = x+y-P_{0,m}$$

(50)

The power allocation problems of SCO$-$AN with perfect channel estimation, imperfect channel and active eavesdropper are similar. Therefore, we use the same algorithm to solve the problem.

3.2. SQP and ISQP Algorithm

μ and λ are Lagrange multipliers. To optimize the problems above, the following conditions must be satisfied:

$$
\begin{aligned}
&\left.\frac{\partial L}{\partial X}\right|_{x=x^*} = 0 &&(a)\\
&\lambda_j \neq 0, &&(b)\\
&u_k \geq 0, &&(c)\\
&u_k g_k(x*) = 0, &&(d)\\
&h_i(x*) = 0 &&i=1 &&(e)\\
&g_j(x*) = 0, &&j=1,2,3,4 &&(f)
\end{aligned}
$$

(51)

(51) are Karush-Kuhn-Tucker conditions (KKT conditions). (a) is a necessary condition when the extreme value of Lagrange function is taken; (b) is a Lagrange coefficient constraint; (c) is an inequality constraint case; (d) is the complementary relaxation condition; (e) and (f) are the original constraints.

The KKT condition is a necessary condition for the optimal solution.

Condition (c) constructs the $L(x,\lambda,\mu)$ function and the condition $L(x,\lambda,\mu) \leq f(x)$ should be satisfied. In $L(x,\lambda,\mu)$, μ is 0, so λ is less than or equal to 0.

A quadratic polynomial is used to approximate $f(x,y)$. By expanding the quadratic polynomial into a positive definite quadratic form, the following quadratic programming subproblem is obtained:

$$
\begin{aligned}
\min \quad & \tfrac{1}{2}d^T\mathbf{B}_k d + \nabla f(x_k,y_k)^T d\\
\text{s.t} \quad & h(x_k,y_k) + \mathbf{A}_k^\varepsilon d - 0\\
& g(x_k,y_k) + \mathbf{A}_k^\Gamma d \geq 0,
\end{aligned}
$$

(52)

where $\mathbf{A}_k^\varepsilon = \nabla h(x_k,y_k)$, $\mathbf{A}_k^\Gamma = \nabla g(x_k,y_k)$, t_k is a positive definite matrix, and d_k is optimal solution of quadratic programming subproblems.

Let x^* denote the KKT point of the optimization constraint problem and $\lambda^*,\mu^* \geq 0$ be its corresponding Lagrange multiplier vectors. For x^*, the following conditions should be satisfied:

1. The Jacobi matrix of $L(x,\lambda,\mu)$ is row full rank.
2. The strict complementary relaxation condition should be satisfied; that is, $g_i(x^*) \geq 0$, $\lambda_i^* \geq 0, g_i(x^*)\lambda_i^* = 0$, and $g_i(x^*)+\lambda_i^* > 0$.
3. A sufficient second-order optimality condition should be satisfied, that is, for any vector $d \neq 0$ that satisfies $\mathbf{A}(x^*)d = 0$, the following condition is satisfied:

$$d^T\mathbf{B}(x^*,y^*,\mu^*,\lambda^*)d > 0,$$

(53)

where $\mathbf{B}(x,y,\mu,\lambda)$ is a positive definite matrix, at the beginning of the iteration, $\mathbf{B}(x,y,\mu,\lambda)$ is usually set as the identity matrix.

If $(x_k,y_k,\mu_k,\lambda_k)$ is close to $(x^*,y^*,\mu^*,\lambda^*)$ sufficiently, the quadratic programming sub-problem of (53) has a local minimum point d^*. The corresponding effective constraint index set is the same as the effective constraint index set of the original problem at (x^*,y^*). Using the KKT conditions, (52) is equivalent to:

$$H_1(d,\mu,\lambda) = \mathbf{B}_k - (\mathbf{A}_k^\varepsilon)^T\mu - (\mathbf{A}_k^\Gamma)^T\lambda + \nabla f(x_k,y_k), \tag{54}$$

$$H_2(d,\mu,\lambda) = h(x_k,y_k) + (\mathbf{A}_k^\varepsilon)^T d, \tag{55}$$

$$\lambda \geq 0, g(x_k,y_k) + \mathbf{A}_k^\Gamma d \geq 0, \lambda[g(x_k,y_k) + \mathbf{A}_k^\Gamma d] = 0, \tag{56}$$

Note that Formula (20) and (23) are linear complementarity problems. We define smooth FB-function:

$$\varphi(\varepsilon,a,b) = a + b - \sqrt{a^2 + b^2 + 2\varepsilon^2}, \tag{57}$$

where $\varepsilon > 0$ is a smooth parameter, and

$$\Phi(\varepsilon,d,\lambda) = (\varphi_1(\varepsilon,d,\lambda), \varphi_2(\varepsilon,d,\lambda) \ldots \varphi_m(\varepsilon,d,\lambda))^T, \tag{58}$$

in (32),

$$\varphi_i(\varepsilon,d,\lambda) = \lambda_i + [g_i(x_k,y_k) + (A_k^\Gamma)_i d] \\ - \sqrt{\lambda_i^2 + [g_i(x_k,y_k) + (A_k^\Gamma)_i d]^2 + 2\varepsilon^2}, \tag{59}$$

where $(A_k^\Gamma)_i$ is the i-th row of $\mathbf{A}_k^\Gamma$. (18) and (19), (21) and (22) are equivalent to

$$H(z) := H(\varepsilon,d,\mu,\lambda) = \begin{pmatrix} \varepsilon \\ H_1(d,\mu,\lambda) \\ H_2(d,\mu,\lambda) \\ \Phi(\varepsilon,d,\lambda) \end{pmatrix} = 0, \tag{60}$$

The Jacobian matrix of (H_z) is

$$H'(z) = \begin{pmatrix} 1 & 0 & 0 & 0 \\ 0 & \mathbf{B}_k & -(\mathbf{A}_k^\varepsilon)^T & -(\mathbf{A}_k^\Gamma)^H \\ 0 & \mathbf{A}_k^\varepsilon & 0 & 0 \\ v & D_2(z)A_k^\Gamma & 0 & D_1(z) \end{pmatrix}, \tag{61}$$

where $v = \nabla_\varepsilon\Phi(\varepsilon,d,\lambda) = (v_1,v_2,\ldots,v_m)^T$ and

$$v_i = -\frac{2\varepsilon}{\sqrt{\lambda_i^2 + [g_i(x_k,y_k) + (A_k^\Gamma)_i d]^2 + 2\varepsilon^2}}, \tag{62}$$

$$\begin{aligned} D_1(z) &= diag(a_1(z),a_2(z),\ldots,a_m(z)), \\ D_2(z) &= diag(b_1(z),b_2(z),\ldots,b_m(z)), \end{aligned} \tag{63}$$

where

$$\begin{aligned} a_i(z) &= 1 - \frac{\lambda_i}{\sqrt{\lambda_i^2 + [g_i(x_k,y_k) + (A_k^\Gamma)_i d]^2 + 2\varepsilon^2}}, \\ b_i(z) &= 1 - \frac{g_i(x_k,y_k) + (A_k^\Gamma)_i d}{\sqrt{\lambda_i^2 + [g_i(x_k,y_k) + (A_k^\Gamma)_i d]^2 + 2\varepsilon^2}}, \end{aligned} \tag{64}$$

here, we make $\gamma \in (0,1)$ and a non-negative functions $\psi(z)$ is

$$\psi(z) = \gamma\|H(z)\| \min\{1, \|H(z)\|\}. \tag{65}$$

Sequence quadratic program (SQP) is an iterative algorithm, the basic idea of SQP is to apply approximate Newton method to the first-order optimality condition of constrained optimization problem. In each iteration step, a quadratic programming problem with the quadratic approximation of Lagrange function as the objective function and the linearization of the original constraint as the constraint condition are solved.

The full SQP is shown as follows:

Algorithm 1 SQP

Step 0: Set β=0.5, σ=0.2, ε=1 $\times$ 10^{-6}, the initial vector $d_0 = (1,1,1)^{\mathrm{T}}$, $\mu_0 = 0$, $\lambda_0 = (0,0,0)^{\mathrm{T}}$, $z_0 = (\varepsilon_0, d_0, \mu_0, \lambda_0)$, $\bar{z}_0 = (\varepsilon_0, 0, 0, 0)$, $i = 0$
Step 1: If $\|H(z_i)\| \leq 0$, stop iteration, else, $\psi_i = \psi(z_i)$, ψ_i is shown in (37), $H(z_i)$ is shown in (32).
Step 2: Solve the equations $H(z_i) + H'(z_i)\Delta z_i = \psi\bar{z}_0$ and then get the solution of the equations: $\Delta z_i = (\Delta\varepsilon_i, \Delta d_i, \Delta\mu_i, \Delta\lambda_i)$
Step 3: Let m be the smallest non-negative integer m that satisfies the following inequality: $H(z_i + \beta^m \Delta z_i) \leq [1 - \sigma(1 - \gamma\varepsilon_0)\beta^m)]\|H(z_i)\|$ where $\alpha_i = \rho^{m_i}$, $z_{i+1} = z_i + \alpha_i \Delta z_i$
Step 4: $i = i + 1$, go to step1

We adopt a improved sequence quadratic program (ISQP) which is based on improvements from sequence quadratic program. At the beginning of the iteration, the initial matrix in (52), $B(x, \mu, \lambda)$ is set as the identity matrix in ISQP. In SQP, the initial matrix is designed as

$$W(x,y,u,\lambda) = \nabla^2(f(x,y)) - \sum_{i=1}^{l} u_i \times \nabla^2(h_i(x,y)) - \sum_{i=1}^{m} \lambda_i \times \nabla^2(g_i(x,y)), \nabla^2(*) \text{ denotes}$$

the Hessian matrix of $(*)$. A second order partial derivative should be calculated in each

iteration of SQP. The complexity of $W(x,y,u,\lambda) = \nabla^2(f(x,y)) - \sum_{i=1}^{l} u_i \times \nabla^2(h_i(x,y)) -$

$\sum_{i=1}^{m} \lambda_i \times \nabla^2(g_i(x,y))$ trivially is much larger than that of $B(x,y,\mu,\lambda)$.

3.3. Complexity Analysis

In this section, we assert the superiority of ISQP by comparing the complexity of the three algorithms: ISQP, SQP, BPA and COCOA [31]. ISQP and SQP have been introduced in detail in previous sections. The BPA algorithm is a traversal algorithm which searches all directions in each iteration and then selects the best direction. The complexity of BPA algorithm is high and the search direction is greatly affected by the step size. BPA is also likely to search in the wrong direction. The complexity of ISQP, SQP, and BPA are shown as follows. The change of angle for BPA is set as 5° so 36 rounds of calculation are needed for just one iteration. The entries in the tables indicate the calculated amount required for one iteration.

In Tables 2–5, N_{H1} denotes derivative of $g_1(x,y)$, and N_{H2} denotes derivative of $g_2(x,y)$. The amounts of computation for derivative of $g_1(x,y)$ in SISO and MIMO are different, so we use N_{H1} in place of the amounts of computation for derivative of $g_1(x,y)$. Similarly, N_{f1} denotes derivative of the objective function, and N_{f2} denotes second derivative of the objective function. Four times of calculation are needed for the second second derivative of the objective function. The objective function is a composite function, so N_{f2} is much larger than N_{f1}.

Table 2. Complexity Analysis for SISO under Perfect Channel Estimation.

Algorithm	Complexity for Each Iteration
ISQP	$16(N_{H1} + N_{H2}) + N_{f1} + 137$
SQP	$122 + 16(N_{H1} + N_{H2}) + N_{f1} + N_{f2}$
BPA	$36\,N_{f2} + 144$

Table 3. Complexity Analysis for SISO under Imperfect Channel Estimation.

Algorithm	Complexity for Each Iteration
ISQP	$16(N_{H1} + N_{H2}) + N_{f1} + 186$
SQP	$157 + 16(N_{H1} + N_{H2}) + N_{f1} + N_{f2}$
BPA	$36\,N_{f2} + 186$

Table 4. Complexity Analysis for MIMO under Perfect Channel Estimation.

Algorithm	Complexity for Each Iteration
ISQP	$16(N_{H1} + N_{H2}) + N_{f1} + 167$
SQP	$142 + 16(N_{H1} + N_{H2}) + N_{f1} + N_{f2}$
BPA	$36\,N_{f2} + 166$

Table 5. Complexity Analysis for MIMO under Imperfect Channel Estimation.

Algorithm	Complexity for Each Iteration
ISQP	$16(N_{H1} + N_{H2}) + N_{f1} + 216$
SQP	$192 + 16(N_{H1} + N_{H2}) + N_{f1} + N_{f2}$
BPA	$36\,N_{f2} + 234$

4. Simulation Results

4.1. Simulation Environment and Discussion

In the simulation experiment for the MIMO communication system, there are two transmitting antennas, two receiving antennas, and two eavesdropping antennas, i.e., $N_A = N_B = N_E = 2$. **H** and **G** are 2×2 Rayleigh fading channels. The distributions of **H** and **G** both have a mean value of 0 and a variance of 0.5. The information-bearing signals are random complex vectors. The transmission power is 10 (i.e., $P = 10$). The Gaussian noise in the channel changes with the SNR. The SNR of **H** increases from 0 to 30, while the SNR of **G** is 5 dB. All Parameters are shown in Table 6.

For the SISO communication system, $N_A = N_R = N_E = 1$ by definition. The information-bearing signals are a random complex number. The SNR of H increases from 0 to 30, while the SNR of G is 5 dB and $P = 10$.

Table 6. Simulation Parameters for MIMO.

Parameter	Value
Number of Transmitting Antennas	2
Number of Receiving Antennas	2
Number of Eavesdropping Antennas	2
Transmission Power	10
Mean of Channel **H** and **G**	0
Variance of Channel **H** and **G**	0.5

4.2. Numerical Simulation and Discussion

Table 7 shows the increase of secrecy capacity after one iteration for SISO under perfect channel estimation. The base value of the iteration step is $\beta = 0.5$. The imperfect channel estimation is not considered here. The secrecy capacity increases the most after one iteration of ISQP, followed closely by SQP. The secrecy capacity of the BPA algorithm has the least increase. The COCOA algorithm, another typical iterative algorithm, is also compared in Tables 7 and 8.

Table 7. Secrecy Capacity Comparison for SISO under Perfect Channel Estimation.

Algorithm	Intial Secrecy Capacity	Optimized Secrecy Capacity ($\beta = 0.5$, One Step)
ISQP	0.3643	0.4243
SQP	0.3643	0.4256
BPA	0.3643	0.3821
COCOA	0.3643	0.4109

Table 8. Complexity Comparison for SISO under Perfect Channel Estimation.

Algorithm	Amount of Computation for Each Iteration	Number of Iterations
ISQP	732	5
SQP	897	5
BPA	6624	13
COCOA	933	7

Table 8 shows the complexity comparison of algorithms for SISO under the perfect channel estimation. BPA leads to the largest calculated amount of computation and the worst optimization result. Such poor performance is owing to the lack of clear search directions for BPA. On the contrary, ISQP has the smallest calculated amount of computation and the best optimization result due to the simplicity of the initial matrix.

According to Tables 7 and 8, the optimization performance of ISQP is similar to that of SQP. However, ISQP requires much less computation. Therefore, we conclude that ISQP is a more effective and thus more desirable algorithm. ISQP and SQP requires fewer iterations than COCOA when $\varepsilon = 1 \times 10^{-6}$. The expression $\varepsilon = 1 \times 10^{-6}$ refers to the two-norm of the gradient rate for BPA and COCOA. Again, BPA requires the most iterations.

Tables 9–11 show the influence of the initial point on the optimization of secrecy capacity. The results are similar across different algorithms. The even distribution of transmission between information-bearing and $SCO-AN$ seems to be the optimal distribution scheme. This result paves the way to an exciting field for future research.

Table 9. Influence of the Initial Point on Secrecy Capacity (ISQP).

Algorithm	(x_0, y_0)	Intial Secrecy Capacity	Optimized Secrecy Capacity
ISQP	$(5, 5)$	0.5612	0.6312
ISQP	$(3, 7)$	0.4785	0.5322
ISQP	$(1, 9)$	0.3821	0.4329

Table 10. Influence of the Initial Point on Secrecy Capacity (SQP).

Algorithm	(x_0, y_0)	Intial Secrecy Capacity	Optimized Secrecy Capacity
SQP	$(5, 5)$	0.5612	0.6375
SQP	$(3, 7)$	0.4785	0.5364
SQP	$(1, 9)$	0.3821	0.4357

Table 11. Influence of the Initial Point on Secrecy Capacity (BPA).

Algorithm	(x_0, y_0)	Intial Secrecy Capacity	Optimized Secrecy Capacity
BPA	$(5, 5)$	0.5612	0.5924
BPA	$(3, 7)$	0.4785	0.4922
BPA	$(1, 9)$	0.3821	0.4012

Figure 4 shows the performance comparison of different algorithms. Figure 4 contains four subfigures, each showing a similar trend in the results. Subfigure (a) shows the optimization performance of SISO under perfect channel estimation versus different SNR. Subfigure (b) shows the optimization performance of SISO under imperfect channel estimation versus different SNR. Subfigure (c) shows the optimization performance of MIMO under perfect channel estimation versus different SNR. Subfigure (d) shows the optimization performance of MIMO under imperfect channel estimation versus different SNR. According to Section 3, BPA requires the most computation and has the worst optimization performance. While SQP and ISQP have similar performance, SQP requires 15% more calculated amounts than ISQP. The optimization performance of COCOA is slightly better than that of BPA, but COCOA is not an excellent iterative algorithm due to the lack of efficacy.

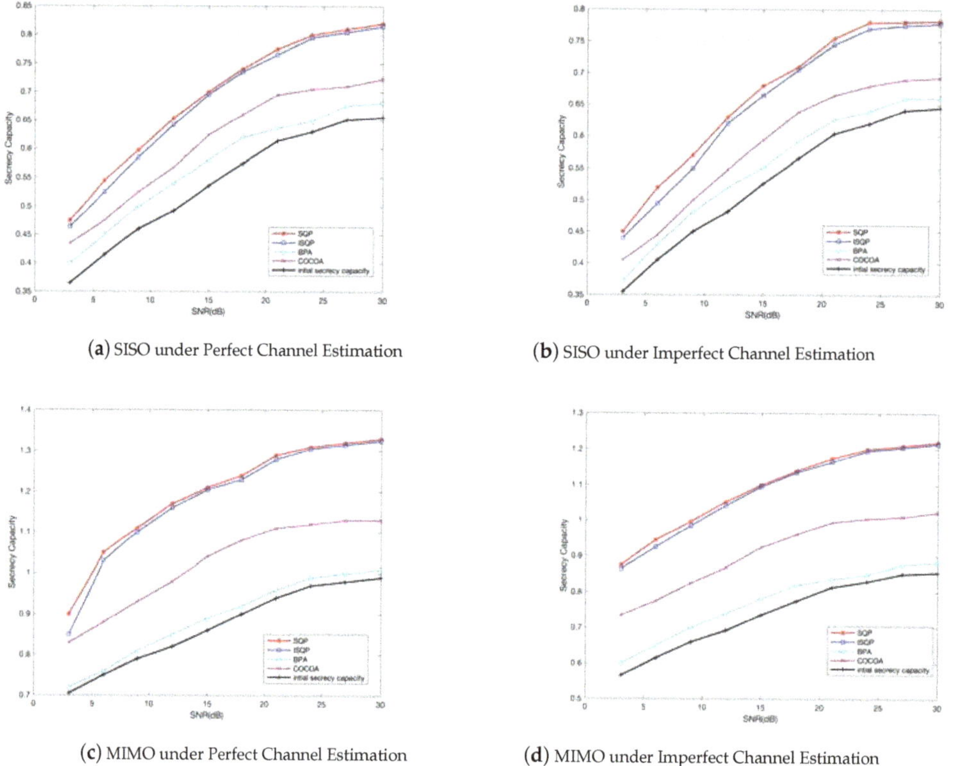

(**a**) SISO under Perfect Channel Estimation

(**b**) SISO under Imperfect Channel Estimation

(**c**) MIMO under Perfect Channel Estimation

(**d**) MIMO under Imperfect Channel Estimation

Figure 4. Comparison of Optimization Performance under Perfect and Imperfect Channel Estimation versus Different SNR.

Figure 5 shows SCO−AN's secrecy capacity with and without allocation versus different SNR. The effects of perfect and imperfect channel estimation are also considered. The SNR of **H** increases from 2 to 30, while the SNR of G is 5 dB. ISQP allocates all the transmission power. In this figure, the solid line shows the lower bound on optimized secrecy capacity. The dashed line shows the lower bound on secrecy capacity without power allocation. The results in Figure 5 are computed according to (6) and (12), and the ISQP algorithm. The results show that the lower bound on secrecy capacity increases with power allocation, implying the high effectiveness of the ISQP algorithm. The secrecy capacity increases with SNR for **H**; that is, a low noise level improves secrecy capacity. The lower bound on the secrecy capacity of SCO−AN decreases when the effect of imperfect channel estimation is taken into consideration. The lower bound on the secrecy capacity with MMSE channel estimation is greater than the lower bound on that with LS channel estimation. We then reach that the higher channel estimation accuracy enhances secrecy capacity.

Figure 6 shows SCO−AN's secrecy capacity with and without active eavesdropper versus different SNR. The effects of different kinds of channel estimation are also considered. The SNR of **H** increases from 2 to 30, while the SNR of G is 5 dB. ISQP allocates all the transmission power. In this figure, the solid line shows the lower bound on optimized secrecy capacity. The results in Figure 6 are computed according to (43), (47) and (48) and the ISQP algorithm. The results show that the lower bound on secrecy capacity increases with power allocation, implying the high effectiveness of the ISQP algorithm. The secrecy capacity increases with SNR for **H**; that is, a low noise level improves secrecy capacity. The lower bound on the secrecy capacity of SCO−AN decreases when the effect of active eavesdropper is taken into consideration. The lower bound on the secrecy capacity with MMSE channel estimation is greater than the lower bound on that with LS channel estimation.

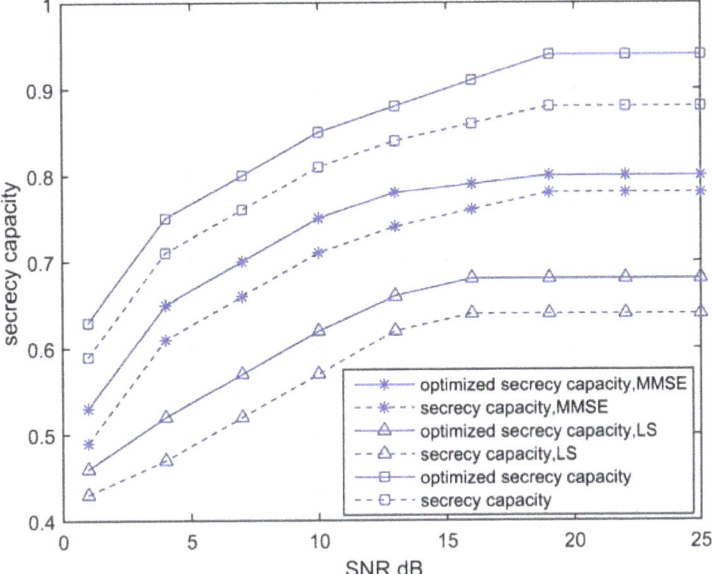

Figure 5. SCO−AN's Secrecy Capacity Before and After Power Allocation versus Different SNR and Channel Estimation Algorithm.

Figure 7 shows SCO−AN's secrecy capacity with active eavesdropper versus different P_E. The effects of different kinds of channel estimation are also considered. The power of P_E increases from 1 to 10, while the P_B is unchanged. ISQP allocates all the transmission power. The results show that the lower bound on secrecy capacity increases with power allocation, implying the high effectiveness of the ISQP algorithm. The secrecy capacity decreases with P_E; that is, a low P_E improves secrecy capacity. The lower bound on the secrecy capacity with MMSE channel estimation is greater than the lower bound on that with LS channel estimation.

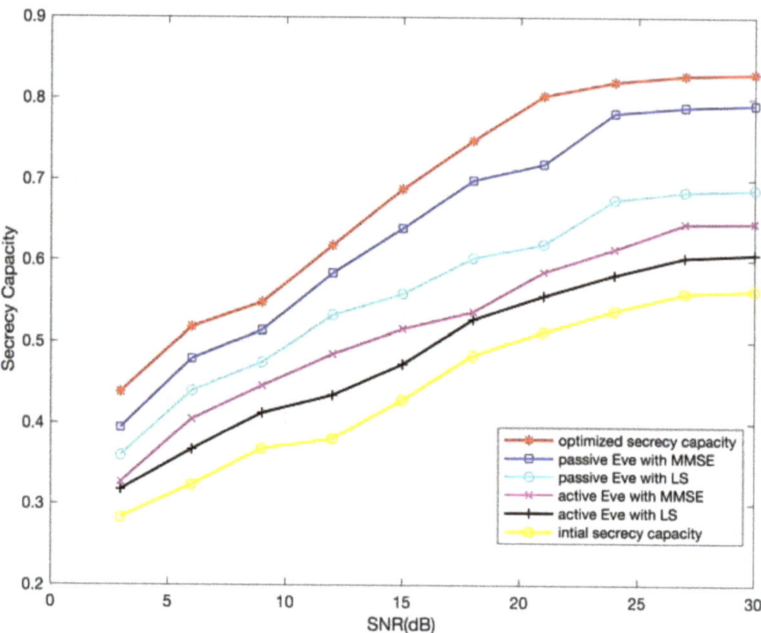

Figure 6. SCO−AN's Secrecy Capacity Before and After Power Allocation versus Different SNR and Channel Estimation Algorithm with and without active eavesdropper.

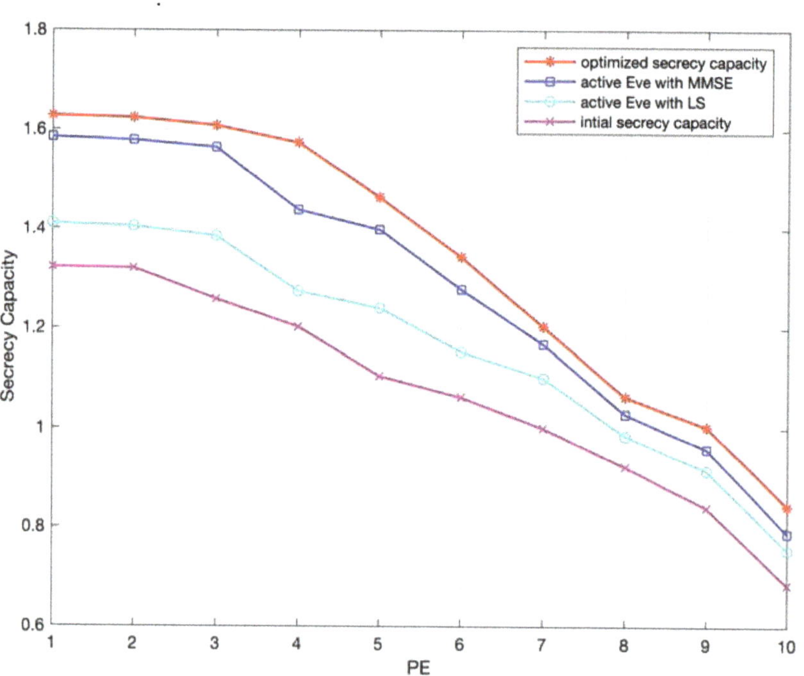

Figure 7. SCO−AN's Secrecy Capacity Before and After Power Allocation versus Different P_E and Channel Estimation Algorithm.

5. Conclusions and Future Work

In this paper, we study the power allocation problems of SCO−AN under perfect and imperfect CSI. First, the power allocation model of SCO−AN with perfect channel estimation is constructed. Then, the effect of the imperfect channel estimation error is examined. The power allocation model of SCO−AN is constructed for the first time in this paper, along with the expression of the imperfect channel estimation's effect on power allocation. The power allocation optimization problem is a crucial contribution to optimizing secrecy capacity under imperfect channel estimation. The power allocation problem's objective function is non-convex, which poses challenges to the solving process. Therefore, we solve this problem by adopting the ISQP algorithm. We compare ISQP with the other three algorithms–SQP, BPA, and COCOA. Although ISQP is slightly worse than SQP in terms of the optimization effect, the ISQP algorithm far exceeds other algorithms. Moreover, ISQP requires the least complex computation. Therefore, we decide to choose the ISQP algorithm. Our simulation results show that the secrecy capacity of SCO−AN wireless communication system increases the most under ISQP algorithm. We then conclude that the ISQP algorithm is the most effective for this purpose.

There is much room for future research. For any optimization problem, there is an upper bound to be reached. What is the upper bound on secrecy capacity for SCO−AN under a specific power? This question lays an exciting background for future research directions. Since 2019, the research on the physical layer security of reflective intelligence surfaces has become a research hotspot. The application of SCO−AN in intelligent reflector technology is one of our future research contents as well. As inspired by many papers, the features of mixing other SCO−AN signals also pose a meaningful research question, such as SCO−AN with interference alignment characteristics and SCO−AN with channel coding characteristics. Among these proposed topics for future studies, we will first study the secrecy capacity's upper bound of SCO−AN.

Author Contributions: Y.G. proposed the framework of the whole algorithm. B.H. handled all the simulations and made all figures and tables in the manuscript. Z.W. was a major contributor in writing the manuscript. All authors have read and agreed to the published version of the manuscript.

Funding: The research in this article is supported by "the National Natural Science Foundation of China" Grant nos. 61571167, 61471142, 61102084 and 61601145).

Institutional Review Board Statement: The study was conducted according to the guidelines of the Declaration of Helsinki, and approved by the Institutional Review Board (or Ethics Committee) of NAME OF INSTITUTE (protocol code XXX and date of approval).

Informed Consent Statement: Informed consent was obtained from all subjects involved in the study.

Data Availability Statement: Data sharing not applicable to this article as no datasets were generated or analysed during the current study.

Conflicts of Interest: The authors declear that they have no competing interests.

Abbreviations

AN	Artificial Noise
SCO−AN	Secrecy Capacity Optimization Artificial Noise
GAN	Global Artificial Noise
MIMO	Multiple-Input Multiple-Output
SQP	Sequence Quadratic Program
CSI	Channel State Information
KKT	Karush–Kuhn–Tucker
SNR	Signal-to-Noise Ratio

References

1. Shannon, C.E. Communication theory of secrecy systems. *Bell Syst. Tech. J.* **1949**, *28*, 142–149. [CrossRef]
2. Wyner, A.D. The wire-tap channel. *Bell Syst. Tech. J.* **1975**, *54*, 1355–1387. [CrossRef]
3. Csiszar, I.; Korner, J. Broadcast channels with confidential messages. *IEEE Trans. Inf. Theory* **2003**, *24*, 339–348. [CrossRef]

4. Negi, R.; Goel, S. Secret communication using artificial noise. In Proceedings of the IEEE Vehicular Technology Conference, Dallas, TX, USA, 25–28 September 2005.

5. Goel, S.; Negi, R. Secret communication in presence of colluding eavesdroppers. In Proceedings of the IEEE Military Communications Conference, Atlantic City, NJ, USA, 17–19 November 2005.

6. Dreschler, W.A.; Verschuure, H.; Ludvigsen, C.; Westermann, S. Icra noises: Artificial noise signals with speech-like spectral and temporal properties for hearing instrument assessment. *Int. Audiol.* **2001**, *40*, 148–157. [CrossRef]

7. Goeckel, D.; Vasudevan, S.; Towsley, D.; Adams, S.; Ding, Z.; Leung, K. Artificial noise generation from cooperative relays for everlasting secrecy in two-hop wireless networks. *IEEE J. Sel. Areas Commun.* **2001**, *29*, 2067–2076. [CrossRef]

8. Rezki, Z.; Ashish, K.; Mohamed, A. On the secrecy capacity of the wiretap channel with imperfect main channel estimation. *IEEE Trans. Commun.* **2014**, *62*, 3652–3664. [CrossRef]

9. Darsena, D.; Gelli, G.; Iudice, I.; Verde, F. Design and performance analysis of channel estimators under pilot spoofing attacks in multiple-antenna systems. *IEEE Trans. Inf. Forensics Secur.* **2020**, *15*, 3255–3269. [CrossRef]

10. Zhou, X.; Maham, B.; Hjorungnes, A. Pilot contamination for active eavesdropping. *IEEE Trans. Wirel. Commun.* **2012**, *11*, 903–907. [CrossRef]

11. Tugnait, J.K. Pilot spoofing attack detection and countermeasure. *IEEE Trans. Commun.* **2018**, *66*, 2093–2106. [CrossRef]

12. Wang, H.M.; Huang, K.W.; Tsiftsis, T.A. Multiple antennas secure transmission under pilot spoofing and jamming attack. *IEEE J. Sel. Areas Commun.* **2018**, *36*, 860–876. [CrossRef]

13. Zhou, X.Y.; Matthew, R.M. Secure transmission with artificial noise over fading channels: Achievable rate and optimal power allocation. *IEEE Trans. Veh. Technol.* **2010**, *59*, 3831–3842. [CrossRef]

14. Ciuonzo, D.; Augusto, A.; Vincenzo, C. Rician MIMO channel-and jamming-aware decision fusion. *IEEE Trans. Signal Process.* **2017**, *65*, 3866–3880. [CrossRef]

15. Cui, M.; Zhang, G.; Zhang, R. Secure wireless communication via intelligent reflecting surface. *IEEE Wirel. Commun. Lett.* **2019**, *8*, 1410–1414. [CrossRef]

16. Zhou, F.; Chu, Z.; Sun, H.; Hu, R.Q.; Hanzo, L. Artificial noise aided secure cognitive beamforming for cooperative MISO-NOMA using SWIPT. *IEEE J. Sel. Areas Commun.* **2018**, *36*, 918–931. [CrossRef]

17. Jiang, Y.; Zou, Y.; Guo, H.; Zhu, J.; Gu, J. Power allocation for intelligent interference exploitation aided physical-layer security in ofdm-based heterogeneous cellular networks. *IEEE Trans. Veh. Technol.* **2020**, *69*, 3021–3033. [CrossRef]

18. Zhao, N.; Yu, F.R.; Li, M.; Leung, V.C. Anti-eavesdropping schemes for interference alignment (IA)-based wireless networks. *IEEE Trans. Wirel. Commun.* **2016**, *15*, 5719–5732. [CrossRef]

19. Cao, Y.; Zhao, N.; Yu, F.R.; Chen, Y.; Liu, X.; Leung, V.C. An anti-eavesdropping interference alignment scheme with wireless power transfer. In Proceedings of the IEEE International Conference on Communication Systems (ICCS), Shenzhen, China, 14–16 December 2016.

20. Wei, L.; Mounir, G.; Bin, C. Secure communication via sending artificial noise by the receiver: Outage secrecy capacity/region analysis. *IEEE Commun. Lett.* **2012**, *16*, 1628–1631.

21. Bai, J.; Dong, T.; Zhang, Q.; Wang, S.; Li, N. Coordinated Beamforming and Artificial Noise in the Downlink Secure Multi-Cell MIMO Systems Under Imperfect CSI. *IEEE Wirel. Commun. Lett.* **2020**, *9*, 1023–1026. [CrossRef]

22. Liao, W.C.; Chang, T.H.; Ma, W.K.; Chi, C.Y. Qos-based transmit beamforming in the presence of eavesdroppers: An optimized artificial-noise-aided approach. *IEEE Trans. Signal Process.* **2011**, *59*, 1202–1216. [CrossRef]

23. Hong, S.; Pan, C.; Ren, H.; Wang, K.; Nallanathan, A. Artificial-noise-aided secure MIMO wireless communications via intelligent reflecting surface. *IEEE Trans. Commun.* **2020**, *68*, 7851–7866. [CrossRef]

24. Zhang, X.; Guo, D.; An, K.; Ma, W.; Guo, K. Secure transmission and power allocation in multiuser distributed massive MIMO systems. *Wirel. Netw.* **2020**, *26*, 941–954. [CrossRef]

25. Ding, Q.; Xi, T.; Lian, Y. Joint Power Allocation Scheme for Distributed Secure Spatial Modulation in High-Speed Railway. *IEEE Syst. J.* **2020**. [CrossRef]

26. Niu, H.; Lei, X.; Xiao, Y.; Liu, D.; Li, Y.; Zhang, H. Power Minimization in Artificial Noise Aided Generalized Spatial Modulation. *IEEE Commun. Lett.* **2020**, *24*, 961–965. [CrossRef]

27. Meng, L.; Wang, Q.; Ji, Z.; Nie, M.; Ji, B.; Li, C.; Song, K. Resource allocation on secrecy energy efficiency for C-RAN with artificial noise. *Wirel. Netw.* **2020**, *26*, 639–650. [CrossRef]

28. Singh, P.; Trivedi, A. NOMA and massive MIMO assisted physical layer security using artificial noise precoding. *Phys. Commun.* **2020**, *39*, 100977. [CrossRef]

29. Gu, Y.; Wu, Z.; Yin, Z.; Zhang, X. The Secrecy Capacity Optimization Artificial Noise: A New Type of Artificial Noise for Secure Communication in MIMO System. *IEEE Access* **2019**, *7*, 58353–58360.

30. Mo, D.; Duarte, M.F. Multi-Branch Binary Modulation Sequences for Interferer Rejection. In Proceedings of the IEEE Statistical Signal Processing Workshop (SSP), Freiburg, Germany, 10–13 June 2018.

31. Bidabadi, S.; Omidi, M.J.; Kazemi, J. Energy efficient power allocation in downlink OFDMA systems using SQP method. In Proceedings of the Iranian Conference on Electrical Engineering, Shiraz, Iran, 10–12 May 2016.

MDPI
St. Alban-Anlage 66
4052 Basel
Switzerland
Tel. +41 61 683 77 34
Fax +41 61 302 89 18
www.mdpi.com

Applied Sciences Editorial Office
E-mail: applsci@mdpi.com
www.mdpi.com/journal/applsci